Ecology Research Handbook

Volume I

Ecology Research Handbook
Volume I

Edited by **Liam Page**

R CALLISTO
REFERENCE

New York

Published by Callisto Reference,
106 Park Avenue, Suite 200,
New York, NY 10016, USA
www.callistoreference.com

Ecology Research Handbook: Volume I
Edited by Liam Page

International Standard Book Number: 978-1-63239-159-9 (Hardback)

Printed in the United States of America.

Contents

Preface

The origins of the word "ecology" can be traced to the year 1866, when it was coined by the German scientist Ernst Haeckel. Ecology is an interdisciplinary field, which is the scientific study of interactions among organisms and their environment. The foundations of this subject were laid by Ancient Greek philosophers such as Hippocrates and Aristotle, who studied natural history. From the 19th century onwards, concepts such as natural selection and adaptation transformed Ecology into a more rigorous discipline.

The concepts of Ecology are premised on ecosystems, which include organisms, the communities they make up, and the non-living components of their environment. The scope of Ecology spans a wide array of interacting levels of organisms, ranging from the micro-level to planetary scale. This subject helps comprehend how the living world interacts. It provides evidence on the interdependence between the inanimate and animate elements of the ecosystem.

Ecology also covers topics of subjects like Biology and Earth Science. Topics of interest to ecologists include diversity, distribution, and population of organisms. The manner in which biodiversity affects ecological function is an important and emerging focus area in ecological studies. A better understanding of ecological systems is crucial in contemporary times, as it allows scientists to predict the consequences of human activity on the environment.

I would like to thank our researchers and writers for sharing their valuable research with us in this book.

Editor

Plants used in traditional beekeeping in Burkina Faso

Schweitzer Paul[1], Nombré Issa[2,3], Aidoo Kwamé[4], Boussim I. Joseph[2]

[1]Laboratoire d'Analyses et d'Ecologie Apicole Centre d'Etudes Techniques Apicole de Moselle, Guenange, France
[2]Institut des Sciences, Ouagadougou, Burkina Faso
[3]Laboratoire de Biologie et Ecologie Végétales UFR Science et Technique Université de Ouagadougou, Ouagadougou, Burkina Faso
[4]International Stingless Bee Centre, Department of Entomology and Wildlife, University of Cape Coast, Cape Coast, Ghana

ABSTRACT

Beekeeping is one of the recommended approaches in the implementation of poverty alleviation programs in rural areas of Burkina Faso. However, plants that are important in beekeeping have not been identified. The use of parts and organs of plants by beekeepers and their methods of harvesting remain unknown. These limit the conservation efforts of these important plants and affect beekeeping development. The study was carried out in the south-central, east-central regions and in Comoé and Boucle of Mouhoun regions of Burkina Faso. The objective of the study was to identify the plants species used by traditional beekeepers, the different uses made of these plant parts and organs and then to discuss the impact of these activities on the survival of the plant resources. An ethno-apiculture survey was conducted in the main apiculture zone of Burkina Faso, using semi-structured interviews. The methodology of botanical coherence or convergence was applied to classify botanical species. Results showed that 35 botanical species were used in traditional beekeeping. The use of plant parts or organs in traditional hives construction represents 55%, attraction of wild swarms in new beehives is 37.50% and use as a torch or as a smoker, 7.50%. The barks are the organs most used. Trees are botanical type most used. The results are not exhaustive and therefore other additional studies need to be carried out. In order to sustain the use of these important plants, their growing in nursery and their planting in the field are recommended.

Keywords: Beekeeping; Melliferous Plants; Pollinating; Biodiversity; Burkina Faso

1. INTRODUCTION

Honeybees are since 2006 victims of colony collapse disorder or (CCD) ([1-3]). Many well intentioned suggestions as to the possible causes of colony losses, including such improbable ideas as mobile telephones, genetically modified crops and nanotechnology, have perhaps overshadowed much more likely explanations such as pests and diseases, pesticides, loss of forage and inappropriate beekeeping practices [2]. Bees are the major pollinators of wild plants and crops in terrestrial ecosystems. Honeybees are known to contribute significantly to the provision of this essential ecosystem service of pollination [4-6]. They are also bio-indicators for environment pollution [7,8] and beekeeping is an effective means to generate monetary incomes that support the livelihood of rural communities. Numerous studies have demonstrated the economic value of honeybees to the agricultural industry of the world [9,10]. In Africa, especially areas in south of Sahara and particularly in Burkina Faso, this phenomenon is not fully known because of the lack of scientific studies [11]. [6] stated that beekeepers and honey hunters are sometimes perceived to cause damages to forests, through the careless use of fire during harvesting and because they kill trees to make beehives. So, traditional beekeeping has been considered as harmful to biodiversity conservation [12]. Others authors differentiate traditional beekeeping from honey hunting as contributing to the increase in honey bee number. The roles of honeybees in providing ecosystem services is a function of their number in the beehive and varies according to the type of beehive used [13].

Burkina Faso has undertaken the modernization of its apiculture since 1987. Studies had been done on the melliferous plants [14,15] and on the plant organs used to attract swarms of the local honeybee *Apis mellifera adansonii* Latreille into newly installed beehives [16]. Traditional beekeeping is widespread in Burkina Faso and their activities understand in the exploitation of plant parts and organs described as "extractivism" may have conservation undertones [17]. Studies on the impact of this activity on colony loss and other effects on the environment have not been carried out. The harvest technologies of plant organs or parts remain unrecognized, limiting the conception of preservation efforts of melliferous plants. This lack of information can moreover constitute a handicap in the development of beekeeping. It is to contribute to raising this constraint in relationship with the lack of information that the present study aims to provide knowledge on the used in traditional beekeeping. It will identify the various uses of the plant parts and organs and discuss their impacts on plant resources sustainability and then suggest solutions for a sustainable management of the identified plants.

2. MATERIAL AND METHODS

2.1. Study Area

The study was carried out in the villages of Nazinga (south-central region), Garango (east-central region), Tiefora (Comoé region) and Dédougou (Boucle of Mouhoun region) (**Figure 1**).

These communities fall within the main beekeeping zones made up of the north and south Soudanian phytogeographical sector of Burkina Faso.

Agriculture (crop cultivation and animal breeding) is the main activity of the population. In all the regions, crops (*Sorghum guineense* Staph., *Zea mays* Linn., *Oryza sativa* Linn., *and Dioscorea dumetotum* (Kunt) Pax.) are dominant.

The vegetation is predominantly savannas with arusticlandscapes dominated by species such *Vitellaria paradoxa* Gaertn, *Tamarindus indica* Linn., *Parkia biglobosa* (Jacq.) Benth, *Lannea microcarpa* Engl. & K. Krause, *Adansonia digitata* Linn., *Faidherbia albida* (Del.) A. Chev.; the groupings of *Anogeissus leiocarpus* (DC.)

Guil. & Perr. and planted species as *Mangifera indica* Linn., *Eucalyptus camaldulensis* Dehnhard, *Azadirachta indica* A. Juss, *Khaya senegalensis* (Desn.) A. Juss, *Anacardium occidentale* Linn., *Borassus aethiopum* Mart., *Psidium guajava* Linn., *Cariaca papaya* Linn., *Annona squamosa* Linn., and *Citrus sinensis* (Linn.) Osbeck.

2.2. Methods

Ethno-apicultural investigations and field observations were carried out using semi-structured inquiry cards on traditional beekeepers who are 25 years of age and possessing colonized traditional beehives. The names of plants used were transcribed into the following local languages: Gourounsi for Nazinga zone; Bissa for Garango zone and Dioula for Tiéfora and Dédougou zones. The plant species scientific identification was made referring to [18]. The plants parts and organs used by beekeepers were identified from responses obtained from at least 10 beekeepers. A total of 103 beekeepers were interviewed.

3. RESULTS

3.1. Different Plant Parts or Organs Used

The results showed that the barks and fibers with 37.5% of utilizations constituted the most organs used (**Figure 1**). The grass, the aerial organs, the thatches of graminaceous and the inflorescences constituted the group of plants aerial part (32.5%). The twigs constituted 12.5%, the fruits and seeds 10%, leave 5% and the tubers were less used (2.5%).

3.2. Plants Used and Their Utilizations

The parts or organs of 35 botanical species were used in the traditional beekeeping practices in Burkina Faso (**Table 1**).

Three kinds of utilization of the plant parts or organs were dominants (**Table 1**). The first concerned the use in the new traditional beehives construction. It repre-

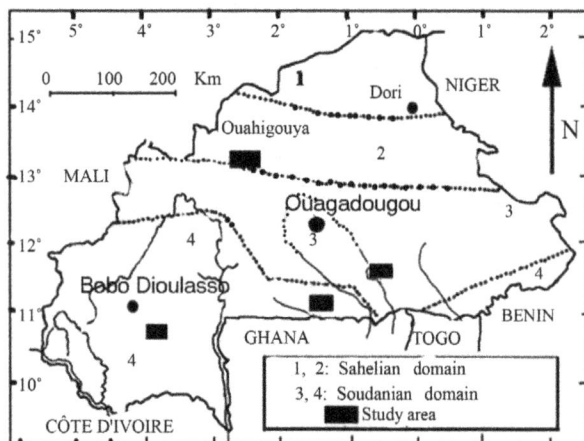

Figure 1. Location of the study area.

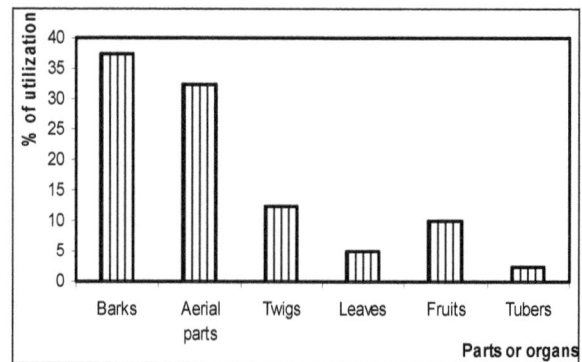

Figure 2. Different plant parts or organs used in traditional beekeeping.

sented 55%. Indeed, beehives can be made with barks, hollowed out tree trunks, plaited straws or twigs.

The second represented the attraction of wild swarms into newly established beehives. It represented 37.50%. Indeed, plant parts or organs can be used as swarm baits substituting the Aristée perfumes, the honeybees' charm or comb foundations used in modern beekeeping.

The third is the use as a torch to light beehives inside or as a smoker during honey harvesting. It represented 7.50%.

3.3. Botanical Type Used

In the biological type, trees were the most used (44%); followed in order by the grass (31%), the shrubs (22%) and the lianas (3%) (**Figure 3**).

Table 1. The use of plant parts and organs in traditional beekeeping in the Burkina Faso.

Scientifique names	Parts or organs used	Utilizations
Acacia seyal Del.	Fruits	Swarms attraction
Combretum glutinosum Perr. ex DC	Twigs	Swarms attraction
Ctenium newtonii Hack.	Thatches	Swarms attraction
Cymbopogon schoenanthus subsp. *proximus* (Hochst. ex A. Rich.) M. & W	Inflorescences	Swarms attraction
Dicoma tomentosa Cass.	Aerial organs	Swarms attraction
Dioscorea dumetorum (Kunth) Pax	Tubers	Swarms attraction
Diospyros mespiliformis Hochst. ex A. DC.	Leaves	Swarms attraction
Guiera senegalensis J. F. Gmel.	Twigs	Swarms attraction
Hyptis spicigera Lam.	Aerial organs	Swarms attraction
Leucas martinicensis (Jacq.) Ait.	Aerial organs	Swarms attraction
Ocimum americanum Linn.	Aerial organs	Swarms attraction
Parkia biglobosa (Jacq.) Benth.	Seeds	Swarms attraction
Piliostigma reticulatum (DC.) Hochst.	Fruits	Swarms attraction
Piliostigma thonningii (Schum.) Milne-Redhead	Fruits	Swarms attraction
Andropogon ascinodis C. B. Clarke	Thatches	Honey harvest
Andropogon gayanus Kunth	Thatches	Honey harvest
Andropogon pseudapricus Stapf	Thatches	Honey harvest
Andropogon ascinodis C. B. Clarke	Thatches	Beehive construction
Andropogon gayanus Kunth	Thatches	Beehive construction
Andropogon pseudapricus Stapf	Thatches	Beehive construction
Borassus aethiopum Mart.	Leaves	Beehive construction
Burkea africana Hook.	Bark	Beehive construction
Daniellia oliveri (Rolfe) Hutch. & Dalz.	Bark	Beehive construction
Detarium microcarpum Guill. & Perr.	Bark	Beehive construction
Feretia apodanthera Del.	Twigs	Beehive construction
Hibiscus asper Linn.	Fibers	Beehive construction
Isoberlinia doka Craib & Stapf	Bark	Beehive construction
Lannea acida A. Rich.	Fibers	Beehive construction
Loudetia togoensis (Pilger) C. E. Hubbard	Thatches	Beehive construction
Piliostigma reticulatum (DC.) Hochst.	Fibers	Beehive construction
Piliostigma thonningii (Schum.) Milne-Redhead	Fibers	Beehive construction
Prosopis africana (Guill. & Perr.) Taub.	Bark	Beehive construction
Pseudocedrela kotschyi (Schweinf.) Harms	Bark	Beehive construction
Pterocarpus erinaceus Poir.	Bark	Beehive construction
Saba senegalensis (A. DC.) Pichon	Twigs	Beehive construction
Fluggea virosa (Roxb. ex Willd.) Baill	Twigs	Beehive construction
Tamarindus indica Linn.	Fibers	Beehive construction
Terminalia avicennioides Guill. & Perr	Bark	Beehive construction
Vitellaria paradoxa Gaertn.	Bark	Beehive construction
Xeroderris stuhlmannii (Taub.) Men	Bark	Beehive construction

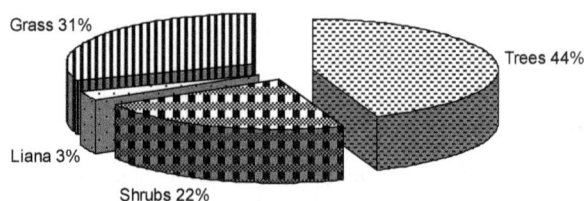

Figure 3. Different plants biological types used in traditional beekeeping.

4. DISCUSSION

In the traditional beekeeping practices, different plants, parts or organs are used by the beekeepers in different ways. The construction of beehives is more important in the utilization of plants and also their parts or organs. The technology used to remove these parts or organs can be negative for the environment because it affects the regeneration and the survival of the plants used. Indeed, according to [17], the cutting down has negative impacts on the individual tree, because, even if it presents a potential of stump rejections, they have only very slight chance to survival. That will appear as habitat degradation and outright destruction and can be the major causal factor in the decline of bees [5].

Often, the plants used were also excellent nectar species and then the loss of trees has negative implications for beekeepers because they lose food and nesting sites for wild bees, materials for building hives and places to keep hives. However, beekeepers must make deliberate and conscious efforts to protect and conserve forests in which their bees forage, despite their dependency on these resources.

There are also positives impacts on traditional beekeeping practices. Indeed, nesting honeybees in appropriate way allows them to increase their number that will increase their role (pollinating, honey production). Often, the traditional beekeepers breed their honeybees even if the hives used are rudimentary; and they only use the smoke of burned thatch to hunt honeybees during the harvest [16]. According to [17], the removal or whatever organ collected is mostly made to secure not only the survival of the exploited individual, but also the regeneration of the resource in a reasonable lapse of time. Furthermore, according to [6], the development of traditional beekeeping based on keeping colonies, to the detriment of the honey hunting can increase the honeybees' number per beehive and even per region, involving thus an increase of their pollinating role. This development minimizes the destructive effects of traditional beekeeping on honeybees on one hand and the environment on the other. Also, according to [16], the utilization of plant organs or parts to attract honeybee swarms in newly established beehives contributes to reduce the installation costs of beekeeping projects development.

According to [6], the technologically modern man has contributed to honeybees declines in Africa. This is evident in that bee diversity and abundance is much greater on crops in areas surrounded by natural vegetation than in ecosystems that have been widely transformed by agriculture and other exotics along with removal of natural vegetation through urbanization.

5. CONCLUSION

Traditional beekeeping contributes greatly to biodiversity conservation in Burkina Faso. Despite the negative effects that are attributed to some of its activities, it allows for the establishment and management of wild swarms of honeybees in appropriated ways for hive products. The pollinating role of honeybees in the ecosystem is therefore enhanced. Plants, parts and organs are used at different levels in this system of apiculture. The effects of the use of plants and their organs in traditional beekeeping practices on the vegetation and the environment remain negligible. Traditional beekeepers therefore sustain the populations of honeybees in the environment which contribute to the essential ecosystem service of pollination and biodiversity conservation. Negative practices of wild honey hunting should be replaced with traditional beekeeping.

REFERENCES

[1]   Chagnon, M. (2008) Causes et effets du déclin mondial des pollinisateurs et les moyens d'y remédier. Fédération Canadienne de la Faune. Bureau régional du Québec.

[2]   Neumann, P., and Carreck, N.L. (2010) Honey bee colony losses. *Journal of Apicultural Research*, **49**, 1-6.

[3]   Kluser, S., Neumann, P., Chauzat, M.-P. and Pettis, J.S. (2010) Global honey bee colony disorder and other threats to insect pollinators. UNEP.

[4]   Vaissière, B., Morison, N. and Carré, G. (2005) Abeilles, pollinisation et biodiversité. *Abeilles & Cie*, **3**, 10-14.

[5]   Brown, M.J.F. and Paxton, R.J. (2009) The conservation of bees: A global perspective, *Apidologie*, **40**, 410-416.

[6]   Eardley, C.D., Gikungu, M. and Schwarz, M.P. (2009) Bee conservation in Sub-Saharan Africa and Madagascar: Diversity, status and threats, *Apidologie*, **40**, 355-366.

[7]   Chauzat, M.-P., Faucon, J.-P., Martel, A.-C., Lachaize, J., Cougoule, N. and Aubert, M. (2006) Les pesticides, le pollen et les abeilles LSA, **216**, 11-12.

[8]   Bogdanov, S. (2006) Contaminants of bee products. *Apidologie*, **37**, 1-18.

[9]   Krell, R. (1996) Value-added products from beekeeping. FAO Agricultural Services Bulletin No. 124.

[10] Bradbear, N. (2010) Le rôle des abeilles dans le dévelop-

pement rural. Manuel sur la récolte, la transformation des produits et services dérivés des abeilles. FAO, Rome, PFNL 19.

[11] Dietemann, V., Pirk, C.W.W. and Crewe, R. (2009) Is there a need for conservation of honeybees in Africa? *Apidologie*, **40**, 285-295.

[12] Fisher, F.U. (1993) L'élevage des abeilles dans l'économie de base de la savane arborée de Miombo au Centre de l'Afrique Australe. *Document du Réseau Forestier Pour le Développement Rural*, **15**, 3-11.

[13] Foucault, B. (2010) Ethonographie de quelques ruches traditionnelles. *Colloque sur les Journées d'échanges sur l'abeille et l'apiculture*, 22-27 Avril 2005, Lille, 121-128

[14] Guinko, S., Guenda, W., Tamini, Z. and Zoungrana, I. (1992) Les plantes mellifères de la zone Ouest du Burkina Faso. *Etudes flor. Vég. Burkina Faso*, **1**, 27-46.

[15] Nombré, I., Schweitzer, P., Sawadogo, M., Boussim, J.I. and Millogo-Rasolodimby, J. (2009) Assessment of melliferous plant potentialities in Burkina Faso. *African Journal of Ecology*, **47**, 622-629.

[16] Nombré, I., Schweitzer, P., Boussim, I.J., Millogo/Rasolomdimby, J. (2009) Plantes utilisées pour attirer les essaims de l'abeille domestique (*Apis mellifera adansonii* Latreille) au Burkina Faso. *International Journal of Biological and Chemical Sciences*, **3**, 840-844.

[17] Aubertin, C., De Castro, A.L., Empenire, L., Lescure, J.-P., Mitja, D. and Pinton, F. (1993) Les activités extractivistes en Amazonie Centrale: Une première synthèse d'un projet multidisciplinaire. ORSTOM/INPA.

[18] Arbonnier, M. (2002) Arbres, arbustes et lianes des zones sèches d'Afrique de l'Ouest. CIRAD, MNHN, UICN.

Parental care in the freshwater crab *Sylviocarcinus pictus* (Milne-Edwards, 1853)*

Bruno Sampaio Sant'Anna, Erico Luis Hoshiba Takahashi, Gustavo Yomar Hattori

Federal University of Amazonas (Amazonas University), Institute of Exact Science and Technology (ICET), Itacoatiara, Brazil;
*Corresponding Author

ABSTRACT

Parental care is a common strategy in many animal groups, to increase survival of the offspring. Here, we report parental care in the freshwater crab *Sylviocarcinus pictus*. A female caught in the Amazon River, Brazil, bore juvenile crabs rather than eggs on her abdomen. Kept in the laboratory, the female retained the juveniles on the abdomen for 17 days, after which the juveniles left the abdomen. A total of 341 juvenile crabs measuring 3.45 ± 0.12 mm were recorded. This pattern of parental care is very important for the maintenance of local populations of *S. pictus*, because if the larvae were released, as occurs in many marine species, they would drift downstream.

Keywords: Freshwater Crab; Parental Care; Reproduction; Trichodactylidae

1. INTRODUCTION

In many animal species, parental care is a common reproductive strategy [1-3]. The many patterns of parental care include biparental, or uniparental by either males or females [1]; that may show manipulation of sex differences in parental care [4] and is energetically costly [5]. Many groups of invertebrates show some form of parental care. For instance, in insects the most rudimenttary form of maternal care is provided by females that incorporate toxins into their eggs, oviposit them in protected places, or cover their eggs with a hard wax-like shell before abandoning them [3]. In arachnids, females of *Bourguyia albiornata* Mello-Leitão 1923 oviposit almost exclusively inside the tube formed by the curled leaves of the bromeliad *Aechmea nudicaulis* (Linnaeus) Grisebach, 1864 [6]. Jawed Hirudinidae deposit desiccation-resistant cocoons on land and many species brood the eggs and young [7].

In decapod crustaceans, parental care is usually restricted to females that carry the eggs in the brood compartment, and care is terminated when the larvae are released into the plankton [8,9]. However, among other examples in crustaceans, [10] presented evidence of a direct link between active brood care and provision of oxygen to the young. For amphipods, [11,12] studied active maternal brooding and juvenile care. The preparation of a nest structure to defend and feed its young in the crab *Metopaulias depressus* Rathbun, 1918 was recorded [9,13].

For freshwater crustaceans, available information about parental care is sparse in comparison with marine species. The species that has been most studied is the crab *M. depressus*, in several aspects, *i.e.*, parental care in an unusual environment [13], protection of larvae from predation by damselfly nymphs [9], maintaining oxygen, pH and calcium levels optimal for the larvae [14] and evolution theory [15]. In freshwater crabs, 15 species have been reported to bear juvenile crabs attached to the female abdomen [16]. The extended brood care was reported in species of all five families of primary freshwater crabs [17]. Here, we record a female of the freshwater crab *Sylviocarcinus pictus* (Milne-Edwards, 1853) with juvenile crabs attached on the abdomen, indicating the existence of parental care in this species.

2. MATERIAL AND METHODS

The female of *S. pictus* was collected by hand, on a bank of the Amazon River (03°08'13.7"S; 58°27'46.8"W) in October 2011 (**Figure 1**). The specimen was placed in a plastic box with aerated water and transported to the laboratory. In the laboratory, the carapace width was

*Research Group on Biology and Production of Amazonic Aquatic Organisms

measured with a caliper (0.05 mm), and the crab was maintained in an aquarium for 17 days.

3. RESULTS AND DISCUSSIONS

This female had a carapace width of 41.5 mm, with 341 juvenile crabs attached to the abdomen (**Figure 2**). The mean size of the juveniles was 3.45 ± 0.12 mm, ranging from 3.10 to 3.66 mm. The juveniles remained on the female's abdomen for 17 days; during this period, they would occasionally leave the female's abdomen for

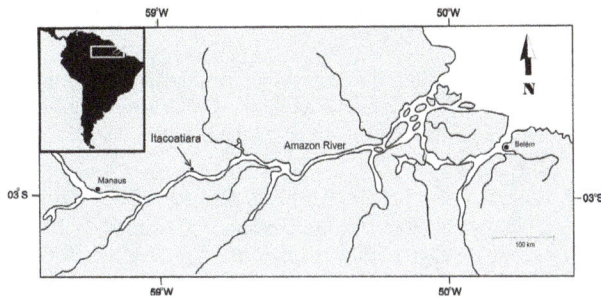

Figure 1. Location of the study area, Itacoatiara (arrow) on the Amazon River, Brazil.

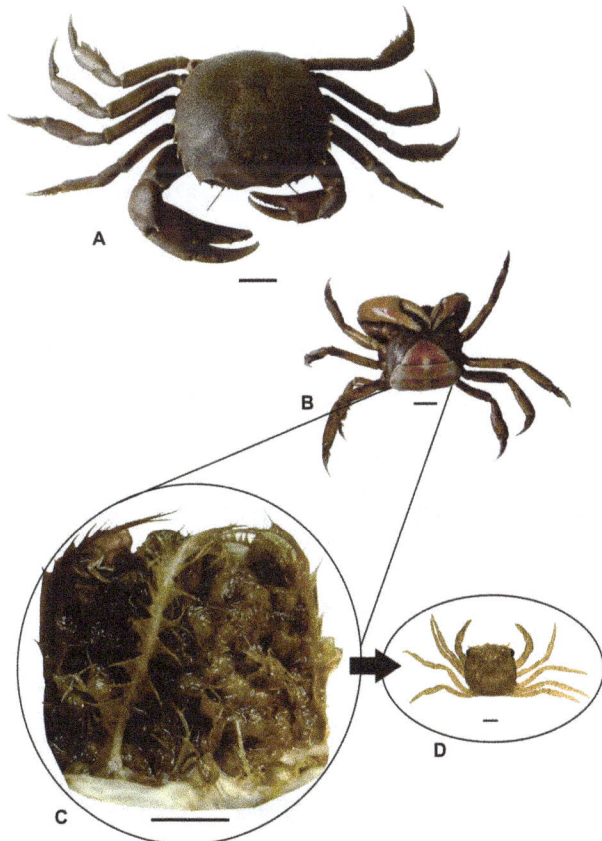

Figure 2. Dorsal view of *Sylviocarcinus pictus* (A), ventral view of *S. pictus* (B), detail of juvenile crabs attached on the female abdomen (C) and juvenile crab (D). Scale bar of the figures A, B and C = 10 mm, and of figure D = 1 mm.

several minutes. After 17 days, all the juveniles permanently left the female's abdomen. *Sylviocarcinus pictus* shows gregarious behavior, and the juveniles can cling to their mother, as also observed in the freshwater crab *Potamon edulis* (*P. fluviatile*) (Latreille, 1818) by [18], this kind of behavior is common for freshwater crabs. According to [20], the juveniles of the potamid crab *Candidiopotamon rathbunae* (de Man, 1914) are essentially independent after their first day of life, but often return to the mother for shelter during the following 2 weeks. The same pattern was observed for the juveniles of *S. pictus*.

In a recent study [16] observed two size groups of juvenile crabs with different carapace morphology, attached on the abdomen of females of the crab *Kingsleya ytupora* Magalhães, 1986, suggesting that the juveniles are attached to females for a prolonged period. In the present study, all juveniles had the same morphology and similar size, and remained on the female's abdomen for 17 days in the laboratory. However, as recorded by [16], we did not observe the hatching process and cannot accurately report the full period of juvenile incubation by females.

According to [11], brood care is called "active" if specific parental activities are directed toward the brood, and "passive" if such specific behavior is lacking. Females of *S. pictus* could be considered "active" in parenmtal activities, since in this freshwater crab the embryonic and larval periods are completed entirely in the egg stage, resulting in hatching of miniature adults [20]; these are considered juveniles, and remain on the abdomen.

Abbreviated larval development is often accompanied by increased parental care. According to [21], in its broadest sense, parental care includes preparation of nests and burrows, production of heavily yoked eggs, care of the eggs, provisioning of the young, and care of the offspring after they reach nutritional independence. Parental care significantly affects the ecological success and evolutionary potential of species by enhancing the survival and fitness of the offspring. In the freshwater caridean shrimp *Dugastella valentina*, [22] observed both the abbreviated development and parental safeguarding until the decapodid stage obviously reduce the risk of being washed away or of being predated upon. Simultaneously, this type of parental care could mean a limited gene flow and hence a high degree of genetic divergence between populations, because of the low dispersal ability of the larvae [22]. In populations of *S. pictus*, the epimorphic development and parental care could produce a similar situation.

This pattern is common in primary freshwater crabs, because in freshwater habitats there are strong selective pressures toward reduction in egg number and increase in egg size, abolishment of free larvae, and extension of brood care until the juvenile stage, resulting in a marked reduction in dispersal and gene flow, and leading to the high degree of endemism and speciation seen in these crustaceans [17]. However, under conditions of rapid habitat destruction, environmental pollution and global warming, slow dispersal of direct developers may become a severe disadvantage, impairing replacement of lost populations and placing the directly developing taxa at a greater risk of extinction than the indirectly developing taxa [17].

4. CONCLUSIONS

The present study records extended parental care in the crab *S. pictus*, contributing to knowledge of the reproduction of freshwater crabs. This pattern of parental care is very important for the maintenance of local populations of *S. pictus*, because if the larvae were released, as occurs in many marine species, they would drift downstream.

REFERENCES

[1]   Webb, J.N., Houston, A.I., McNamara, J.M. and Székely, T. (1999) Multiple patterns of parental care. *Animal Behaviour*, **58**, 983-993.

[2]   Cockburn, A. (2006) Prevalence of different modes of parental care in birds. *Proceedings of the Royal Society B*, **273**, 1375-1383.

[3]   Scott, M.P. (2009) Parental care. In: Resh, V.H. and Cardé, R.T., Eds., *Encyclopedia of Insects*, 2nd Edition, Academic Press, New York, 751-753.

[4]   Wring, J. and Cuthill, I. (1989) Manipulation of sex differences in parental care. *Behavioral Ecology and Sociobiology*, **25**, 171-181.

[5]   Lardies, M.A., Cotoras, I.S. and Bozinovic, F. (2004) The energetics of reproduction and parental care in the terrestrial isopod *Porcellio laevis*. *Journal of Insect Physiology*, **50**, 1127-1135.

[6]   Machado. G. and Oliveira, P. (2002) Maternal care in the neotropical harvestman *Bourguyia albiornata* (Arachnida; Opiliones): Oviposition site selection and egg protection. *Behaviour*, **139**, 1509-1524.

[7]   Kutschera, U. and Wirtz, P. (2001) The evolution of parental care in freshwater leeches. *Theory in Bioscience*, **120**, 115-137.

[8]   Hazlett, B.A. (1983) Parental behaviour in decapod Crustacea. In: Rebach, S. and Dunham, D.W., Eds., *Studies in Adaptation. The Behaviour of Higher Crustacea*, John Wiley, New York, 171-193.

[9]   Diesel, R. (1992) Maternal care in the bromeliad crab, *Metopaulias depvessus*: Protection of larvae from predation by damselfly nymphs. *Animal Behaviour*, **43**, 803-812.

[10] Baeza, J.A. and Fernández, M. (2002) Active brood care in *Cancer setosus* (Crustacea: Decapoda): The relationship between female behaviour, embryo oxygen consumption and the cost of brooding. *Functional Ecology*, **16**, 241-251.

[11] Dick, J.A., Faloon, S.E. and Elwood, R.W. (1998) Active brood care in an amphipod: Influences of embryonic development, temperature and oxygen. *Animal Behaviour*, **56**, 663-672.

[12] Thiel, M. (1998) Extended parental care in marine amphipods. I. Juvenile survival without parents. *Journal of Experimental Marine Biology and Ecology*, **227**, 187-201.

[13] Diesel, R. (1989) Parental care in an unusual environment: *Metopaulias depressus* (Decapoda: Grapsidae), a crab that lives in epiphytic bromeliads. *Animal Behaviour*, **38**, 561-575.

[14] Diesel, R. and Schuh, M. (1993) Maternal care in the bromeliad crab *Metopaulias depressus* (Decapoda): Maintaining oxygen, pH and calcium levels optimal for the larvae. *Behaviour Ecology and Sociobiology*, **32**, 11-15.

[15] Schubart, C. D, Diesel, R. and Hedges, S. B. (1998) Rapid evolution to terrestrial life in Jamaican crabs. *Nature*, 393, 363-365.

[16] Wehrtmann, I.S., Magalhães, C., Hernáez, P. and Mantelatto, F.L. (2010) Offspring production in three freshwater crab species (Brachyura: Pseudothelphusidae) from the Amazon region and Central America. *Zoologia*, **27**, 965-972.

[17] Vogt, G. (2013) Abbreviation of larval development and extension of brood care as key features of the evolution of freshwater Decapoda. *Biological Reviews*, **88**, 81-116.

[18] Pace, F., Harris, R.R. and Jaccarini, V. (1976) The embryonic development of the Mediterranean freshwater crab, *Potamon edulis* (*P. fluviatile*) (Crustacea, Decapoda, Potamonidae). *Journal of Zoology*, **180**, 93-106.

[19] Liu, H.C. and Li, C.W. (2000) Reproduction in the freshwater crab *Candidiopotamon rathbunae* (Brachyura: Potamidae) in Taiwan. *Journal of Crustacean Biology*, **20**, 89-99.

[20] Von Sternberg, R., Cumberlidge, N. and Rodriguez, G. (1999) On the marine sister groups of the freshwater crabs (Crustacea: Decapoda: Brachyura). *Journal of Zoological Systematics and Evolutionary Research*, **37**, 19-38.

[21] Clutton-Brock, T.H. (1991) The evolution of parental care. Princeton University Press, Princeton, 1-352,

[22] Cuesta, J.A., Palacios-Theil, E., Drake, P. and Rodríguez, A. (2006) A new rare case of parental care in decapods. *Crustaceana*, **79**, 1401-1405.

Biodiversity and secretion of enzymes with potential utility in wastewater treatment

Susanne Facchin[1,2], Priscila Divina Diniz Alves[2], Flávia de Faria Siqueira[1],
Tatiana Moura Barroca[1], Júnia Maria Netto Victória[2], Evanguedes Kalapothakis[1*]

[1]Departamento de Genética, Instituto de Ciências Biológicas, Universidade Federal de Minas Gerais, Belo Horizonte, Brazil;
[*]Corresponding Author
[2]Phoneutria Biotecnologia e Serviços Ltda, Belo Horizonte, Brazil

ABSTRACT

The main organic contaminants in municipal wastewater are proteins, polysaccharides, and lipids, which must be hydrolyzed to smaller units. A high concentration of oil and grease in wastewater affects biological wastewater treatment processes by forming a layer on the water surface, which decreased the oxygen transfer rate into the aerobic process. Microbial proteases, lipases, amylases, and celullases should play essential roles in the biological wastewater treatment process. The present study aimed to isolate lipase- and other hydrolytic enzyme-producing microorganisms and assess their degradation capabilities of fat and oil wastewater in the laboratory. We also evaluated microbial interactions as an approach to enhance lipolytic activity. We place emphasis on lipase activity because oil and grease are not only environmental pollutants, but also form an undesirable tough crust on pipes of sewage treatment plants. Thirty-five lipolytic microorganisms from sewage were identified and assessed for hydrolytic enzyme profiles. Lipases were characterized in detail by quantification, chain length affinity, and optimal conditions for activity. The good stability of isolated lipases in the presence of chemical agents, thermal stability, wide range of pH activity and tolerance, and affinity for different lengths of ester chains indicates that some of these enzymes may be good candidates for the hydrolysis of organic compounds present in wastewater. A combination of enzymes and fermenting bacteria may facilitate the complete hydrolysis of triglycerides, proteins, and ligno-cellulose that normally occur in the wastes of industrial processes. This study identifies enzymes and microbial mixtures capable of digesting natural polymeric materials for facilitating the sewage cleaning process.

Keywords: Lipase; Esterase; Wastewater; Sewage Treatment; Lipolytic Microorganisms; Amylase; Protease; Cellulase

1. INTRODUCTION

For more than a century, biological wastewater treatment has been used to minimize anthropogenic damage to the environment. Oil and Grease (O&G) are the major problems and contaminants in biological wastewater treatment processes. Because of their nature, O&G form a layer on the water surface and decrease the oxygen transfer rate into an aerobic process [1]. These contaminants are mainly discharged from restaurants, food industries, and households [1,2]. Proteins and polysaccharides must also be hydrolyzed to smaller units by extracellular enzymes in a municipal wastewater treatment plant [3,4].

The composition and activity of the microbial community within a wastewater treatment plant play a substantial role in the efficiency and robustness of the purification process [5]. The efficiency of conventional biological processes in wastewater treatment is reduced by the high concentrations of O&G in effluents [6]. In the activated sludge process, high levels of O&G lead to a reduction of biological activity of the flocs due to the difficulty of oxygen and substrate to penetrate the floc due to the oil film formation around it [7]. Moreover, in the case of anaerobic digestion, excessive amounts of O&G inhibit the action of acetogenic and methanogenic bacteria [6,8,9]. The Brazilian National Council on the Environment (CONAMA) established the maximum level of mineral oil concentration allowed for effluent in water bodies at 20 mg/l, and the maximum level of vegetable oils and animal fats to 50 mg/l in Article 34, Resolution number 357 established on March 17, 2005 [10].

Traditional approaches to treat oily effluents include

gravity separation, dissolved air flotation (DAF), de-emulsification, coagulation, and flocculation. Free oil is removed from wastewater by gravity separation; however, this process cannot remove small oil droplets and emulsions. Oil that adheres to the surface of solid particles can be removed by particle settling [11]. DAF uses solubilized air to increase buoyancy of the smaller oil droplets and improve separation. In addition, emulsified oil is removed by chemical or thermal de-emulsifying processes, or both [11]. Wastewater containing emulsified oil is heated to reduce the viscosity, accentuate density differences, and weaken the interfacial films stabilizing the oil phase. Thereafter, acidification and the addition of a cationic polymer neutralize the negative charges, and elevation of pH to an alkaline level induces flocculation of inorganic salts. Flocs with adsorbed oil are separated and the sludge is dewatered [11]. In this context, the usage of lipolytic microorganisms in wastewater treatment could eliminate this pretreatment process [12]. Chigusa et al. [13] showed that the percentage of fat in wastewater treated with a mixed culture of nine lipase-producing yeast strains decreased by 94%.

Oily effluents can also be pretreated as an approach to conform with the CONAMA resolution, but the achievement of such pretreatment processes depends on the costs of the enzyme [14]. Many industrial processes require breakdown of solids and the prevention of fat blockage or filming in waste systems before the wastewater can be released into the sewage system. This can be accomplished 1) by degradation of organic polymers with a commercial mixture of lipase, cellulase, protease, amylase, and inorganic nutrients; or 2) by sewage treatment, cleaning of holding tanks, septic tanks, grease traps, and other systems. WW07P is produced by Environmental Oasis Ltd. and contains a range of high-performance microorganisms adapted for use in the biological treatment of wastewater containing high fat and oils. It also contains surfactants capable of liquefying heavy fat deposits, thereby assisting in their biodegradation [15].

Lipases and esterases constitute a large category of ubiquitous enzymes expressed by many organisms. Carboxylesterases (EC 3.1.1.1) have broad substrate specificity toward esters and thioesters. Esterases that hydrolyze long-chain acylglycerols (containing more than 10 carbon atoms) are termed lipases (EC 3.1.1.3) and can be considered lipolytic and esterolytic enzymes [16]. Most lipases are water-soluble enzymes that hydrolyze ester bonds of water-insoluble substrates [17]. Therefore, lipases act at the interface between a substrate phase and an aqueous phase, in which the enzyme is dissolved [18]. It is often necessary to combine two or more lipases in order to release a glycerol molecule, since all three acyl chains of a triacylglycerol molecule are rarely released

by a single lipase [17].

The α-amylases (E.C.3.2.1.1) are enzymes that hydrolyze starch molecules to generate progressively smaller polymers composed of glucose units [19]. Today, a large number of microbial amylases have almost completely replaced the chemical hydrolysis of starch. The main advantage of using microorganisms for the production of amylase is the ability to bulk produce the enzyme and the easy manipulation of microbes to achieve enzymes with desired characteristics. Moreover, the stability of microbial amylases are higher than those of plant and animal [20].

Microbial proteases are used in waste treatment from various food-processing industries and household activities to solubilize proteinaceous waste and reduce the biological oxygen demand of aquatic systems [21,22]. Hydrolytic enzymes, such as lipases, amylases, and proteases, have a promising application in wastewater treatment of candy, ice cream, dairy, and meat industries. Enzymes for wastewater treatment do not require purification and thus should present a low production cost [23]. These characteristics have led to an increasing interest in enzyme production technology and the search for new microorganisms with a diverse ability to produce enzymes [24-27].

Microbial cellulases are widely used in the paper, wine, animal feed, and textile industries as well as for biofuels production, food processing, olive oil and carotenoid extraction, and waste management [28]. The wastes generated from agricultural fields and agroindustries contain a large amount of unutilized cellulose, thereby causing environmental pollution. Today, these wastes are utilized to produce valuable products, such as enzymes, sugars, biofuels, chemicals, and others [29-33].

Biosurfactants are amphiphilic molecules that possess both polar and nonpolar domains that have effective surface-active properties. There are two main types of these molecules: 1) those that reduce surface tension at the air-water interface (biosurfactants), and 2) those that reduce the interfacial tension between immiscible liquids or at the solid-liquid interface (bioemulsifiers) [34,35]. Biosurfactants usually exhibit an emulsifying capacity, but bioemulsifiers do not necessarily reduce the surface tension [34,35]. These molecules are microbial synthesized, and the different types of biosurfactants include lipopeptides synthesized by many species of Bacillus, glycolipids synthesized by Pseudomonas and Candida sp., phospholipids synthesized by Thiobacillus thiooxidans, and polysaccharidelipid complexes synthesized by Acinetobacter sp. [36-38].

The emulsification of lipids through the breakdown of lipid droplets favors the occurrence of hydrolysis, since the water-soluble lipolytic enzymes have greater surface contact with the substrate to be hydrolyzed.

Natural bio-surfactants exhibit low toxicity, biodegradability, and ecological acceptability, which provide an alternative to chemically-prepared conventional surfactants. They can be produced from various substrates but are often generated from renewable resources, such as vegetable oils as well as distillery and dairy wastes [39]. They are applicable for environmental protection and management, bioremediation of soil [40], crude oil recovery, cleanup of hydrocarbon contaminated groundwater, and enhanced oil recovery [41], antimicrobial agents in healthcare [36] or in a wide variety of industrial processes involving emulsification, foaming, detergency, wetting, dispersing, or solubilization [42].

Lipase-producing microorganisms are also found in fat and oil contaminated sources. Thus, they originate from dairy, household, and biotechnology industry wastewaters. The intense competition for limited carbon sources may result in the evolution of novel genes and/or novel biochemical pathways in the specialized environment of wastewaters [43]. Therefore, in this study we isolated lipase and other hydrolytic enzyme-producing microorganisms and assessed their fat and oil degradation capabilities.

2. MATERIALS AND METHODS

2.1. Selection of Lipase-Producing Microorganisms Using Tributyrin as a Growth Substrate

Our group currently has a microbial stock composed of more than 1100 lipolytic microorganisms that were obtained randomly. Among this library, 35 strains were selected for this study that had been originally isolated from four different sewage tanks. Samples were obtained from wastewaters at 2 farm houses (A and C), a dairy industry (B), and a biotechnological industry (D). Samples or its dilutions in sterile water were spread over Spirit Blue agar (Himedia, Mumbai/India) supplemented with 3% (v/v) tributyrin emulsion (20% v/v tributyrin; 0.2% v/v tween 80). After incubation at 25°C for 1 - 7 d, representative lipolytic colonies of each morphological type were isolated and purified in the same media. The strains were maintained in LB medium (10.0 g/l peptone, 5.0 g/l yeast extract, and 5.0 g/l sodium chloride) supplemented with 50% (v/v) fetal bovine serum and kept at −80°C.

2.2. Determination of the Degradative Enzymatic Profile of Organic Compounds between the Lipolytic Selected Microorganisms

Lipolytic strains were inoculated in 2 ml of LB broth in a 96-well plate. After 24 h at 25°C and 30 hz agitation, 2 μl of culture were inoculated in the follow media: 1) tributyrin-agarose [1% (w/v) agarose; 50 mM Tris-HCl pH 6,8; 1 mM CaCl2; 0.6% (v/v) tributyrin emulsion—to detect lipase]; 2) casein-agarose [1% (w/v) agarose, 1 mM CaCl2, 10% (v/v) of a casein solution in 1X PBS pH 7.4—to detect caseinase activity]; 3) corn starch-agarose [1% (w/v) agarose; 50 mM Tris-HCl pH 6.8, 1 mM CaCl2, 0.5% (w/v) corn starch—to detect amylase]; 4) carboxymethylcellulose-agarose [1% (w/v) agarose, 50 mM Tris-HCl pH 6.8, 1 mM CaCl2, 0.5% (w/v) carboxymethylcellulose—to detect cellulase modified from Akhtar et al. [44]; and 5) gelatin media [2 ml media/ assay tube: 50 mM Tris-HCl pH 6.8, 1 mM CaCl2, 10% (w/v) gelatin—to detect gelatin-specific proatease], the formulation of all those media were adapted from Vuong et al. [45]. After 24 h incubation at 25°C, enzymatic activities were detected by assessing for the presence of a clear halo in the media tests listed above except for gelatin media (see below). For media 1, the result was obtained by direct observation. Media 2 - 4 required revelation prior to a final analysis of the results as follows: 2) 2 min incubation with 1 N HCl solution to precipitate remaining casein; 3) 2 min incubation with 2% iodine solution to colorize remaining starch; 4) 30 min incubation with Congo red [0.25% (w/v) in 0.1 M Tris-HCl, pH 8,0) followed by 5 minutes in a destaining solution (0.5 M NaCl, 0.1 M Tris-HCl, pH 8.0) [46]. Microorganisms that produced a gelatin-specific protease were capable of liquefying the gelatin medium in media 5.

To evaluate the emulsifying capacity of the isolates, 200 μl of a culture with optical density (OD 600 nm) 0.5 were inoculated into 35 ml of LB broth supplemented with 2% soybean oil. The culture was then incubated for 35 d at 37°C and 250 rpm agitation. When the emulsification of the oil took place in the media, it acquired a milky appearance and consistency.

2.3. Characterization of Extracellular Lipase

2.3.1. Production of Extracellular Enzymatic Extract

Cells (500 μl) were grown in LB medium for 24 h under 30 hz agitation at 25°C and then seeded on the surface of a sterile dialysis membrane (12,000 Da) that had an equal diameter as a petri dish. The membrane was then placed on Spirit Blue agar containing 0.6% (v/v) tributyrin and incubated at 25°C for a period of 1 - 7 d (see results section) [47]. The membranes were washed with 2 ml of buffer (10 mM Tris-HCl pH 8.4 and 40 mM NaCl). Thereafter, all experimental steps were conducted on ice. The cell suspension was centrifuged at 25,000 g for 20 min at 4°C and filtered through a 0.22 μm filter. Then, 4 μl of enzyme extract was applied to the tributyrin-agarose (media one described previously) and incubated for 24 h at 25°C in order to verify lipase activity.

2.3.2. Determination of Optimal ph and Temperature of Action

Optimal pH was tested using 2 ml of freshly prepared 0.3% tributyrin broth [50 mM Tris-HCl—evaluated at pH values of 4.3, 6.8, 9.8, and 12.3—1 mM $CaCl_2$, and 1.5% (v/v) tributyrin emulsion] and 75 µl of enzymatic extract in a 1 cm cuvette. The optimal temperature was tested using 2 ml of the same broth at the optimal pH and 75 µl of enzymatic extract. Analyzed temperatures were 4°C, 25°C, 37°C, and 50°C. The reduction in OD was measured at 800 nm in a Shimadzu double beam spectrophotometer (UV-ISO-02, Kyoto, Japan) using a negative control as a standard at 0, 3, and 21 h of incubation.

2.3.3. Quantification of Lipase Activity in P-Nitrophenyl Esters

The lipase assay was performed by measuring the increase in the absorbance at 405 nm in a Thermo Plate microplate reader (TP-reader) caused by the release of p-nitrophenol after hydrolysis of p-nitrophenyl-butyrate (C4), decanoate (C10), and palmitate (C16) at 25°C for 15 min at pH 8.0 as previously described [48], but modified by adding 10 mM $CaCl_2$ to solution B. An enzyme-free control was used as the reference. One unit of lipase (U) was defined as the amount of enzyme that releases 1 µmol p-nitrophenol per min under the assay conditions.

2.3.4. Test of Lipase Activity in the Presence of Chemical Agents and Thermal Resistance

For further characterization, thermal and chemical resistance was evaluated using p-NPB as a substrate. Solution B described above was added with one of following chemical compounds: 0.25% (v/v) H_2O_2, 0.1% (v/v) NaClO, 0.1% (v/v) liquid detergent, or 10 mM EDTA. For thermal resistance, extracellular enzyme extracts were incubated for 30 min at 50°C and residual activity was also evaluated using p-NPB as a substrate.

2.4. Identification of Microorganisms

2.4.1. DNA Extraction

Genomic DNA was prepared from a loopful of cells grown in LB agar for 24 h. The cell pellet was resuspended in 250 µl of solution I (50 mM glucose, 25 mM Tris-HCl pH 8.0, and 10 mM EDTA). The cells were lysed by adding 25 µl of solution II [200 mM NaOH and 1% (w/v) SDS], and mixed for 5 min. Then, 500 µl of solution I and 2.5 µl of RNAse A (10 mg/ml) was added and incubated for 2 h at 37°C. This methodology was adapted from alkaline lysis first described by Birnboim & Doly [49]. DNA was then purified with phenol-chloroform using a standard laboratory protocol and after precipitation, DNA was resuspended in 30 µl of TE (10 mM Tris-HCl pH 8.0 and 1 mM EDTA).

2.4.2. Ribosomal RNA Gene Amplification

Bacterial isolates were identified by sequencing rDNA. The PCR reaction was performed as previously described [50] using the primers 8F (5'-AGAGTTTGATYMTGG-CTCAG-3') [51] and 907R (5'-CCGTCAATTCMTTT-RAGTTT-3') [52]. Fungal identification was performed as previously described [53] by sequencing D1D2 of 26S rDNA using primers NL1 (5'-GCATATCAATAAGCGG-AGGAAAAG) and NL4 (5'-GGTCCGTGTTTCAAGA-CGG). The sequences of PCR products were analyzed using standard protocols with a dideoxy nucleotide dye terminator (Big Dye vs. 3.1—Applied Biosystems, CA, USA) and Genetic Analyzer 3130 (Applied Biosystems, CA, USA). All 16S and 26S rRNA gene sequences were checked for quality, aligned, and analyzed with Codon-Code Aligner v.3.7.1 (CodonCode Corp., Centerville, MA, USA). All sequences were compared with reference sequences in the Ribosomal Database Project (RDP) using Sequence Match and sequences in GenBank using BLASTN.

2.5. Induction of Lipase Production and Synergistic Effect between Different Strains

Pre-inoculum of five different isolates was induced with tributyrin or left untreated (group control) to evaluate if microbial metabolism is increased or not by this triglyceride (pre-induction). Then, 100 µl of pre-inoculum from each isolate were inoculated in LB broth containing 10 µl alamar blue in four different treatments groups: 1) no supplementation, 2) supplemented with 2% tributyrin, 3) supplemented with 2% soybean oil, or 4) supplemented with 2% soybean oil emulsified with sterile bacterial extract contained lipase to evaluate possible synergism between different strains. Monitoring the percentage of alamar blue reduction indicated the conditions in which the culture demonstrated higher metabolism. A calculation of standard deviation indicated the differences between the assessed values and the average. Two-way ANOVA with Bonferroni correction post test was performed using GraphPad Prism version 5.03 for Windows [54].

3. RESULTS AND DISCUSSION

Our lipolytic microbial stock was selected by the presence of a halo around colonies when wastewaters were spread over Spirit Blue agar supplemented with 3% (v/v) tributyrin emulsion. Because of this triglyceride is formed by a glycerol and three four-carbon chains, esterases were selected preferably over lipases. However, because any lipase can also be classified as an esterase, some also showed lipolytic activity among the selected microorganisms.

The 16S/28S rDNA sequence analysis provides molecular identification of isolates. The microorganism collection site, identification, surfactant production [ability to emulsify 2% (v/v) soybean oil in culture medium], and enzymatic profile against agarose tributyrin, agarose casein, gelatin media, agarose corn starch, and agarose carboxymethylcellulose are shown in **Table 1**. Collection

Table 1. Collection site, identification, GenBank accession numbers and enzymatic profile of isolates.

Collection site	n°	Ribossomal Data Project (RDP) identification	Lipase	Gelatinase	Caseinase	Amilase	CM Cellulase	Time required for emulsifying 2% (v/v) soybean oil*
Grease trap/ São Sebastião das Águas Claras/MG	A1	*Terribacillus* sp. (GenBank accession number KC310807)	++	-	-	-	-	not
	A2	Unclassified Saccharomycetales (GenBank accession number KC310808)	++	-	++	-	-	partial
	A3	*Bacillus pumilus* (GenBank accession number KC310809)	++	-	++	-	++	5 days
	A4	*Bacillus pumilus* (GenBank accession number KC310810)	+	-	++	-	++	10 days
	A5	*Terribacillus* sp. (GenBank accession number KC310811)	++	-	-	-	-	7 days
	A6	*Bacillus pumilus* (GenBank accession number KC310812)	++	-	++	-	++	7 days
	A7	*Bacillus pumilus* (GenBank accession number KC310813)	+	-	++	-	++	7 days
	A8	*Bacillus* sp. (*B. megaterium/B. flexus*) (GenBank accession number KC310814)	+	++	-	-	-	not
	A9	*Staphylococcus* spp. (*S. sciuri/S. xylosus*) (GenBank accession number KC310815)	++	++	++	-	-	not
	A10	family Flavobacteriaceae - (*Empedobacter brevis* ou *Wautersiella flasenii*) (GenBank accession number KC310816)	+++	++	++	-	-	partial
Dairy waste/ Crucilândia/MG	B1	*Bacillus megaterium* (GenBank accession number KC310817)	+	++	-	-	-	partial
	B2	*Bacillus* spp. (*B. subtilis* subsp. *subtilis/B. amyloliquefaciens*) (GenBank accession number KC310818)	+	+++	++	++	++	not
	B3	*Lysinibacillus* spp. (*L. sphaericus/L. fusiformis*) (GenBank accession number KC310819)	++	-	-	-	-	7 days
	B4	*Lysinibacillus* spp. (*L. sphaericus/L. fusiformis*) (GenBank accession number KC310820)	++	-	-	-	-	10 days
	B5	*Lysinibacillus* spp. (*L. sphaericus/L. fusiformis*) (GenBank accession number KC310821)	++	-	-	-	++	7 days
	B6	*Bacillus cereus* (GenBank accession number KC310822)	++	-	-	-	++	not
	B7	*Bacillus subtilis* (GenBank accession number KC310823)	+	++	-	++	++	not
	B8	*Bacillus subtilis* (GenBank accession number KC310824)	+	++	++	++	++	not
	B9	*Bacillus* spp. (*B. simplex/B. macroides/B. frigotolerans*) (GenBank accession number KC310825)	+	++	-	-	-	partial
	B10	Mutualistic association *Bacillus sp.* (GenBank accession number KC310826) + *Staphylococcus epidermidis* (GenBank accession number KC310827)	+	-	-	-	-	7 days
	B11	*Bacillus cereus* (GenBank accession number KC310828)	++	-	++	-	++	not
	B12	*Acetobacter pasteurianus* (GenBank accession number KC310829)	+	-	-	-	-	3 days
	B13	*Acetobacter pasteurianus* (GenBank accession number KC310830)	+	-	-	-	-	7 days
Grease trap/ Crucilândia/ MG	C1	family Enterobacteriaceae (*Serratia* sp.; *Yersinia* sp.) (GenBank accession number KC310831)	++	+++	+++	-	-	14 days
	C2	*Pseudomonas* spp. (*P. corrugata/P. fluorescens/Marinobacter arcticus*) (GenBank accession number KC310832)	++	++	++	-	-	7 days
	C3	family Enterobacteriaceae (*Enterobacter aerogenes/Leclercia adecarboxylata/Pantoea agglomerans*) (GenBank accession number KC310833)	+	-	-	-	-	7 days
	C4	*Bacillus* spp. (*B. subtilis/B. amyloliquefaciens*) (GenBank accession number KC310834)	+	+++	++	++	++	7 days
	C5	*Bacillus subtilis* (GenBank accession number KC310835)	+	+++	++	++	+++	7 days
Biotechnology industry sewage/ Belo Horizonte/ MG	D1	*Bacillus megaterium* (GenBank accession number KC310836)	++	++	+++	++	-	not
	D2	*Pseudomonas* spp. (*P. rhodesiae/P. putida/P. fluorescens*) (GenBank accession number KC310837)	+++	+++	+++	-	-	14 days
	D3	*Pseudomonas* spp. (*P. taetrolens/P. putida/P. umsongensis/P. fluorescens*) (GenBank accession number KC310838)	+++	+++	+++	-	-	partial
	D4	family Enterobacteriaceae (*Serratia fonticola/Rahnella* sp.) (GenBank accession number KC310839)	+	-	-	-	-	10 days
	D5	*Pseudomonas* spp. (*P. rhodesiae/P. fluorescens/P. putida*) (GenBank accession number KC310840)	++	-	-	-	-	7 days
	D6	family Enterobacteriaceae (*Serratia fonticola/Rahnella* sp.) (GenBank accession number KC310841)	+	-	-	-	-	partial
	D7	*Pseudomonas* spp. (*P. rhodesiae*; *P. putida*) (GenBank accession number KC310842)	+++	+++	+++	-	-	partial

(-) no enzymatic activity; (+) low enzymatic activity (colony size/halo size < 2); (++) median enzymatic activity (colony size/halo size $2 \leq x < 5$); (+++) strong enzymatic activity (colony size/halo size ≥ 5); *maximum experimental time was 36 days.

sites A and B presented a predominance of members of the *Bacillaceae* family and at sites C and D, we observed *Enterobacteriaceae*, *Pseudomonas*, and *Bacillus* spp. At site D, *Pseudomonas* sp. was prevalent.

The extracellular bacterial lipases are of commercial importance, as thousands of lipase units can be produced from only several liters of culture medium [55]. Bacterial lipases are mostly extracellular and are greatly influenced by nutritional and physicochemical factors, such as temperature, pH, nitrogen and carbon sources, presence of lipids, inorganic salts, agitation, and dissolved oxygen concentration [55]. Lipases are mostly inducible enzymes and are thus generally produced in the presence of a lipid source, or any other inducer, such as triacylglycerols, fatty acids, hydrolysable esters, tweens, bile salts, and glycerol [55-57]. However, their production is significantly influenced by other carbon sources, such as sugars, sugar alcohol, polysaccharides, whey, casamino acids, and other complex sources [58,59]. Therefore, 10 ml of extracellular extract were produced for lipase characterization of each of the 35 isolates. Despite the use of an inductor at this step, the lipase activity of two isolates (A9—*Staphylococcus* spp.- and B9—*Bacillus* spp.) could not be recovered, even with tributyrin induction (**Table 2**). In a grease trap ecosystem, the coexistence of microbial strains could supply the nutritional needs due to partial degradation of other biopolymers naturally present in wastewater. Moreover, competition for nutrients and microbial interaction could also lead to the induction of lipase expression so that it may remains active during the first few subcultures. We found that each strain required a different incubation period to express extracellular lipase (**Table 2**). The exact incubation period required to express lipase over the dialysis membrane was monitored by parallel incubation in spirit blue agar supplemented with tributyrin. It needs to be pointed out that the optimum growth condition was not determined for each strain separately.

Table 2. Collection site, incubation period over dialysis membrane, optimal pH, optimal temperature, and quantification in lipase units with p-NPB (C4), p-NPD (C10), and p-NPP (C16).

Collection site	n°	Incubation period over dialysis membrane	pH*	Temperature* (°C)	In enzimatyc units**		
					p-NPB C4	p-NPD C10	p-NPP C16
grease trap/São Sebastião das Águas Claras/MG	A1	6 days	12.3	25	0.96	1.01	0.96
	A2	24 hours	12.3	5 to 25 (25)***	2.21	5.45	1.09
	A3	24 hours	6.8 to 9.8 (9.8)***	25 to 37 (25)	5.30	2.08	0.82
	A4	24 hours	6.8 to 12.3 (9.8)	25	0.13	0.12	0.02
	A5	4 days at 25°C + 1 day at 4°C	6.8 to 9.8 (9.8)	5 to 37 (5)	2.50	0.95	0.20
	A6	24 hours	6.8 to 12.3 (9.8)	25	2.40	1.01	0.44
	A7	24 hours	6.8 to 9.8 (9.8)	25	0.00	0.41	0.00
	A8	24 hours	9.8	5 and 37	15.34	0.92	0.08
	A9	activity could not be recovered					
	A10	6 days	9.8 to 12.3 (12.3)	50	9.97	9.18	0.49
Dairy waste/Crucilândia/MG	B1	24 hours	12.3	5 to 25	8.74	0.21	0.29
	B2	48 hours	9.8 to 12.3 (12.3)	5 to 25 (5)	5.17	0.23	0.19
	B3	48 hours	12.3	37	0.73	0.46	0.30
	B4	48 hours	12.3	37 to 50	1.21	1.19	0.39
	B5	24-48 hours	4.3 to 6.8 and 12.3 (12.3)	50	0.64	1.65	1.01
	B6	24-48 hours	12.3	37 to 50	0.44	0.92	0.68
	B7	6 days	9.8 to 12.3	5 to 25 (5)	3.07	0.19	0.33
	B8	6 days	12.3	37 to 50	1.09	0.08	0.05
	B9	activity could not be recovered					
	B10	24 hours	4.3	5 to 25 (25) and 50	0.24	0.04	0.05
	B11	24-48 hours	12.3	37 to 50	0.31	0.83	0.52
	B12	6 days	9.8 to 12.3 (12.3)	5 to 25 (5)	1.87	0.33	0.02
	B13	6 days	9.8 to 12.3 (9.8)	5 to 50 (25 to 37)***	6.15	1.74	0.78
grease trap/ Crucilândia/MG	C1	24 hours	6.8 to 12.3 (6.8 to 9.8)***	5 to 37	6.43	28.06	10.72
	C2	6 days	6.8 to 12.3 (9.8)	5 to 37 (25 to 37)	4.11	8.36	2.43
	C3	6 days at 25°C + 1 day at 4°C	9.8 to 12.3	5 to 37 (37)	0.48	0.18	0.16
	C4	48 hours	12.3	37 to 50 (50)	4.41	0.66	0.28
	C5	48 hours	12.3	37 to 50	5.39	1.11	0.49
Biotechnology industry sewage/ Belo Horizonte/MG	D1	24 hours	9.8 to 12.3	5 and 37	0.30	0.22	0.15
	D2	24 hours	4.3 to 6.8 (6.8)	5 to 37 (37)	3.77	16.68	11.29
	D3	24 hours	6.8 to 9.8 (9.8)	25 to 37 (37)	22.83	16.80	1.73
	D4	6 days at 25°C + 1 day at 4°C	4.3 and 9.8 to 12.3 (9.8 to 12.3)	5 to 37 (5)	4.93	6.50	1.21
	D5	24 hours	12.3	37 to 50	1.65	2.28	0.38
	D6	6 days	4.3 to 6.8 (6.8)	5 and 37	1.85	10.14	3.10
	D7	24 hours	4.3 to 9.8 (4.3 to 6.8)	5 to 37 (37)	9.61	28.60	16.59

*Range of values in which the extract retains more than 70% of its activity; **One unit of lipase (U) is defined as the amount of enzyme that releases 1 μmol p-nitrophenol per min, in the assay conditions; ***Numbers between parentheses indicate the value in which the enzyme showed higher activity—it sometimes indicates a range of values.

After confirmation of lipase activity, the optimal pH and temperature of the extracts were defined (**Table 2**). Among the extracts, mesophilic lipases with an optimal pH in the basic range were prevalent, but psychrophiles and acidophilus were also observed. In general, bacterial lipases have optimal activity at neutral or alkaline pH [60-63]. Lipases from *Bacillus* species are active over a broad pH range (pH 3-12) [64], and our findings indicated that lipase extracts produced by *Bacillus* species often presented more than one optimal pH value. However, in contrast to the findings of the previous study, lipases secreted by ours isolates belonging to the *Bacillus* genus showed less activity at high temperatures, with optimal activity at mesophilic temperatures. Only lipases secreted by *B. cereus* were more thermotolerant, reaching optimal activity at 50°C. Some of our lipase extracts showed thermal stability up to 50°C and retained more than 70% of activity after thermal treatment for 30 min, including A5 (*Terribacillus* sp.), A10 (family Flavobacteriaceae), B3, B4 and B5 (*Lysinibacillus* spp.), B7, B8, and C5 (*Bacillus subtilis*), B10 (Mutualistic association of *Bacillus sp.* and *Staphylococcus epidermidis*), B11 (*Bacillus cereus*), C2, D2, D3, D5, and D7 (*Pseudomonas* spp.), and D1 (*Bacillus megaterium*) (**Table 3**). The thermal resistance of lipases from *Bacillus* and *Pseudomonas* has already been described [65-68].

Table 3. Percentage of lipase residual activity after various treatments.

Collection site	n°	[1]0.25% (v/v) H_2O_2	[2]0.1% (v/v) NaClO	[3]0.1% (v/v) Liquid detergent	[4]10 mM EDTA	[5]Thermal resistance for 30 min at 50°C
grease trap/São Sebastião das Águas Claras/MG	A1	101%	56%	0%	5%	36.9%
	A2	93%	95%	110%	59%	65.0%
	A3	117%	8%	40%	88%	21.5%
	A4	96%	0%	17%	78%	42.7%
	A5	57%	103%	78%	12%	77.4%
	A6	97%	1%	80%	80%	18.3%
	A7	82%	17%	28%	87%	45.8%
	A8	136%	116%	156%	110%	5.4%
	A9	Activity could not be recovered				
	A10	86%	0%	0%	34%	96.9%
Dairy waste/Crucilândia/MG	B1	102%	111%	101%	66%	7.3%
	B2	22%	0%	101%	56%	54.3%
	B3	14%	11%	11%	0%	81.8%
	B4	67%	10%	3%	384%	124.1%
	B5	63%	11%	43%	0%	73.6%
	B6	64%	109%	88%	28%	56.5%
	B7	60%	0%	118%	47%	72.9%
	B8	23%	36%	0%	22%	76.3%
	B9	Activity could not be recovered				
	B10	42%	0%	167%	0%	98.6%
	B11	0%	103%	28%	18%	94.6%
	B12	25%	18%	74%	29%	44.3%
	B13	41%	28%	7%	6%	26.1%
grease trap/Crucilândia/MG	C1	102%	110%	126%	0%	68.1%
	C2	103%	102%	125%	35%	100.2%
	C3	131%	52%	89%	0%	61.8%
	C4	0%	0%	0%	0%	51.5%
	C5	41%	0%	20%	28%	75.3%
Biotechnology industry sewage/ Belo Horizonte/MG	D1	53%	0%	122%	41%	77.4%
	D2	95%	45%	70%	0%	71.0%
	D3	44%	9%	128%	0%	94.5%
	D4	111%	0%	1008%	2%	49.2%
	D5	47%	18%	192%	10%	85.1%
	D6	72%	0%	11%	81%	20.7%
	D7	70%	0%	54%	0%	75.7%

Results are shown for lipase activity after incubation in the presence of [1]0.25% (v/v) H_2O_2, [2]0.1% (v/v) NaClO, [3]0.1% (v/v) liquid detergent, [4]10 mM EDTA, or [5]after thermal treatment for 30 min at 50°C. Hydrolysis of p-NPB (C4) was used for all measurements. Controls with no treatment were used for comparison.

Bacterial lipases generally have optimal activity in the temperature range of 30˚C - 60˚C; however some reports have shown that bacterial lipases exist with optimal activity at both low and high temperature ranges [60,61,63, 69]. Lipases quantification using pNP-butyrate (C4), decanoate (C10), and palmitate (C16) was done in a fixed condition for all enzymes, at 25˚C and pH 8.0 (**Table 2**). The majority of lipase extracts were active on short chain esters, which was expected since the selection was made with tributyrin. In addition, these findings are consistent with the fact that all lipases are also esterases and harbor esterolytic activity. However, isolates A2, A10, B4, B5, B6, C1, C2, D2, D3, D4, D5, D6, and D7 showed a greater affinity for long chain esters, therefore characterizing a "true" lipase extract, since a "true" lipase hydrolyses esters with more than 10 carbon atoms (**Table 2**).

There are three categories of microbial lipases: nonspecific, regiospecific, and fatty acid-specific [55]. Non-specific lipases act randomly on triacylglyceride molecules, which results in the complete breakdown tofatty acid and glycerol. Regiospecific lipases are 1, 3-specific lipases that hydrolyze only primary ester bonds, which is observed in lipases produced by some *Bacillus* species. The third group, fatty acid-specific lipases, comprise those with a pronounced fatty acid preference [55]. Despite showing activity on tributyrin-agarose, the isolated A7 (*Bacillus pumillus*) was inactive when evaluated for hydrolysis of pNP-ester with 4 and 16 carbons and exhibited poor activity for pNP-ester with 10 carbon atoms. This can be explained by the fact that some lipases have affinity for triacylglycerols, exhibiting no or little activity against mono- and diglycerides [17,70]. Another possibility is that A7 isolate was originally capable of producing more than one type of esterase. However, over the course of the passages *in vitro*, the strain began to only express the esterase that hydrolyzes the ester of 10 carbon atoms. This behavior has been observed in some species *in vitro*. In addition, it is possible that our system was not sensitive enough to detect the reduced expression after several passages in culture media.

Some biochemical similarities can be observed between microorganisms with the same identification. *Lysinibacillus* spp. isolates showed many similarities, including the expression of active lipases within 48 h of incubation, having an optimal pH that was alkaline and an optimal temperature of 37˚C of higher, and the absence of gelatinase, caseinase, or amylase activity. Twenty strains belonged to the *Bacillaceae* family, of which 15 belonged to the *Bacillus* genus. All strains identified as *B. megaterium* were gelatin-specific protease producers and were unable to produce cellulase. They expressed extracellular lipase after 24 h of incubation over a dialysis membrane and their lipases had optimal activity at a pH in the alkaline range. In addition, these strains are mesophilic/psicrophilic and are more active against esters with only 4 carbon atoms. Among the five strains identified as *B. subtilis*, all were amylase and cellulase producers. Five were also capable of producing gelatin-specific proteases, but only 4 strains were able to produce casein-specific proteases as well. All five *B. subtilis* lipases had better activity at an alkaline pH and against 4 carbon esters. Among the five strains identified as *Pseudomonas* spp. all were amylase and cellulase producers, and 4 were still capable of producing gelatin- and casein-specific proteases All *Pseudomonas* lipases presented considerable activity at 37˚C and against 10 carbon atom esters, indicating the presence of at least one type of lipase (**Tables 1** and **2**).

For preliminary assessment of the potential use of lipases in sewage treatment, the thermal and chemical resistance of the enzymes was evaluated using p-NPB as a substrate (**Table 3**). An abundance of cleaning products and other chemical compounds are released daily from grease traps. None of the extracts maintained more than 70% residual activity in all analyzed conditions. Isolate A2 (unclassified Saccharomycetales) exhibited the best results for all conditions combined and maintained more than 50% of residual activity in the presence of H_2O_2, NaClO, liquid detergent, EDTA, and thermal treatment. Other extracts that showed good results included B6 (from *Bacillus cereus*) and B12 (from *Acetobacter pasteurianus*). They showed residual activity in greater than 20% of all of the conditions. In addition to the A2 extract, a combination of extracts could also be potentially advantageous for the development of a biological method for cleaning O&G from grease traps and sewage treatment plant equipment. The strong inhibition caused by 0.1% (v/v) detergent could be due to inactivation of the enzyme as a result of a disruption of its tertiary structure. When dishes, clothes or floor are washed, high concentrations of detergents and soaps are released into the sewer over a short period of time. Therefore, the choice of a stable enzyme is an important aspect for sewage treatment. Therefore, we analyzed the ability of isolates to produce multiple degradative enzymes and found that isolates B2, B8, C4, C5, D1, D2, and D7 presented a wide range of ability in utilizing biopolymers commonly present in wastewater (**Table 3**). These isolates all produced proteases and at least one other hydrolytic enzyme besides lipase. Among the 35 strains, more than 70% were able to produce at least two enzymes or more. This multiple approach allows for the use of a small number of isolates in sewage treatment, since each one individually has the ability to secrete more than one enzyme at a time. This can also reduce the requirement for nutritional supplementation of the system. Among the extracellular lipases produced from the 35 strains, approximately 45%

were resistant to 0.25% (v/v) H_2O_2, approximately 20% were resistant to 0.1% (v/v) of NaClO or 10 mM EDTA, and more than 50% were resistant to 0.1% (v/v) liquid detergent (**Table 3**).

Bioremediation techniques *in situ* include the introduction of different strains of live microorganisms to wastewater at various stages of its treatment. Almost all known methods for sludge treatment introduce microbial strains in the log phase of growth. These microbes are in active phase of multiplication, however their action requires time to degrade the substrats [71]. Estera *et al.* [72] previously developed a method for reducing the time for degradation, which includes first providing an enzyme mixture capable of digesting natural polymeric materials, and only the adding at least one species of fermenting bacteria to the system that is able to ferment the resulting suspension. Dash *et al.* [71] described a composition for the treatment of wastewater to remove pollutants that was comprised of a synergistic composition of microbes, enzymes, and cofactors/nutrients. The microbes in the composition were selected *Pseudomonas aeruginosa, Pseudomonas fluorescens, Pseudomonas putida, Pseudomonas desmolyticum, Coriolus versicolour, Lactobacillus* sp., *Bacillus subtilis, Bacillus cereus, Staphylococcus* sp., and *Phanerochaete chrysosporium*, alone or in combination. The enzymes produced include proteases, lipases, amylases, glucose oxidases, and others. This composition exhibits synergy and effectively removes pollutants from the wastewater. Enzymes act by dissociating the molecules to simpler forms, and microbes utilize these intermediates in their metabolism, which results in the complete degradation of the pollutants in the wastewater. Microbes will grow faster due to the increasing availability of intermediates and therefore will produce more enzymes that can further degrade the pollutants. Thus, enzymes and microbes are interdependent and work together to facilitate faster degradation of the pollutant molecules [71].

Diverse microorganisms are able to hydrolyze different types of oil. However, the biodegradation process can be lengthy due to the low water solubility of oil [73]. In natural or induced conditions, many microorganisms are able to produce emulsifying agents, which minimizes the time required for biodegradation of O&G by enhancing hydrophobic substrate bioavailability [74]. Cell-bound esterase synthesis has been recently reported in association with the generation of surface active substances, indicating the coupled function of emulsification with lipolytic activity [75,76]. Biosurfactants increase the uptake of microorganisms when grown on insoluble substrates and also increase the efficiency of bioremediation [77].

As shown by Gautam *et al.* [78], Saharam *et al.* [79], and Pattanathu *et al.* [80], several species belonging to the genera *Pseudomonas* are capable of producing different classes of biosurfactants. In our study, we found that various isolates belonging to those genera and Bacilaceae class (A3, A4, A5, A6, A7, B3, B4, B5, B10, C2, C4, C5, D2, and D5) were able to emulsify 2% soybean oil, although we did not identify the class of biosurfactant produced (**Table 1**).

Acetobacter pasteurianus is an acetogenic bacterial species normally associated with wine production and spoilage [81]. It produces acetic acid due to the incomplete oxidation of a carbon source into CO_2 [81]. In our study, isolates B12 and B13 were associated with wastewater and lipase production. These enzymes preferentially hydrolyzed triglycerides with esters of 4 carbon atoms and were active in the basic pH range. Despite these similarities, they showed different chemical and thermal resistance, indicating that they are most likely different enzymes. In addition, these two strains were not able to produce any other type of hydrolytic enzyme or biosurfactant among those evaluated.

According to the literature, emulsification of lipids would favor its hydrolysis, since the water-soluble lipolytic enzymes have greater surface contact with the substrate to be hydrolyzed due to the breakdown of lipid droplets. To evaluate the behavior of some of our isolates, five strains with different emulsification and hydrolysis profiles were subjected to metabolic quantification by reduction of alamar blue, in four different conditions: 1) LB broth culture media, 2) LB broth with 2% tributyrin (triglyceride of 4 carbons), 3) LB with 2% soybean oil, and (iv) LB with 2% lipid emulsion (**Figure 1**). The initial curve of alamar blue reduction indicated the best volume of pre-inoculum and the optimal period of incubation in the presence of the reagent for each strain. The graphs in **Figure 1** show that the presence of an emulsified lipid in the culture medium did not increased the metabolic rates of any of the microorganisms, and rather the metabolic rate was reduced. However, previous induction of lipase production by the addition of tributyrin to pre-inoculum media proved effective in raising the metabolism in all experimental conditions for the A3 and B1 strains. The only situation in which the induction was not efficient was for strain B13, wherein the pre-inoculum that was not induced was more metabolically efficient in all experimental conditions. However, because lipase production may be influenced by the carbon source used in the induction process, the absence of lipase production for strain B13 in the presence of tributyrin is justifiable, since no other inducer was evaluated. The metabolism of strains B1 and B12 was markedly increased in presence of tributyrin compared to the presence of soybean oil, and these findings were in agreement with the hydrolysis of p-nitrophenol esters. Strains B12 and B13 were also more efficient at hydrolyzing

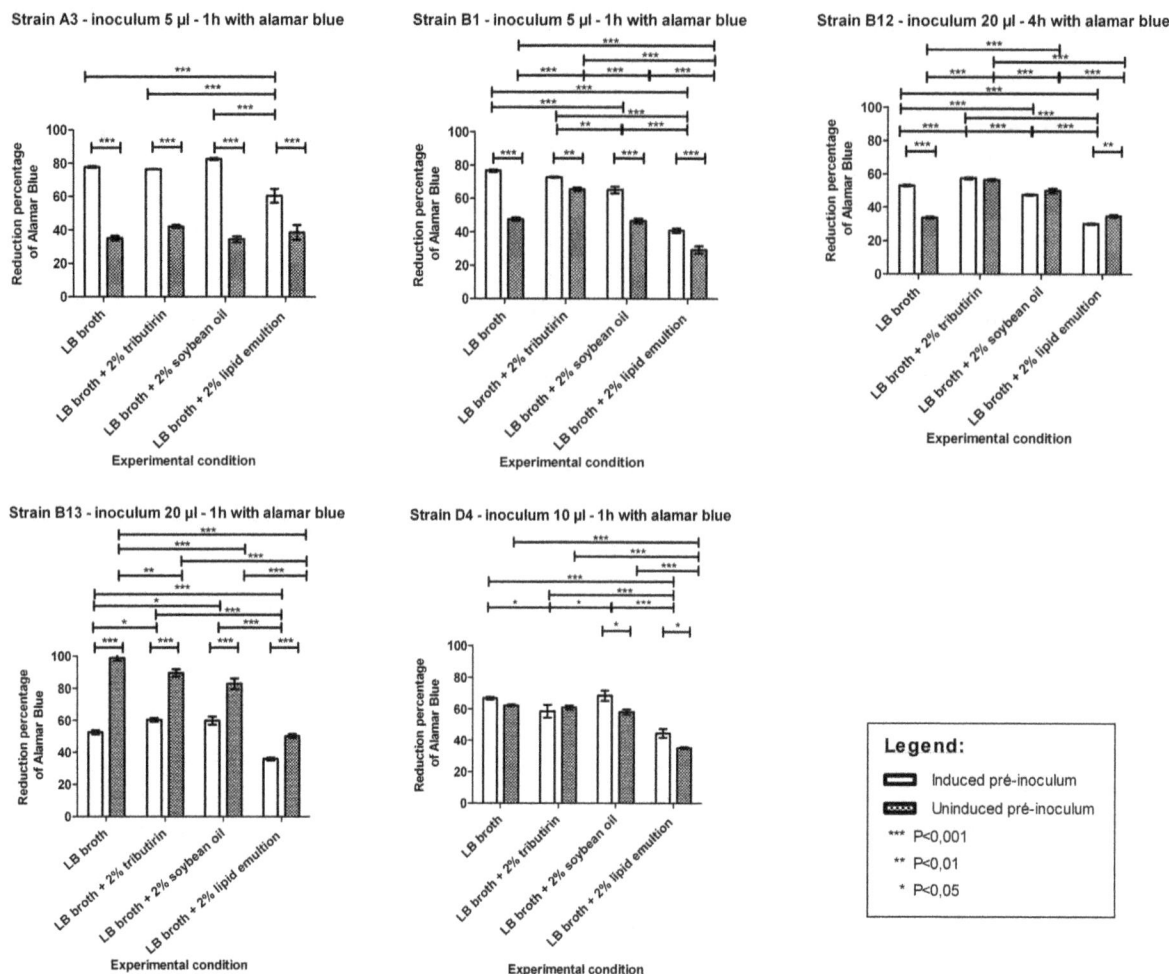

Figure 1. Metabolic status of five different strains when induced or non-induced pre-inocula were challenged against a simple trigliceride, a complex mixture of triglicerides, or an oily emulsion of a complex mixture of triglicerides. Quantification of metabolism was performed by determining the percent reduction of 10% (v/v) alamar blue reagent.

p-nitrophenol butyrate and more metabolically active in LB media containing 2% tributyrin. Similar results were observed for strain D4, which was more efficient in hydrolyzing p-nitrophenol decanoate and more metabolically active in LB media containing 2% soybean oil (**Figure 1**).

Cellulose is the most common organic polymer. It is the most prevalent material in waste from agriculture and the most abundant renewable biopolymer on Earth [82]. A promising strategy for utilization of this energetic renewable source is microorganism-mediated hydrolysis of discarded lignocellulose, followed by fermentation of the resulting compound, which produces the desired metabolites or biofuel [82]. Among our selected lipolytic microorganisms, 34.3% presented also hydrolytic activity against CM cellulose. Parmar et al. [83] showed that a mixture of hydrolytic enzymes, such as cellulases, proteases, and lipases, in equal proportion by weight, reduced total suspended solids (TSS) by 30% - 50% and improved sedimentation of solids in sludge. An increase in

solid reduction was observed with increasing enzyme concentration. That reduction occurred due to the hydrolysis of residual polymers, proving that enzymatic synergism can effectively reduce the organic matter in industrial wastewater pretreatment plants.

4. CONCLUSION

Cultures isolated from the vast diversity of microorganisms provide a major source of biological material for industrial biocatalysts and other environmental applications. Lipases and esterases obtained in this study presented different resistances and affinities. Enzymes were characterized with the aim of identifying suitable candidates for use in wastewater treatment. The good stability of isolated lipases in the presence of chemical agents, thermal stability, wide range of pH activity and tolerance, and affinity for different lengths of ester chains indicates that some of these enzymes may be good candidates for the hydrolysis of organic compounds and polymers present in the wastewater of diverse industries. As bacterial enzymes are highly robust, being active over a wide range of pH and temperature and possessing a diverse range of substrate specificity, they could easily be used in pretreatment sludge processes, since they possess adequate resistance of some chemical elements and can be produced at a low cost. The absence of purification requirements contributes to the cost reduction of using these enzymes for sewage treatment. In addition, it is possible that a combination of two or more enzymes may facilitate the process of complete hydrolysis of triglycerides, proteins, and lignocellulose that normally occurs in the wastes of industrial processes. However, careful selection of the strains to be used in sewage treatment is essential, because it may be possible to use fewer strains to achieve the same purpose, since several strains showed the capability of producing two or more enzymes. This will ensure that the hydrolysis of all compounds commonly discarded in wastewater will be sufficiently achieved. In order to give a more real reflection of degradation capabilities of those microorganisms, further studies will use a simulation condition of common sewage as culture to test the degradation capabilities, and also focus on optimizing hydrolysis conditions with the aim of using combined enzyme/microbial strategies for improving industrial wastewater treatment processes.

5. ACKNOWLEDGEMENTS

We thank CNPQ for financial support, projects number: 580311/2008-2, 560912/2010-2, 551113/2011-1, 300721/2012-9.

REFERENCES

[1] Becker, P., Koster, D., Popov, M.N., Markossian, S., Antranikian, G. and Markl, H. (1999) The biodegradation of olive oil and treatment of lipid-rich wool wastewater under aerobic thermophilic condition. *Water Research*, **33**, 653-660.

[2] Stoll, U. and Gupta, H. (1997) Management strategies for oil and grease residues. *Waste Management & Research*, **15**, 23-32.

[3] Cadoret, A., Conrad, A. and Block, J.C. (2002) Availability of low and high molecular weight substrates to extracellular enzymes in whole and dispersed activated sludges. *Enzyme and Microbial Technology*, **31**, 179-186.

[4] Sheng, G.P. and Yu, H.Q. (2006) Characterization of extracellular polymeric substances of aerobic and anaerobic sludge using three-dimensional excitation and emission matrix fluorescence spectroscopy. *Water Research*, **40**, 1233-1239.

[5] Wagner, M., Loy, A., Nogueira, R., Purkhold, U., Lee, N. and Daims, H. (2002) Microbial community composition and function in wastewater treatment plants. *Antonie van Leeuwenhoek*, **81**, 665-680.

[6] Perle, M., Kimchie, S. and Shelef, G. (1995) Some biochemical aspects of the anaerobic degradation of dairy wastewater. *Water Research*, **29**, 1549-1554.

[7] Lefebvre, X. Paul, E. and Mauret, M. (1998) Kinetic characterization of saponified domestic lipid residues aerobic biodegradation. *Water Research*, **32**, 3031-3038.

[8] Vidal, G., Carvalho, A., Méndez, R. and Lema, J.M. (2000) Influence of the content in fats and proteins on the anaerobic biodegradability of dairy wastewaters. *Bioresource Technology*, **74**, 231-239.

[9] Masse, L., Kennedy, K.J. and Chou, S. (2001) Testing of alkaline and enzymatic hydrolysis pretreatments for fat particles in slaughterhouse wastewater. *Bioresource Technology*, **77**, 145-155.

[10] CONAMA 357—Conselho Nacional do Meio Ambiente. (2005) Legislação Ambiental Federal, resolution 357.

[11] Cheryan, M. and Rajagopalan, N. (1998) Membrane processing of oil streams. Wastewater treatment and waste reduction. *Journal of Membrane Science*, **151**, 13-28.

[12] Bhumibhamon, O., Koprasertsak, A. and Funthong, S. (2002) Biotreatment of high fat and oil wastewater by lipase producing microorganisms. *Kasetsart Journal Natural Science*, **36**, 261-267.

[13] Chigusa, S., Hasegawa, T., Yamamoto, N. and Watanabe Y. (1996) Treatment of wastewater form oil manufacturing plant by yeast. *Water Science and Technology*, **34**, 51-58.

[14] Alberton, D., Mitchell, D.A., Cordova, J., Peralta-Zamora, P. and Krieger, N. (2010) Production and application of *R. microsporus* lipases. *Food Technology and Biotechnology*, **48**, 28-35.

[15] Environmental Oasis Ltd. (2012) WW07P—Grease removal and food processing.

[16] Wong, H. and Schotz, M.C. (2002) The lipase gene family. *Journal of Lipid Research*, **43**, 993-999.

[17] Gilham, D. and Lehner, R. (2005) Techniques to measure lipase and esterase activity *in vitro. Methods*, **36**, 139-147.

[18] Bussamara, R., Fuentefria, A.M., Oliveira, E.S., Broetto, L., Simcikova, M., Valente, P., Schrank, A. and Vainstein, M.H. (2010) Isolation of a lipase-secreting yeast for enzyme production in a pilot-plant scale batch fermentation. *Bioresource Technology*, **101**, 268-275.

[19] Windish, W.W. and Mhatre, N.S. (1965) Microbial amylases. In: Wu, W., Ed., *Advances in Applied Microbiology*, Elsevier Inc., Houston, 273-304.

[20] Tanyildizi, M.S., Ozer, D. and Elibol, M. (2005) Optimi-

zation of -amylase production by *Bacillus* sp. using response surface methodology. *Process Biochemistry*, **40**, 2291-2296.

[21] Gupta, R., Beg, Q.K. and Lorenz, P. (2002) Bacterial alkaline proteases: Molecular approaches and industrial applications. *Applied Microbiology and Biotechnology*, **59**, 15-32.

[22] Ichida, J.M., Krizova, L., LeFevre, C.A., Keener, H.M., Elwell, D.L. and Burtt Jr., E.H. (2001) Bacterial inoculum enhances keratin degradation and biofilm formation in poultry compost. *Journal of Microbiological Methods*, **47**, 199-208.

[23] Rigo, E., Rigoni, R.E., Lodea, P., Oliveira, D., Freire, D.M.G. and Luccio, M. (2008) Application of different lipases as pretreatment in anaerobic treatment of wastewater. *Environmental Engineering Science*, **25**, 1243-1248.

[24] Hu, W.C., Thayanithy, K. and Forster, C.F. (2002) A kinetic study of the anaerobic digestion of ice-cream wastewater. *Process Biochemistry*, **37**, 965-971.

[25] Mongkolthanaruk, W. and Dharmsthiti, S. (2002) Biodegradation of lipid-rich wastewater by a mixed bacterial consortium. *International Biodeterioration & Biodegradation*, **50**, 101-105.

[26] Cavalcanti, E.A.C., Gutarra, M.L.E., Freire, D.M.G., Castilho, L.R. and Sant'Anna Jr., G.L. (2005) Lipase production by solid-state fermentation in fixed-bed bioreactors. *Brazilian Archives of Biology and Technology*, **48**, 79-84.

[27] Leal, M.C.C.R., Freire, D.M.G., Cammarota, M.C. and Sant'Anna Jr., G.L. (2006) Effect of enzymatic hydrolysis on anaerobic treatment of dairy wastewater. *Process Biochemistry*, **41**, 1173-1178.

[28] Kuhad, R.C., Rishi, G. and Singh, A. (2011) Microbial cellulases and their industrial applications. *Enzyme Research*, **2011**, Article ID: 280696.

[29] Kuhad, R.C., Gupta, R. and Khasa, Y.P. (2010) Bioethanol production from lignocellulosic biomass: An overview. In: Lal B, Ed., *Wealth from Waste*, Teri Press, New Delhi, 53-106.

[30] Karmakar, M. and Ray, R.R. (2011) Current trends in research and application of microbial cellulases. *Research Journal of Microbiology*, **6**, 41-53.

[31] Gupta, R., Mehta, G., Khasa, Y.P. and Kuhad, R.C. (2011) Fungal delignification of lignocellulosic biomass improves the saccharification of cellulosics. *Biodegradation*, **22**, 797-804.

[32] Gupta, R., Khasa, Y.P. and Kuhad, R.C. (2011) Evaluation of pretreatment methods in improving the enzymatic saccharification of cellulosic materials. *Carbohydrate Polymers*, **84**, 1103-1109.

[33] Gupta, R., Sharma, K.K. and Kuhad, R.C. (2009) Separate hydrolysis and fermentation (SHF) of *Prosopis juli-*

flora, a woody substrate, for the production of cellulosic ethanol by *Saccharomyces cerevisiae* and *Pichia stipitis-*NCIM 3498. *Bioresource Technology*, **100**, 1214-1220.

[34] Batista, R.M., Rufino, R.D., Luna, J.M., Souza, J.E.G. and Sarubbo, L.A. (2010) Effect of medium components on the production of a biosurfactant from *Candida tropicalis*, applied to the removal of hydrophobic contaminants in soil. *Water Environment Research*, **82**, 418-425.

[35] Luna, J.M., Rufino, R.D., Campos-Takaki, G.M. and Sarubbo, L.A. (2012) Properties of the biosurfactant produced by *Candida sphaerica* cultivated in low-cost substrates. *Chemical Engineering Transactions*, **27**, 67-72.

[36] Moussa, T.A.A., Ahmed, A.M. and Abdelhamid, S.M.S. (2006) Optimization of cultural conditions for biosurfactant production from *Nocardia amarae*. *Journal of Applied Sciences Research*, **11**, 844-850.

[37] Nitschke, M. and Pastore, G.M. (2006) Production and properties of a surfactant obtained from *Bacillus subtilis* grown on cassava wastewater. *Bioresource Technology*, **97**, 336-341.

[38] Anyanwu, C.U. (2010) Surface activity of extracellular products of a *Pseudomonas aeruginosa* isolated from petroleum contaminated soil. *International Journal of Environmental Sciences*, **1**, 225-235.

[39] Abouseoud, M., Maachi, R. and Amrane, A. (2007) Biosurfactant Production from olive oil by *Pseudomonas fluorescens*. In: Méndez-Vilas, A., Ed., *Communicating Current Research and Educational Topics and Trends in Applied Microbiology*, Formatex, Badajoz, 340-347.

[40] Van Dyke, M.I., Lee, H. and Trevors, J.T. (1991) Application of microbial surfactants. *Biotechnology Advances*, **9**, 241-252.

[41] Ron, E.Z. and Rosenberg, E. (2001) A review of natural roles of biosurfactants. *Environmental Microbiology*, **3**, 229-236.

[42] Desai, J.D. and Banat, I.M. (1997) Microbial production of biosurfactants and their commercial potential. *Microbiology and Molecular Biology Reviews*, **61**, 47-64.

[43] Bramucci, M., Kane, H., Chen, M. and Nagarajan, V. (2003) Bacterial diversity in an industrial wastewater bioreactor. *Applied Microbiology and Biotechnology*, **62**, 594-600.

[44] Akhtar, N., Ghauri, M.A., Iqbal, A., Anwar, M.A. and Akhtar, K. (2008) Biodiversity and phylogenetic analysis of culturable bacteria indigenous to Khewra Salt Mine of Pakistan and their industrial importance. *Brazilian Journal of Microbiology*, **39**, 143-150.

[45] Vuong, C., Gotz, F. and Otto, M. (2000) Construction and characterization of an *agr* deletion mutant of *Staphylococcus epidermidis*. *Infection and Immunity*, **68**, 1048-1053.

[46] Ruegger, M.J.S. and Tauk-Tornisielo, S.M. (2004) Atividade da celulase de fungos isolados do solo da Estação Ecológica de Juréia-Itatins, São Paulo, Brasil. *Brazilian*

Journal of Botany, **27**, 205-211.

[47] Christen, G.L. and Marshall, R.T. (1984) Selected properties of lipase and protease of *Pseudomonas fluorescens* 27 produced in four media. *Journal of Dairy Science*, **67**, 1680-1687.

[48] Fakhreddine, L., Kademi, A., Ait-Abdelkader, N. and Baratti, J.C. (1998) Microbial growth and lipolytic activities of moderate thermophilic bacterial strains. *Biotechnology Letters*, **20**, 879-883.

[49] Birnboim, H.C. and Doly, J. (1979) A rapid alkaline extraction procedure for screening recombinant plasmid DNA. *Nucleic Acids Research*, **7**, 1513-1523.

[50] Pontes, D.S., Pinheiro, F.A., Lima-Bittencourt, C.I., Guedes, R.L,, Cursino, L., Barbosa, F., Santos, F.R., Chartone-Souza, E. and Nascimento, A.M. (2009) Multiple antimicrobial resistance of gram-negative bacteria from natural oligotrophic lakes under distinct anthropogenic influence in a tropical region. *Microbial Ecology*, **58**, 762-772.

[51] Felske, A., Rheims, H., Wolterink, A., Stackebrandt, E. and Akkermans, A.D.L. (1997) Ribosome analysis reveals prominent activity of an uncultured member of the class Actinobacteria in grassland soils. *Microbiology*, **143**, 2983-2989.

[52] Lane, D.J. (1991) 16S/23S rDNA sequencing. In: Stackebrandt, E. and Goodfellow, M., Eds., *Nucleic Acid Techniques in Bacterial Systematics*, John Wiley &Sons, New York, 115-148.

[53] Lachance, M.A., Bowles, J.M., Starmer, W.T. and Barker, J.S.F. (1999) Kodamaea kakaduensis and Candida tolerans, two new ascomycetous yeast species from Australian Hibiscus flowers. *Canadian journal of microbiology*, **45**, 172-177.

[54] GraphPad Software (2009) Prism 5 for Windows: Version 5.03.

[55] Gupta, R., Gupta, N. and Rathi, P. (2004) Bacterial lipases: An overview of production, purification and biochemical properties. *Applied Microbiology and Biotechnology*, **64**, 763-781.

[56] Sharma, R., Chisti, Y. and Banerjee, U.C. (2001) Production, purification, characterization and applications of lipases. *Biotechnology Advances*, **19**, 627-662.

[57] Rathi, P., Saxena, R.K. and Gupta, R. (2001) A novel alkaline lipase from *Burkholderia cepacia* for detergent formulation. *Process Biochemistry*, **37**, 187-192.

[58] Ghanem, E.H., Al-Sayeed, H.A. and Salch, K.M. (2000) An alkalophilic thermostable lipase produced by a new isolate of *Bacillus alcalophilus*. *World Journal of Microbiology and Biotechnology*, **16**, 459-464.

[59] Rashid, N., Shimada, Y., Ezaki, S., Atomi, H. and Imanaka, T. (2001) Low-temperature lipase from psychrotrophic *Pseudomonas* sp. strain KB700A. *Applied and Environmental Microbiology*, **67**, 4064-4069.

[60] Dharmsthiti, S. and Luchai, S. (1999) Production, purification and characterization of thermophilic lipase from *Bacillus* sp. THL027. *FEMS Microbiology Letters*, **179**, 241-246.

[61] Lee, O.-W., Koh, Y.-S., Kim, K.-J., Kim, B.-C., Choi, H.-J., Kim, D.-S., Suhartono, M.T. and Pyun, Y.-R. (1999) Isolationa and characterization of a thermophilic lipase from *Bacillus thermoleovorans* ID-1. *FEMS Microbiology Letters*, **179**, 393-400.

[62] Kanwar, L. and Goswami, P. (2002) Isolation of a *Pseudomonas* lipase produced in pure hydrocarbon substrate and its applications in the synthesis of isoamyl acetate using membrane-immobilized lipase. *Enzyme and Microbial Technology*, **31**, 727-735.

[63] Sunna, A., Hunter, L., Hutton, C.A. and Bergquist, P.L. (2002) Biochemical characterization of a recombinant thermoalkalophilic lipase and assessment of its substrate enantioselectivity. *Enzyme and Microbial Technology*, **31**, 472-476.

[64] Bradoo, S., Saxena, R.K. and Gupta, R. (1999) Two acidothermotolerant lipases from new variants of *Bacillus* spp. *World Journal of Microbiology and Biotechnology*, **15**, 87-91.

[65] Hassan F, Shah AA, and Abul-Hameed A. (2006) Influence of culture conditions on lipase production by *Bacillus* sp. FH5. *Annals of Microbiology*, **56**, 247-252.

[66] Bora, L. and Kalita, M.C. (2007). Production and optimization of thermostable lipase from a thermophilic *Bacillus* sp. LBN 4. *The Internet Journal of Microbiology* 4.

[67] Zhang, J.W. and Zeng, R.Y. (2008) Molecular cloning and expression of a cold-adapted lipase gene from an Antarctic deep sea psychrotrophic bacterium *Pseudomonas* sp. 7323. *Marine Biotechnology*, **10**, 612-621.

[68] Kumar, S., Kikon, K., Upadhyay, A., Kanwar, S.S. and Gupta, R. (2005) Production, purification, and characterization of lipase from thermophilic and alkaliphilic *Bacillus coagulans* BTS-3. *Protein Expression and Purification*, **41**, 38-44.

[69] Litthauer, D., Ginster, A. and Skein, E.V.E. (2002) *Pseudomonas luteola* lipase: a new member of the 320-residue *Pseudomonas* lipase family. *Enzyme and Microbial Technology*, **30**, 209-215.

[70] Xu, X. (2000) Production of specific-structured triacylglycerols by lipase-catalyzed reactions: a review. *European Journal of Lipid Science and Technology*, **102**, 287-303.

[71] Dash, S.S., Subramani, R. and Kompala, D.S. (2011). A method for rapid treatment of wastewater and a composition thereof. World Intellectual Property Organization (WIPO), Geneva.

[72] Estera, S.D., Lund, S., Olof, N. and Helsingborg, S. (2006). Method for digestion of sludge in water purification. US Patent No. 20060086659, PCT No. PCT/SE03/01436.

[73] Snape, I., Ferguson, S., Harvey, P.M. and Riddle M. (2006) Investigation of evaporation and biodegradation of fuel spills in Antarctica: II extent of natural attenuation at Casey station. *Chemosphere*, **63**, 89-98.

[74] Banat, I.M., Makkar, R.S. and Cameotra, S.S. (2000) Potential commercial applications of microbial surfactants. *Applied Microbiology and Biotechnology*, **53**, 495-508.

[75] Bach, H., Bedichevsky, Y. and Gutnick, D. (2003) An exocellular protein from the oil-degrading microbe *Acinetobacter venetianums* RAG-1 enhances the emulsifying activity of the polymeric bioemulsifier emulsan. *Applied and Environmental Microbiology*, **69**, 2608-2615.

[76] Mathur, C., Prakash, R., Ali, A., Kaur, J., Cameotra, S.S. and Prakash, N.T. (2010) Emulsification and hydrolysis of oil by *Syncephalastrum racemosum*. *Defence Science Journal*, **60**, 251-254.

[77] Saimmai, A., Rukadee, O., Onlamool, T., Sobhon, V. and Maneerat, S. (2012) Isolation and functional characterization of a biosurfactant produced by a new and promising strain of *Oleomonas sagaranensis* AT18. *World Journal of Microbiology Biotechnology*, **28**, 2973-2986.

[78] Gautam, K.K. and Tyagi, V.K. (2006) Microbial surfactants: a review. *Journal of Oleo Science*, **55**, 155-166.

[79] Saharan, B.S., Rahu, R.K. and Sharma, D. (2011) A review on biosurfactants: Fermentation, current developments and perspectives. *Genetic Engineering and Biotechnology Journal*, **2011**, GEBJ-29.

[80] Pattanathu, K.S.M., Rahman, K.S.M. and Gakpe, E. (2008) Production, characterisation and applications of biosurfactants—Review. *Biotechnology*, **7**, 360-370.

[81] Prieto, C., Jara, C., Mas, A. and Romero, J. (2007) Application of molecular methods for analysing the distribution and diversity of acetic acid bacteria in Chilean vineyards. *International Journal of Food Microbiology*, **115**, 348-355.

[82] Sukumaran, R.K., Singhania, R.R. and Pandey, A. (2005) Microbial cellulases—Production, applications and challenges. *Journal of Scientific & Industrial Research*, **64**, 832-844.

[83] Parmar, N., Singh, A. and Ward, O.P. (2001) Enzyme treatment to reduce solids and improve settling of sewage sludge. *Journal of Industrial Microbiology and Biotechnology*, **26**, 383-386.

4

Seasonal nekton assemblages in a flooded coastal freshwater marsh, Southwest Louisiana, USA

Sung-Ryong Kang[1*], Sammy L. King[2]

[1]School of Renewable Natural Resources, Louisiana State University Agricultural Center, Baton Rouge, USA;
[*]Corresponding Author
[2]U.S. Geological Survey, Louisiana Cooperative Fish and Wildlife Research Unit, School of Renewable Natural Resources, Louisiana State University Agricultural Center, Baton Rouge, USA

ABSTRACT

Marsh flooding and drying may be key factors affecting seasonal nekton distribution and density because habitat connectivity and water depth can impact nekton accessibility to the marsh surface. Recent studies have characterized freshwater nekton assemblages in marsh ponds; however, a paucity of information exists on the nekton assemblages in freshwater emergent marshes. The principal objectives of this study are to characterize the seasonal nekton assemblage in a freshwater emergent marsh and compare nekton species composition, density, and biomass to that of freshwater marsh ponds. We hypothesize that 1) freshwater emergent marsh has lower taxa richness than freshwater marsh ponds; and 2) freshwater emergent marsh has a lower seasonal density and biomass than freshwater marsh ponds. Mosquitofish _Gambusia affinis_ and least killifish _Heterandria formosa_ were abundant species in both habitats while some abundant species (e.g., banded pygmy sunfish _Elassoma zonatum_) in freshwater ponds were absent in freshwater emergent marsh. Our data did not support our first and second hypotheses because taxa richness, seasonal density and biomass between freshwater emergent marsh and ponds did not statistically differ. However, freshwater emergent marsh was dry during the summer months and thus supports no fish species during this period. Additional long-term research on the effects of flow regime in the freshwater marsh on nekton assemblages would potentially improve our understanding of nekton habitat requirements.

Keywords: Freshwater Emergent Marsh; Freshwater Pond; Nekton Assemblage; Hydrologic Connection

1. INTRODUCTION

Regional-scale patterns in the distribution of organisms result primarily from species responses to their physical environment because dominant abiotic variables are thought to act like a physiological sieve [1,2]. Marsh flooding and drying are likely to be key factors affecting seasonal nekton distribution and density because habitat connectivity and water depth can determine nekton accessibility to the marsh surface [3-7]. Moreover, flow regime plays a profound role in the lives of fish through its effect on critical life events (e.g., reproduction, spawning, larval survival, recruitment) [8-13]. In this sense, lateral hydrologic connectivity between coastal freshwater emergent-herbaceous marsh (adjacent to ponds and channels; hereafter termed "freshwater emergent marsh, FEM") and ponds during flooding may increase nekton density in the freshwater emergent marsh while nekton density in ponds may decrease due to nekton movement from ponds to the freshwater emergent marsh. However, shallow water depths may not provide equal access for all nekton (e.g., larger species) thereby restricting some nekton taxa from the freshwater ponds. Also, ponds that have a relatively longer hydroperiod and longer hydrologic connectivity to permanent water bodies may have relatively higher nekton density and biomass than the freshwater emergent marsh. For example, several studies suggest that a low degree of connection with adjacent waterways support relatively few organisms due to limited recruitment [14] and severe environmental conditions (e.g., salinization, drying [15-17]).

In freshwater habitats, low dissolved oxygen (DO) also creates stressful conditions for many species [18]. However, relatively abundant species (e.g., mosquitofish) in freshwater marsh are adapted to low DO. [19] documented that mosquitofish reached the greatest abundance in habitats with relatively low DO (e.g., 2 mg/L), high submerged aquatic vegetation (SAV) cover, and low salinity (e.g., <0.5 ppt). Thus, nekton assemblages in freshwater emergent marshes and ponds may have similar dominant species even though freshwater emergent marshes exhibit severe environmental conditions (e.g., drying).

The extent of coastal marsh loss in many parts of the world has intensified efforts to develop marsh management and conservation strategies that include habitat value assessment for nekton [20-23]. [24] characterized freshwater nekton assemblages in marsh ponds, however, a paucity of information exists on nekton assemblages in freshwater emergent marshes compared to assemblages in freshwater marsh ponds. A clear understanding of the similarity and differences between freshwater emergent marsh and marsh ponds would enhance our understanding of nekton habitat requirements in freshwater marshes as well as the effects of anthropogenic activities, such as habitat conversion (e.g., freshwater emergent marsh to pond), on their distribution. The principal objectives of this study are to characterize the seasonal nekton assemblage in a freshwater emergent marsh and compare nekton species composition, density, and biomass to that of freshwater marsh ponds. We hypothesize that 1) freshwater emergent marsh has lower taxa richness than marsh ponds; and 2) freshwater emergent marsh has a lower seasonal density and biomass than marsh ponds.

2. STUDY AREA AND METHODS

2.1. Study Area

This study was conducted in White Lake Wetlands Conservation Area (WLWCA, 29°52'N, 92°31'W, **Figure 1**) in the Chenier Plain of southwestern Louisiana. WLWCA, a 28,719 ha freshwater marsh, is bounded on the south by White Lake (28.2 km north of the Gulf of Mexico). Dominant vegetation is maidencane (*Panicum hemitomon* Schultes) and bulltongue arrowhead (*Sagittaria lancifolia* Linnaeus). We used marsh vegetation (*i.e.*, freshwater marsh: *Panicum hemitomon*, [25]) to define our marsh types because vegetation does not respond to daily salinity fluctuations [25,26]. Salinity (*i.e.*, freshwater marsh: 0.1 - 3.4 ppt) was also a major consideration of our decision to select marsh types.

2.2. Data Collection

In November 2008, we deployed continuous water

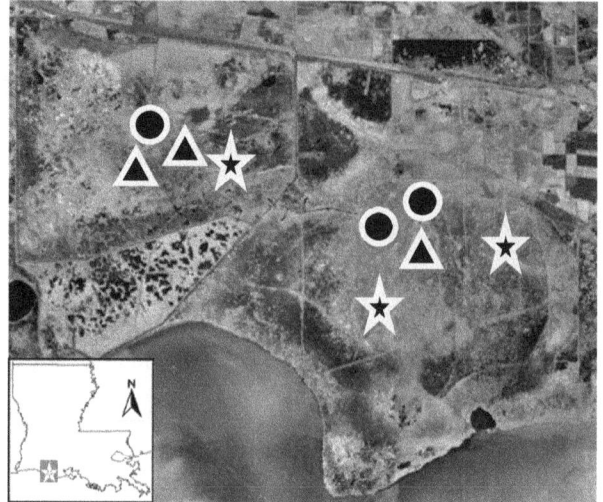

Figure 1. White lake wetlands conservation area is located in the Chenier plain of southwestern Louisiana. Stars (FEMs), triangles (PCPs), and circles (TCPs) are our sampling points in the marshes.

level recorders in freshwater emergent marshes (*i.e.*, 100 m from channel or pond margin) and ponds to measure water depth 6 times per day. Water depths were validated by comparing water level recorder readings to discrete monthly water depths obtained with a meter stick adjacent to the recorder; both water depths were always within 1 cm of each other. We then determined flooding depth and duration based on the criteria that daily water depth (DWD) > 0. We also deployed a staff gage at the border between the pond and freshwater emergent marsh to measure disconnection of surface water and connected water depth (CWD). CWD was the water depth at the border between the pond and the freshwater emergent marsh when the pond is connected with surface water to the channel or surrounding marsh (marginal zone of the pond).

To determine nekton characteristics, we sampled freshwater emergent marshes (*i.e.*, 100 m from channel or pond margin) seasonally from March 2009 to February 2010. Seasons were defined as: 1) Spring (March-May); 2) Summer (June-August); 3) Fall (September-November); 4) Winter (December-February). A 1-m^2 aluminum-sided throw trap (mesh size: 3 mm), similar to that described by [28], was tossed at three random points in each sampling plot within the freshwater emergent marsh (4 sampling sites) and ponds (*i.e.*, 3 permanently connected ponds [PCP: permanently connected by a channel during all seasons], 3 temporarily connected ponds [TCP: temporarily connected by surface water to the surrounding marsh but not permanently connected to a channel], [24]). Sweeps with a 1 m wide bar seine (3 mm mesh size) were used to remove the nekton from the trap. Five consecutive sweeps without collecting organisms were completed before the trap was considered free of nekton. Fish and decapod crustaceans were frozen and returned to the laboratory where they were sorted and identified to species or to the lowest possible taxon. All nekton were weighed to the nearest 0.001 g wet-weight to determine biomass ($g \cdot m^{-2}$).

2.3. Statistical Analysis

Data are reported as mean ± standard error (SE), and significance level was chosen at $\alpha = 0.05$. Analyses of variance (ANOVA) and T-test (Proc Mixed, Version 9.3, Cary, SAS Institute, North Carolina) were used to test for statistical differences in environmental variables and nekton density and biomass by season. We used one-way ANOVA for each response variable that included environmental variable and nekton density. We conducted a one-way ANOVA with one fixed effect. Significant one-way ANOVA effects were tested using post-hoc comparisons of Tukey adjusted least squared means. For ANOVA analyses, data were tested for normality with the Shapiro-Wilks test. In the event that the residuals were not normally distributed, the data were log-transformed. Linear regression (Proc Mixed, Version 9.3, SAS Institute, North Carolina) was used to examine the potential relationship between nekton assemblage characteristics (*i.e.*, density, biomass) and environmental factors.

3. RESULTS

In the freshwater emergent marshes, summer was the driest period (flooded days: 23/92 days) and winter was the wettest period (flooded days: 90/90 days). DWD ranged from 31.7 ± 0.54 cm (mean ± SE; winter) to 1.3 ± 0.41 cm (summer). DWD differed among all seasons ($F_{3,12} = 190.55$, $p < 0.01$). CWD ranged from 40.2 ± 2.14 cm (winter, PCP) to 2.9 ± 1.02 cm (summer, PCP).

We recorded 439 individuals of 11 taxa in 60 samples in the freshwater emergent marsh. Seasonal nekton density (organisms/m^2) ranged from 14.7 ± 5.37 (mean ± SE; winter) to 0 (summer, **Figure 2**). However, nekton density within freshwater emergent marsh did differ among spring, fall, and winter ($F_{2,9} = 0.52$, $p = 0.61$). Nekton biomass (g wet wt/m^2) ranged from 4.9 ± 0.95 (winter) to 0 (summer). Nekton biomass had similar seasonal patterns as nekton density ($F_{2,9} = 2.47$, $p = 0.14$). No statistically significant relationships were observed between environmental variables and nekton density/biomass in the freshwater emergent marsh. Relatively abundant species were mosquitofish (spring: 58%, fall: 29%, winter: 23%), least killifish (spring: 34%, fall: 30%, winter: 24%), and swamp dwarf crawfish (spring: 7%, fall: 30%, winter: 34%).

In the freshwater marsh ponds, we recorded 22 nekton

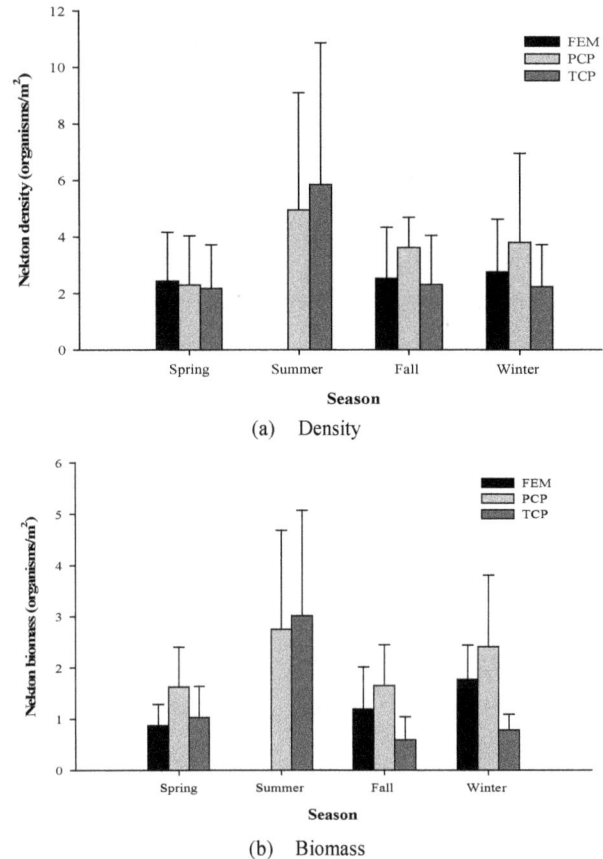

(a)   Density

(b)   Biomass

Figure 2. Seasonal nekton density (log(x + 1) transformed organisms/m^2 (±SE), (a) and biomass (log(x + 1) transformed g wet wt/m^2, (b) in throw trap samples by different habitat types in freshwater marsh from March 2009 to February 2010.

taxa in 90 samples. Nekton density and biomass between freshwater emergent marsh and ponds did not differ for any season (**Table 1**). A total of 22 taxa were found in ponds and 11 taxa in the freshwater emergent marsh; no unique species were observed in the freshwater emergent marsh. Freshwater emergent marsh and ponds shared some abundant species (*i.e.*, mosquitofish, least killifish) but some abundant species (*i.e.*, banded pygmy sunfish, golden topminnow *Fundulus chrysotus*) in freshwater ponds were absent in the freshwater emergent marsh.

4. DISUSSION

The present study considered the hypothesis that freshwater emergent marsh would have lower nekton taxa richness than freshwater marsh ponds due to seasonal isolation of surface water from other water bodies, such as ponds/channels, and relatively shallow CWD. As habitats become spatially reduced, the contact among species may intensify and/or harsh abiotic conditions may develop; in either case, some species may go locally extinct [29]. In addition, the relatively shallow flooded water depth (<32 cm) in freshwater emergent marsh may

Table 1. Mean species nekton density (organisms/m^2 (±SE)) in throw trap samples by different habitat types in freshwater marsh from March 2009 to February 2010.

	FEM	PCP	TCP
Banded pygmy sunfish		4.1 (1.84)	1.0 (0.57)
Bantam sunfish	0.1 (0.06)	1.5 (0.49)	0.0 (0.03)
Bayou killifish		0.1 (0.06)	
Bluegill	0.1 (0.06)	0.7 (0.63)	
Creek chubsucker		0.0 (0.03)	0.1 (0.06)
Golden topminnow	0.2 (0.16)	2.1 (1.52)	1.1 (0.97)
Grass pickerel	0.0 (0.04)	0.1 (0.11)	
Grass shrimp	0.0 (0.04)	9.5 (4.65)	1.6 (0.54)
Least killifish	2.6 (0.88)	19.1 (12.61)	17.7 (15.27)
Mosquitofish	3.2 (1.23)	12.8 (9.82)	69.5 (65.66)
Northern starhead topminnow		0.2 (0.11)	0.1 (0.06)
Pirate perch		0.0 (0.00)	
Rainwater killifish	0.0 (0.02)	1.4 (1.37)	
Redspotted sunfish		0.2 (0.07)	
Red swamp crawfish	0.2 (0.20)	0.1 (0.06)	
Sailfin molly	0.4 (0.30)	2.9 (2.80)	0.3 (0.24)
Sheepshead minnow		0.6 (0.57)	
Spotted bass		0.1 (0.06)	
Swamp darter		0.0 (0.03)	
Swamp dwarf crawfish	2.3 (1.15)	1.2 (0.73)	0.3 (0.21)
Warmouth		0.0 (0.00)	
Yellow bullhead		0.0 (0.03)	

restrict accessibility of large predator species (e.g., bantam sunfish, bluegill). Our data did not support our first hypothesis as taxa richness between freshwater emergent marsh and ponds did not statistically differ, although no fish taxa used the emergent marsh in summer because of lack of water. Similarly, [27] noted that nekton taxa in intermediate marsh ponds included most of the nekton taxa in flooded intermediate freshwater emergent marsh (88% same species). This finding suggests that nekton in freshwater emergent marsh is a nested subset of those in freshwater marsh ponds.

[24] noted that nekton density in freshwater marsh ponds was negatively correlated with CWD and this relationship appears to be related to flooding of the adjacent freshwater emergent marsh. When freshwater emergent marsh is flooded (i.e., lateral hydrologic connectivity), some nekton species will migrate from ponds to the marsh, resulting in decreased nekton density in ponds

[30]. We hypothesized that freshwater emergent marsh had lower nekton density and biomass to that of freshwater marsh ponds, but our results indicate that they did not statistically differ. High variability in nekton density within the freshwater emergent marsh and ponds suggests that nekton in freshwater emergent marsh are patchily distributed. Despite the high variability and limited temporal availability, the freshwater emergent marsh is still an important and widely distributed habitat for nekton.

Individual species responses to habitat attributes (e.g., vegetation cover) may be predicted in the context of their life history-environment relationships [31]. Our results indicated that common pond inhabitants (i.e., mosquitofish, least killifish) were common in the freshwater emergent marsh. This finding is similar to previous studies that found relatively higher population densities of mosquitofish and least killifish in shallow water with thick vegetation, low DO and salinity [19,32,33]. Some abundant species in freshwater ponds, however, were not caught in freshwater emergent marsh as expected. We expected banded pygmy sunfish and golden topminnow to have relatively higher density in freshwater emergent marsh because they prefer shallow water with macrophytes [34-35]. [36] noted vegetation structural complexity may affect nekton habitat use in SAV (e.g., pond) and the freshwater emergent marsh. Differences in the structural complexity of vegetation between habitat types may have been responsible for the absence of banded pygmy sunfish and the relatively low density of golden topminnow. These findings suggest that some abundant species in freshwater emergent marsh and ponds may be well adapted to low DO with high vegetation cover.

Dry conditions are common in wetlands and are an important part of the hydrological cycle. Variation in life history traits of nekton seems to be correlated with hydrologic condition (i.e., flooding duration). [37] noted that flow regime adaptations range from behaviors that result in the avoidance of individual floods or droughts, to life-history strategies that are synchronized with long-term flow patterns. In addition, [13] noted that many fish species in highly variable flow regimes have evolved life history strategies that ensure strong recruitment. Our results of high variability in nekton density within the freshwater emergent marsh indicate that nekton is patchily distributed. Furthermore, we observed that variability in flooding is common among years during the same season. During spring sampling, the freshwater emergent marsh was flooded, providing ample access to the marsh by nekton. However, during March to May 2010 (spring period), dry conditions prevailed and the marsh remained unflooded (unpublished data, no nekton sample). Strictly from a nekton perspective, our results suggest that anthropogenic activities such as marsh management that increases or decreases duration of lateral hydrologic connection between freshwater emergent marsh and adjacent water bodies can potentially alter nekton habitat value (*i.e.*, non-suitable, less suitable, suitable) in freshwater marsh.

Previous studies [38,39] noted that the natural flow regime has a profound influence on the biodiversity of aquatic ecosystems (e.g., streams, rivers and their floodplain wetlands). Several interrelated flow characteristics influence nekton assemblages in aquatic systems at different temporal and spatial scales; no single flow characteristic is responsible. [13] noted that it is difficult to resolve which attributes of the altered flow regime are directly responsible for observed impacts. Similarly, in our study, it is unclear as to what hydrologic characteristics are most important in structuring nekton communities in freshwater marsh. Additional long-term research on the effects of flow regime in the freshwater marsh on nekton assemblages would potentially improve our understanding of nekton habitat requirements.

5. ACKNOWLEDGEMENTS

This project was supported by a Louisiana Department of Wildlife and Fisheries and U.S. Fish and Wildlife Service State Wildlife Grant with support also from the International Crane Foundation. We thank M. La Peyre, J. A. Nyman, R. Keim, A. Rutherford, and S. Piazza for their critical insights. The authors would like to acknowledge the field and laboratory contributions of J. Linscombe, R. Cormier, M. Huber, and A. Williamson. In addition, we extend gratitude to M. Kaller for statistical assistance. We appreciate the comments of two anonymous reviewers, whose suggestions improved this manuscript. Collections were made under Louisiana State University AgCenter Animal Care and Use protocol (#AE2008-012). Any use of trade, firm, or product names is for descriptive purposes only and does not imply endorsement by the U.S. Government.

REFERENCES

[1] Remmert, H. (1983) Studies and thoughts about the zonation along the rocky shores of the Baltic. *Zoologica*, **22**, 121-125.

[2] Martino, E.J. and Able, K.W. (2003) Fish assemblages across the marine to low salinity transition zone of a temperate estuary. *Estuarine, Coastal and Shelf Science*, **56**, 969-987.

[3] Whoriskey, F.G. and Fitzgerald, G.J. (1989) Breeding-season habitat use by sticklebacks (Pisces: Gasterosteidae) at Isle Verte, Quebec. *Canadian Journal of Zoology*, **67**, 2126-2130.

[4] Szedlmayer, S.T. and Able, K.W. (1993) Ultrasonic telemetry of age-0 summer flounder, *Paralichthys dentatus*, movements in a southern New Jersey estuary. *Copeia*, **1993**, 728-736.

[5] Lake, P.S. (2003) Ecological effects of perturbation by drought in flowing waters. *Freshwater Biology*, **46**, 1161-1172.

[6] Humphries, P. and Baldwin, D.S. (2003) Drought and aquatic ecosystem: An introduction. *Freshwater Biology*, **48**, 1141-1146.

[7] Minello, T.J., Rozas, L.P. and Baker, R. (2012) Geographic variability in salt marsh flooding patterns may affect nursery value for fishery species. *Estuaries and Coasts*, **35**, 501-514.

[8] Welcomme, R.L. (1985) River fisheries. Food and Agriculture Organization of the United Nations, FAO Fisheries Technical Paper 262.

[9] Junk, W.J., Bayley, P.B. and Sparks, R.E. (1989) The flood-pulse concept in river-floodplain systems. In: Dodge, D.P., Ed., *Proceedings of the International Large River Symposium (LARS), Canadian Journal of Fisheries and Aquatic Sciences Special Publication* 106, NRC research press, Ottawa, 110-127.

[10] Copp, G.H. (1990) Effect of regulation on 0+ fish recruitment in the Great Ouse, a lowland river. *Regulated Rivers: Research and Management*, **5**, 251-163.

[11] Sparks, R.E. (1995) Need for ecosystem management of large rivers and floodplains. *BioScience*, **45**, 168-182.

[12] Humphries, P., King, A.J. and Koehn, J.D. (1999) Fish, flows and floodplains: Links between freshwater fishes and their environment in the Murray-Darling River system, Australia. *Environmental Biology of Fishes*, **56**, 129-151.

[13] Bunn, S.E. and Arthington, A.H. (2002) Basic principles and ecological consequences of altered flow regimes for aquatic biodiversity. *Environmental Management*, **30**, 492-507.

[14] Rozas, L.P. and Minello, T.J. (1999) Effects of structural marsh management on fishery species and other nekton before and during a spring drawdown. *Wetlands Ecology and Management*, **7**, 121-139.

[15] Dunson, W.A., Friacano, P. and Sadinski, W.J. (1993) Variation in tolerance to abiotic stresses among sympatric salt marsh fish. *Wetlands*, **13**, 16-24.

[16] Rowe, C.L. and Dunson, W.A. (1995) Individual and interactive effects of salinity and initial fish density on a salt marsh assemblage. *Marine Ecology Progress Series*, **128**, 271-278.

[17] Gascon, S., Boix, D., Sala, J. and Quintana, X.D. (2008) Relation between macroinvertebrate life strategies and habitat traits in Mediterranean salt marsh ponds (Emporda wetlands, NE Iberian Peninsula). *Hydrobiologia*, **597**, 71-83.

[18] McKinsey, D.M. and Chapman, L.J. (1998) Dissolved oxygen and fish distribution in a Florida spring. *Environmental Biology of Fishes*, **53**, 211-223.

[19] Hubbs, C. (1971) Competition and isolation mechanisms in the *Gambusia affinis* X *G. heterochir* hybrid swarm.

Texas Memorial Museum Bulletin, **19**, 1-46.

[20] Cattrijsse, A., Makwaia, E.S., Dankwa, H.R., Hamerlynck, O. and Hemminga, M.A. (1994) Nekton communities of an intertidal creek of a European estuarine brackish marsh. *Marine Ecology Progress Series*, **109**, 195-208.

[21] Hampel, H. (2003) Factors influencing the habitat value of tidal marshes for nekton in the Westerschelde estuary. Ph.D. Dissertation, University of Gent, Belgium.

[22] Cattrijsse, A. and Hampel, H. (2006) European intertidal marshes: A review of their habitat functioning and value for aquatic organisms. *Marine Ecology Progress Series*, **324**, 293-307.

[23] La Peyre, M.K., Gossman, B. and Nyman, J.A. (2007) Assessing functional equivalency of nekton habitat in enhanced habitats: Comparison of terraced and unterraced marsh ponds. *Estuaries and Coasts*, **30**, 526-536.

[24] Kang, S.R. and King, S.L. (2013) Effects of hydrologic connectivity and environmental variables on nekton assemblage in a coastal marsh system. *Wetlands*, 33, 321-334

[25] Chabreck, R.H. and Nyman, J.A. (2005) Management of coastal wetlands. In: Braun, C.E., Ed., *Techniques for Wildlife Investigations and Management*, The Wildlife Society, Bethesda, 839-860.

[26] Visser, J.M., Sasser, C.E., Chabreck, R.H. and Linscombe, R.G. (1998) Marsh vegetation types of the Mississippi river deltaic plain. *Estuaries*, **21**, 818-828.

[27] Rozas, L.P. and Minello, T.J. (2010) Nekton density patterns in tidal ponds and adjacent wetlands related to pond size and salinity. *Estuaries and Coasts*, **33**, 652-667.

[28] Kushlan, J.A. (1981) Sampling characteristics of enclosure fish traps. *Transactions of the American Fisheries Society*, **110**, 557-562.

[29] Fernandes, R., Gomes, L.C., Pelicice, F.M. and Agostinho, A.A. (2009) Temporal organization of fish assemblages in floodplain lagoons: The role of hydrological connectivity.

Environmental Biology of Fishes, **85**, 99-108.

[30] Minello, T.J. (1999) Nekton densities in shallow estuarine habitats of Texas and Louisiana and the identification of essential fish habitat. *American Fisheries Society Symposium*, **22**, 43-75.

[31] Olden, J.D., Poff, N.L. and Bestgen, K.R. (2006) Life-history strategies predict fish invasions and extirpations in the Colorado River basin. *Ecological Monographs*, **76**, 25-40.

[32] Douglas, N.H. (1974) Freshwater fishes of Louisiana. Claitor's Publishing Division, Baton Rouge.

[33] Chervinski, J. (1983) Salinity tolerance of the mosquito-fish, *Gambusia affinis* (Baird and Girard). *Journal of Fish Biology*, **22**, 9-11.

[34] Shute, J.R. (1980) *Fundulus chrystotus*, Golden topminnow. In: Lee, D.S., Gilbert, C.R., Hocutt, C.H., Jenkins, R.E., McAllister, D.E. and Stauffer J.R. *Atlas of North American Freshwater Fishes*, North Carolina Museum of Natural Sciences Publication, Raleigh, 510.

[35] Moriarty, L.J. and Winemiller, K.O. (1997). Spatial and temporal variation in fish assemblage structure in Willage Creek, Hardin County Texas. *Texas Journal of Science*, **49**, 85-110.

[36] Castellanos, D.L. and Rozas, L.P. (2001) Nekton use of submerged aquatic vegetation, marsh, and shallow unvegetated bottom in the Atchafalaya River delta, a Louisiana tidal freshwater ecosystem. *Estuaries*, **24**, 184-197.

[37] Lytle, D.A. and Poff, N.L. (2004) Adaptation to natural flow regimes. *Trends in Ecology and Evolution*, **19**, 94-100.

[38] Poff, N.L., Allan, J.D., Bain, M.B., Karr, J.R., Prestegaard, K.L., Richter, B.D., Sparks, R.E. and Stromberg, J.C. (1997) The natural flow regime. *BioScience*, **47**, 769-784.

[39] Hart, D.D. and Finelli, C.M. (1999) Physical-biological coupling in streams: The pervasive effects of flow on benthic organisms. *Annual Review of Ecology and Systematics*, **30**, 363-395.

5

Patterns of nest placement of lappet faced vulture (*Torgos tracheliotos*) in Lochinvar National Park, Kafue Flats, Zambia

Chansa Chomba[1*], Eneya M'simuko[2], Vincent Nyirenda[3]

[1]School of Agriculture and Natural Resources, Disaster Management Training Centre, Mulungushi University, Kabwe, Zambia;
*Corresponding Author
[2]School of Natural Resources, Copperbelt University, Kitwe, Zambia
[3]Department of Research, Zambia Wildlife Authority, Chilanga, Zambia

ABSTRACT

This study assessed the nesting patterns of lappet faced vulture in Lochinvar National Park, on the Kafue flats, Zambia. Road drives and foot patrols were used to identify and take GPS coordinates of lappet faced vulture nests. The main objectives of the study were: 1) to obtain basic breeding information of lappet-faced vulture in Lochinvar National Park and the Kafue Flats in general, 2) to determine size of the breeding population in the National Park, 3) to document distribution of the nesting sites, 4) to facilitate development of a monitoring programme that would secure the nesting sites from human disturbance, and 5) to determine availability of suitable nesting sites and major threats that may interfere with breeding. Tree species on which nests were found were identified and height of nest above ground was estimated. A total of 22 nests were recorded with 5 (23%) being lappet faced vulture nests on seven species of trees. The mean height for nest placamenet was 10 m above ground. Host tree physiognomy, size and height were important characteristics in nest placement. Large trees of 10 m and above are critical in facilitating nest placement and must be protected in identified breeding sites.

Keywords: Raptors; Nest Placement; Tree Height; Lochinvar; Kafue Flats; Habitat

1. INTRODUCTION

Vultures are very special type of raptors. They locate carcasses by scanning the ground while soaring high in the air [1]. While they soar, they also observe each other, so that when one identifies a carcass and descends, others would follow. They are large and robustly built (**Figure 1**), and due to their large size, they mainly depend on soaring flight and often congregate in large numbers at a carcass. After identifying potential food, they usually perch on nearby trees to be sure that the animal is dead, and this is why it is important to have large trees with branches large enough to support their weight. From the perch, one or few brave birds will take the lead to the carcass and soon others will follow. Like other raptors, vultures are on top of the food chain and as such play an important role in the functioning of ecosystems. Regarding its size, the lappet-faced vulture is the largest of the three dark vulture species, the other two being hooded (*Necrosyrtes monachus*) and white headed (*Aegypius occipitalis*) (**Figure 1**). It also ranks as the longest and largest winged vulture in its range behind the closely related cinereous vulture (*Aegypius monachus*) [1], although some co-occurring *Gyps* vultures tend to be heavier on average, especially the cape vulture (*Gyps coprotheres*) and Eurasian griffon (*Gyps fulvus*). Its heavy yellowish hooked bill and bare red head with its consipicuous skin folds are a major diagnostic feature among other raptors (**Figure 2**).

The Lappet-faced vulture or Nubian vulture (*Torgos tracheliotos*) is mostly African Old World vulture belonging to order Accipitriformes, which also includes eagles, kites, buzzards and hawks, but its size and shape of beak are unmistakenly larger than other raptors (**Figure 2**). It is the only member of the genus *Torgos*. It was formerly considered monotypical, but has now been separated into two subspecies: 1) the nominate race which is found almost throughout Africa, and 2) sub-

Figure 1. Lappet faced vulture (*Torgos tracheliotus*) is one of the largest vultures reaching about 100 cm (39 inches) in length, its massive size of the bird enables it to catch small mammals and some time to scare off other raptors from carrion (Source: Micrsoft Encarta 2009).

black vulture
(Coragyps atratus)

lappet-faced vulture
(Torgos tracheliotus)

common buzzard
(Buteo buteo)

Steller's sea eagle
(Haliaeetus pelagicus)

hook-billed kite
(Chondrohierax uncinatus)

gyrfalcon
(Falco rusticolus)

Figure 2. Lappet faced vulture with bear head and massively built and hooked bill which are major diagnostic features which sets it apart in appearance from other raptors (Source: Encyclopaedia Britannica 2010).

species T. t. *negevensis* occurring in the Negev desert of Sinai differs considerably in appearance from African vultures [2].

The distribution of the species in Africa is intermittent being absent from much of the central and western parts of the continent and recorded to be declining elsewhere in its range. It is recorded to nest in: Senegal, Mali, Burkina Faso, Niger, Chad, Sudan, southeastern Egypt,

Ethiopia, Somalia, Kenya, Tanzania, Uganda, Rwanda, easternmost part of the Democratic Republic of the Congo, parts of Zambia, Malawi, Mozambique, Swaziland, northeastern South Africa, Zimbabwe, Botswana, Namibia, the Gambia, Guinea, Ivory Coast, Benin, the Central African Republic, southern Angola and possibly in Mauritania and Nigeria. Across the Red Sea, the species nests in Arabia, Yemen, Oman and the United Arab Emirates [2-4].

With regard to foraging, its scavenging habits of feeding on carcasses are fairly well understood. It finds its food by sight or by watching other vultures. More so than many other African vultures that find carrion on their own and start tearing through the skin. The lappet faced vulture is the most powerful and aggressive of the African vultures, and other vultures will usually cede a carcass if the Lappet-faced decides to assert itself. This is often beneficial to the less powerful vultures because, with their powerful beak, bare head and knotty muscles, they can tear through the tough hides of large mammals that the other raptors cannot penetrate (**Figure 2**), although hyenas (*Crocuta crocuta*) are more efficient in this regard. The bald head is advantageous, because a feathered head would become spattered with blood and other fluids, and thus be difficult to keep clean. However, pioneering in the opening up of carcasses before other species come in, has the potential to expose the species to mainly chemical poisoning, particularly in areas of Africa where poisonous chemicals are used in poaching of wild animals or control of livestock predators. Usually after opening the carcass, it frequently hangs around the edges of the throngs at large carcasses, waiting until the other vultures are done to feed on remnant skin, tendons and other coarse tissues that the others will not eat, which in itself, is an important feeding behavioural component that facilitates ecological separation and avoids direct competition and hence underscoring their importance in the food chain. Big game animals, up to the size of elephant (*Loxodonta africana*), with tough skin are preferred as carrion, since they provide the most subsistence at a sitting of up to a full crop of 1.5 kg (3.3 lb) of meat [5,6].

Regarding breeding, records show that in southern Africa it takes place from May to January. Nests consist of a pile of neatly formed sticks, and are large measuring 120 - 220 cm (47 - 87 inches) across and 30 - 70 cm (12 - 28 inches) deep. They are often lined with green leaves, as well as animal hair and skins [2]. Nests are almost always placed in the main fork or top of tree species such as *Acacia*, *Balanites* and *Terminalia*, at 5 to 15 m (16 to 49 ft) off the ground. The clutch contains 1 or 2 eggs, which are incubated by both parents over a period of 54 to 56 days. The young fledge at 124 to 135 days old, although they can be dependent on their parents for up to an age of 1 year or more, sometimes forcing parents to only nest in alternate years. There is a single remarkable record of a Lappet-faced vulture pair successfully raising a white-headed vulture. The Lappet-faced Vultures do not usually breed until it is around 6 years of age [2]. Their social behaviour is that of being generally solitary birds which do not nest in cohesive colonies as do many smaller vultures, with one tree or area usually only having 1 to 2 nests in it, though sometimes up to 10 nests have been recorded in one area. The home range of a Lappet-faced Vulture is usually at least 8 to 15 km (5.0 to 9.3 miles) [6,7].

On the aspect of distribution and population status, the species is believed to have decreased perceptibly. Declining in Sahel and several parts of their southern, northern and western distribution in Africa. The declines are almost entirely due to human activities, including disturbances from habitat destruction and cultivation, disturbances at the nesting site to which the species is reportedly quite sensitive, and ingestion of pesticides, which are usually set out for jackals (Canis spp) and other small mammalian carnivores [8]. Cattle, which have replaced natural prey over much of the range, are now often sold off, rather than abandoned, due to the proliferation of markets and abattoirs and rarely left to die and be consumed by vultures [8]. Lappet-faced vultures are also sometimes victims of direct persecution, including shooting and the use of strychnine and other poisons. In Namibia, 86 Lappet-faced vultures were poisoned at once through a group of cattle carcasses, because the farmers erroneously believed they were killing and eating the cattle. In some cases the poisoning is done by poachers, who fear the presence of vultures will alert authorities to their activities, particularly the illegal killings of protected species [8]. They are considered Vulnerable at species level, with an estimated world population of 8500 individuals [8].

With low global population estimated at 8500, and possibly 1000 pairs (almost 3000 individuals) in southern Africa [4], the species was ranked as vulnerable and put on the IUCN red data list in 2009. Breeding records for Lochinvar National Park were earlier recorded in the 1970s by Osborne [9], and no survey has since been done until 2009 when a preliminary survey was done by Tokura Wataru, Volunteer Biologist-Lochinvar National Park, [10]. This paucity of data on the species has a potential of exposing it to habitat loss and other unkown negative effects that arising from poor management of the habitat. It was for this reason that this study was undertaken to identify nesting sites and safeguard them from human disturbance.

The objectives of the study were as follows: 1) To obtain basic breeding information of Lappet-faced vulture in Lochinvar National Park and the Kafue Flats in general; 2) To determine size of the breeding population in the National Park; 3) To document distribution of nesting sites and 4) To develop a monitoring programme that would secure and safeguard nesting sites from human disturbance.

2. MATERIALS AND METHODS

Study Site

The survey was undertaken in Lochinvar National Park (**Figures 3(a)** and **(b)**), mainly in the termitaria and floodplain area in the month of September 2011-2012 following the same methods used by Tokura Wataru [10]. The survey was done mostly in the open woodland or grassland and in the shrub Dichrostachys cinerea which forms sparse thicket around Chunga Camp.

Between three and five observers walked around the study site (**Figure 3(b)**), and searched for possible lappet-faced vulture's nesting sites. We defined the possible Lappet-face vulture's nest based on the earlier survey done by Wataru that vulture's nests are: 1) bowl-shaped, 2) built near the top of a tree, and 3) its size is larger than one meter in diameter. When a nest was located, GPS coordinates were taken, using GPS receiver (GPS60, Germin Ltd.). The tree species hosting the nest, and visual estimation of height of the tree were recorded. The nest was only recorded as belonging to lappet faced vulture when the species was physically seen on the nest. The observation was done by naked eye with an aid of ×10 wide angle Bushnell binoculars. Pictures were taken using a 560× digital zoom Sony camera. Since the survey was carried out in September which is dry season, and most trees are defoliated, visility was high, making it possible for the observer to view up to 100 m by the naked eyes, and as far as 200 m with the aid of a pair of binoculars in the open woodland or grassland.

3. RESULTS

3.1. Selection of Tree Species for Placement of Nests

The nests encountered during the survey were placed on selected tree species as follows: Acacia nigrescens 5 (23% of total), Acacia seyal 6 (27% of total), Faidherbia albida 6 (27% of total), Acacia gerrardii 2 (9% of total), Combretum imberbe 1 (5% of total), Commiphora spp 1 (5% of total) and one unkown species 1 (5% of total) (**Table 1**). The mean height above ground for nest placemenet was 10 m.

3.2. Number and Location of Nests

A total of 22 nests were found in the area surveyed. Of the 22 nests seen, lappet faced vultures were only physically seen in five (23%) which were confirmed as their nests (**Figures 4(a)-(e)**). It is likely that there were more lappet vulture nests as birds may not lay and incubate eggs at the same time, or some of them may have already

(a)

(b)

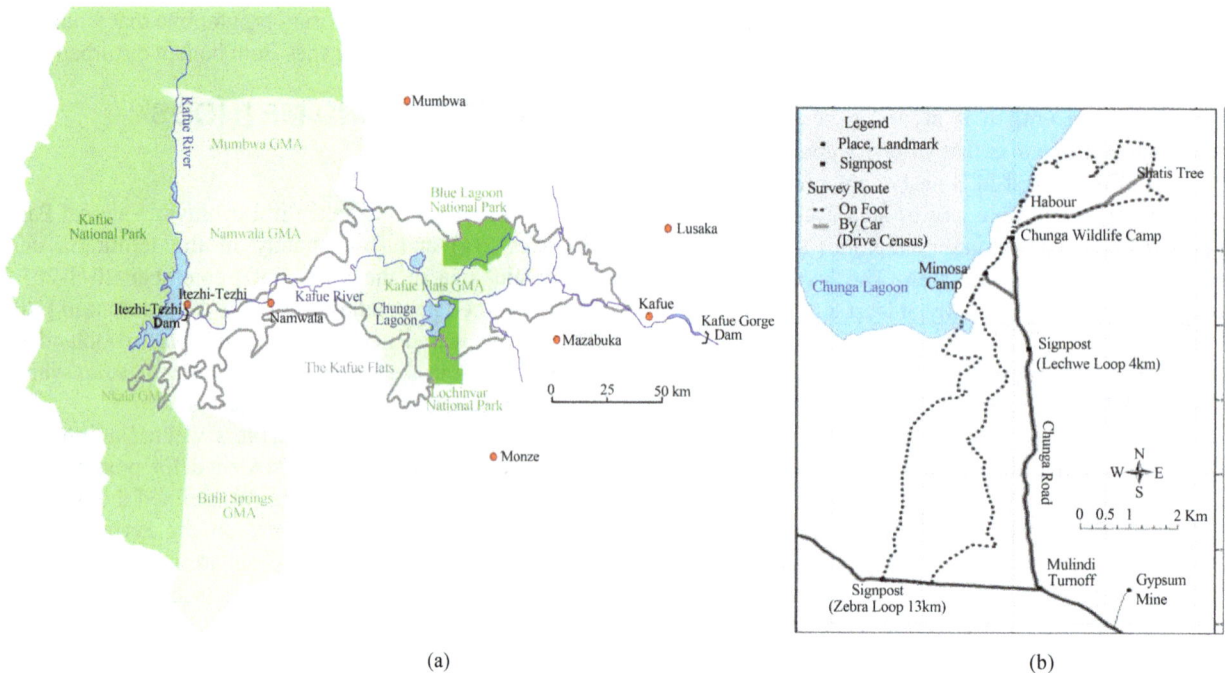

Figure 3. (a) Location of study area, Lochinvar National Park on the Kafue Flats; (b) Details of the study routes in the study area, Zambia 2012.

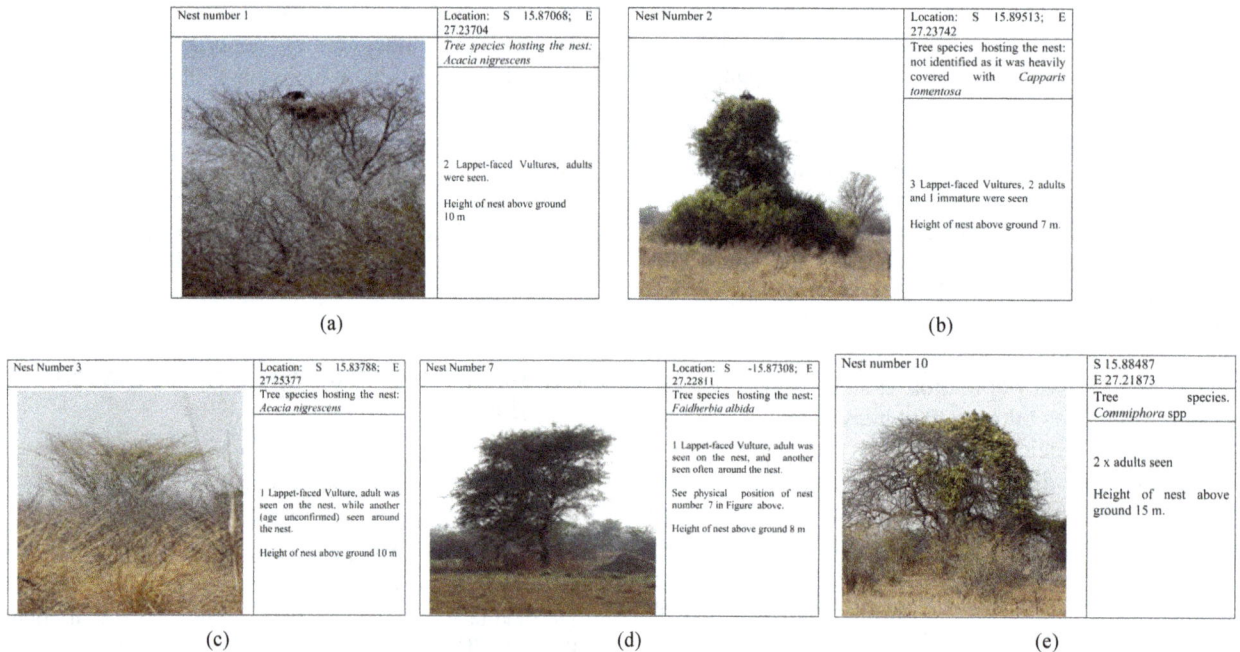

(a)

(b)

(c)

(d)

(e)

Figure 4. (a)-(e) Tree species and physiognomy of trees selected for placement of raptor nests, Lochinvar National Park, Kafue Flats, Zambia, 2012.

hatched at the time of this study.

3.3. Nature of Vegetation Community for Nest Placement

Nests recorded for lappet faced vulture were in the flood plain and termitaria areas of the National Park,

where density of woody plant species were in fact sparse plant species was sparse. The vultures built large (Mean often more than 100 cm in diameter) (**Table 1**). Nests were made from small branches near the top of the tree (**Figures 4(a)-(e)**).

Regarding plant phenology; tree species were the only ones used for nest placement and not bushes, climbers,

Table 1. Nests recorded during the survey, location, host tree species and possible species of bird, Lochinvar National Park, 2012 Zambia.

No.	Species using nest	GPS location		Tree species		Observation	Estimated size of nest	Signs on ground	Other
		S	E	Name	Height (m)				
1	Lappet faced vulture	−15.87068	27.23704	*Acacia nigrescens*	10	2× adults			
2	Lappet faced vulture	−15.89513	27.23742	Unidentified, heavily coverd with *Capparis tomentosa*	7	2× adults; 1 immature			
3	Lappet faced vulture	−15.83788	27.25377	*A. nigrescens*	10	1× Adult; 1× young one			
4	African fish eagle	−15.86369	27.22386	*F. albida*	18	None			African Fish Eagle
5	African fish eagle	−15.87256	27.22023	*F. albida*	12	Immature African fish eagle			
6	Unknown	−15.87287	27.22018	*F. albida*	12	None	Nest smaller than Nos. 4, 5.		
7	Lappet-faced Vulture	−15.87308	27.22811	*F. albida*	8	1× adult; 1× young one.			
8	Unknown	−15.88146	27.22296	*F. albida*	14	None			
9	Unknown	−15.88226	27.22555	*F. albida*	7	None	50 cm		
10	Lappet-faced Vulture	−15.88487	27.21873	*Commiphora spp*	15	2× adults			Tree covered by climber, *Capparis tomentosa*.
11	Unknown	−15.91651	27.23074	*A.nigrescens*	7	None	120 cm		Nest partially mended with new twigs
12	Unknown	−15.91005	27.22960	*Acacia seyal*	5	None	120 cm	Feathers	Nest partially mended
13	Unknown	−15.88321	27.22855	*A. seyal*	9	None	90 cm	Droppings	
14	Martial Eagle?	−15.83214	27.24106	*A. seyal*	8	1× immature Martial Eagle	100 × 120 cm	Fresh droppings, feathers, and Kafue Lechwe fur	
15	Unknown	−15.82961	27.24220	*A. seyal*	10	None	100 cm	Droppings	
16	Unknown	−15.82784	27.24591	*A. gerrardii*	10	None	100 cm	Droppings;egg shell, fur of Kafue Lechwe fur, feathers	Two nests on the same tree
17	Unknown	−15.83319	27.24826	*A. seyal*	10	Dead chicks seen, but not identified		None	
18	Unknown	−15.83458	27.24742	*Combretum imberbe*	8	None	120 cm	Feathers; droppings	
19	Tawny Eagle	−15.83010	27.24948	*A. gerrardii*	10	1× Tawny Eagle, adult			
20	Unknown	−15.82502	27.25286	*A. seyal*	8	None	100 cm	Droppings, feathers.	
21	Unknown	−15.83856	27.26115	*A. nigrescens*	8	None	100 cm	Droppings, feathers	
22	Unknown	−15.83842	27.25690	*A. nigrescens*	7	None	120 cm	Droppings	
Total					**213**		**1130**		
n					**22**		**11**		
Mean					**10 (9.68)**		**103 (102.72)**		

anthill tops, kopjes, or other raised features. Of the tree species used for nest placement, only three were commonly used. Placement height varied from 7 - 15 meters with overall mean of 10 m (**Table 1**). Some tall trees were however, not chosen for nest placement, probably because branches were too feeble to support the weight of a large nest and incubating pair. For instance, *Acacia sieberana* was very common in the study area, but no nest were built on it. The species had widely branching crown which provided no suitable folks for nest placement and nest stability during incubation.

4. DISCUSSION

4.1. Size of Breeding Population in Lochinvar National Park

In this study, potential lappet faced vulture nests were assumed to be nest numbers: 11, 12, 13, 18, 21, and 22 (**Figure 4**), with their physical dispersion shown in **Figure 5**. This is a minimum figure, as some of the remaining 17 nests in which the species were not observed could likely belong to lappet faced vulture.

Moreover, the study area did not cover the whole floodplain and termitaria zones in the National Park. Therefore, total breeding population could probably be higher than what was recorded in this study.

In the 1970s, for instance [10], eight breeding pairs were recorded in Lochinvar National Park, but such results could not be compared with the present study, because the methodology and area covered during the sur-

vey were not indicated. Based on these estimates however, it can be stated that the breeding population of Lappet-faced vulture in Lochinvar National Park, has probably not decreased significantly since 1970s. A follow-up study would be required to draw a logical and objective conclusion on the matter.

4.2. Monitoring Programme

We recommend a comprehensive monitoring programme for the nesting sites to safeguard the species from further decline as it is currently classified as vulnerable under IUCN's red data list. In addition, wild ungulate populations which are the main source of food need monitoring, as their decline would also negatively affect the species. Monitoring of large trees which are the potential nesting sites should be incorporated in the Park Ecologist's routine functions. Human disturbance in potential and confirmed breeding sites should be minimized so that the birds do not abandon their nests. As earlier reported [7], tall trees of the height exceeding 10 m are critical for placement of raptor nests and these should be protected. Confirmed breeding sites should be zoned as low visitor use zones, because frequent and uncontrolled visitation may lead to nest abandonment. Construction of infrastructure including permanent roads should take into account the need to maintain large trees for raptor nest placements.

This study has established that mature trees ≥10 m in height in areas with minum human disturbance are critical to successful breeding of raptors on the Kafue Flats, Zambia.

5. ACKNOWLEDGEMENTS

We wish to thank the Regional Manager Mrs. Marina Sibbuku for allowing the researchers to operate in the National Park without interruption, Mr Wataru Tokura for initiating the survey in 2009. Mr. Benjamin Wishikoti for his skills in nest identification and other readers that contributed through constructive criticisms.

REFERENCES

[1] Oberprieler, U. and Cillie' B. (2009) The raptor guide of Southern Africa. Game Parks Publishing, Pretoria.

[2] Hardy, E. (1947) The northern lappet faced vulture in Palestine—A new record for Asia. *Auk*, **64**, 471-472.

[3] BirdLife International (2007) Haliaeetus vocifer. *IUCN Red List of Threatened Species.* International Union for Conservation of Nature.

[4] BirdLife International (2009) Haliaeetus vocifer. *IUCN Red List of Threatened Species.* International Union for Conservation of Nature, Gland.

[5] Hadoram, S. (1987) Field chracaters of the Negev lappet

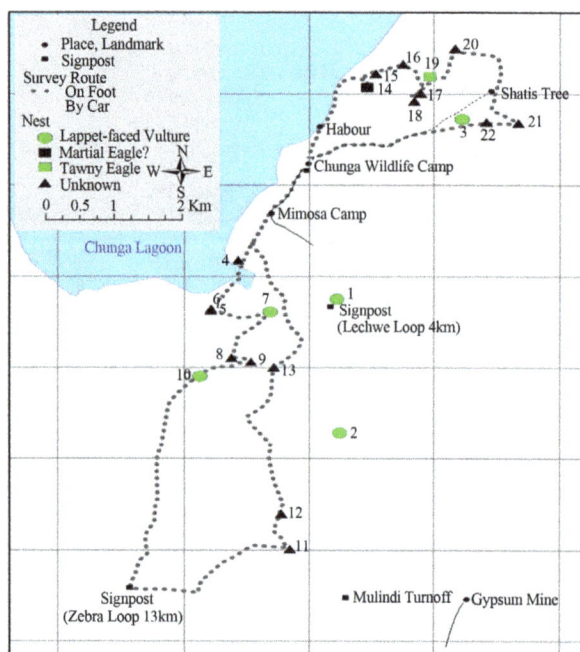

Figure 5. Physical location of active and confirmed lappet faced vulture nests and other nests in Lochinvar National Park, Kafue Flats, Zambia.

faced vulture. *Proccedings of the 4th International Iden-tification Meeting*, Eilat, 1-8 November 1986, 8-11.

[6] Ferguson-Lees, J. and Christie, D.A. (2001) Raptors of the world. Houghton Mifflin Company, New York.

[7] Chomba, C. and Msimuko, E. (2013) Nesting patterns of raptors; White backed vulture (*Gyps africanus*) and African fish eagle (*Haliaeetus vocifer*) in Lochinvar Na-tional Park, on the kafue flats, Zambia. *Open Journal of Ecology*, **3**, 325-330.

[8] BirdLife International (2012) Torgos tracheliotos. IUCN 2012. IUCN Red List of Threatened Species.

[9] Leonard, P. (2005) Important bird areas in Zambia. Zam-bian Ornithological Society, Lusaka.

[10] ZAWA (2010) Report on the operations of the JICA Volunteer Biologist Mr. Wataru Tokura, to the Director Research. Zambia Wildlife Authority, Chilanga. Unpub-lished Report.

6

Comparative analysis of magnetic fields, low temperatures and their combined action on growth of some conditionally pathogenic and normal humane microflora

Olena V. Derev'yanko[1], Oleksandr I. Raichenko[1], Vladimir S. Mosienko[2],
Vladimir O. Shlyakhovenko[2], Yuri V. Yanish[2], Olena V. Karnaushenko[2]

[1]Frantsevych Institute for Problems of Materials Science of NASU, Kyiv, Ukraine; *Corresponding Author
[2]Kavetsky Institute of Experimental Patology, Oncology and Radiobiology of NASU, Kyiv, Ukraine

ABSTRACT

Growth dynamic of bacterial population after influence of magnetic field and cryoaction has been studied. *In vitro* **experiments were performed on** *Staphylococcus aureus* **(202),** *Staphylococcus aureus* **(wild)** *Micrococcus lysodeicticus* **and** *Pseudomonas aeruginosa.* **Cryotreatment: it has been demonstrated that both** *Staphylococcus aureus* **(202) and** *Staphylococcus aureus* **(wild)** *Micrococcus lysodeicticus* **and** *Pseudomonas aeruginosa* **respond to the impact of cryoaction by pronounced growth retardation followed by rapid increase in biomass increment of cryo-resistant clones selected during the experiment. Magnetic influence: it has been shown that neither alternating magnetic field (MF) (induction 30 mT) nor constant MF (induction 52 mT) applied separately from deep freezing have not induced any changes in further growth of both** *Staphylococcus aureus* **(202) and** *Staphylococcus aureus* **(wild) cultures. Combined action: Deep freezing (up to −45˚C) achieved within 70 sec compared with the action of constant MF significantly stimulated (P < 0.05) growth of** *Staphylococcus aureus* **(202) and** *Staphylococcus aureus* **(202). Combined influence of cryoaction and alternating MF had no substantial effect on growth.**

Keywords: Biological Objects; Magnetic Field; Cryotreatment; Bacteria

1. INTRODUCTION

Development of anthropogenic civilization led to new problems. The influence on the living organisms of physical and chemical factors of external environment, products of chemical and radiation contamination, vibrations of acoustic and subacoustic range, in particular, electromagnetic fields, and others take important place.

A modern man is under the mentioned influences not only by passive appearance, using a consumer electronics or transport. The physical factors affected a patient directly: during radial therapy and chemotherapy, in oncologic medical practice, in particular, by using criosurgery methods which acquire all of greater value in clinic and cosmetology.

It is impossible to forget in this sense, that the human's body practically everywhere has the normal microflora (on the whole 10^{14} bacteria and other microorganisms), here and there changed a pathological process or direct infection during a sharp disease. Microorganisms accompany us during all of life (and some time after). They are involved in the production links of microbiological and food industry. Therefore researches directed on suppression of harmful and stimulation of useful microbes are actual problems.

2. MATERIALS AND METHODS

Combined influence of low temperatures and magnetic field was performed with using of bacterial culture suspensions. Representatives of the microflora, occurring in postoperative wounds with septic complicated healing, and normal microflora from human large intestine were used as test objects [1,2]. We carried out our experiments with two opportunistic cocci (*Staphylococcus aureus*, strain 202 and *Staphylococcus aureus*, strain wild) as well as nonpathogenic spore-forming microorganism, *Bacillus cereus*, and blue pus bacillus *Pseudomonas aeruginosa*. Normal microflora was represented by *Lactobacillus delbrueckii*. Bacteria were suspended in distilled water in compliance with all requirements of sterility.

The value of magnetic field used in experiments was 50 mT for constant MF and 30 mT for alternating MF. Temperature range from −20°C to −45°C was used in the experiments in vitro. Magnetic field was applied at both freezing and thawing stages. Exposition of magnetic field applied was in the range of 120 - 900 sec.

To affect microorganisms by MF, the samples of bacterial suspension (volume 0.5 - 1 ml) in Eppendorf tubes of total volume of 1.5 ml were placed between the poles of electromagnet. In case of control and in the experiments when cryoaction was studied only, electromagnet was switched off. Sterile applicator, which brings liquid nitrogen, was immersed in the sample to a depth of 10 mm measured from the top of the meniscus of bacterial suspension.

Cryoaction on bacterial suspensions consisted of their cooling to −45°C and storing for 70 sec. Thawing procedure took place in free regime at room temperature and lasted approximately 10 min each time.

In case of separate action of MF on bacteria, two regimes of magnetization (120 sec or 900 sec) were applied. Since there was no difference in results, exposition of both alternating and constant MF in further experiments was chosen to be 120 sec. In case of combined influence of MF and cryoaction, MF was applied only during freezing of the samples plus another 50 sec after thawing with gradual elevation of temperature inside the tube.

After thawing, samples exposed to cryoaction or combined influence of MF and deep freezing, were divided into aliquots and then suspensions of both stains of *Staphylococcus aureus* and *Bacillus cereus* were sown in standard Petri dishes onto solid nutrient medium (ribopeptone agar, RPA). With regard to *Bacillus cereus*, non-frozen samples (control and samples exposed to MF

only) were subjected to the same procedure. As a rule, results of experiments were calculated on a day after sowing of material, and if it is necessary, to 7 consecutive days.

Staphylococci and *Bacillus* were cultivated at 37°C, *Lactobacillus* was cultivated at 39°C (cultivation was carried out on liquid manufactured nutrient medium MRS). The latter culture was diluted with distilled water at a ratio of 1:25 before spectrophotometrical measurement.

Growth rate of bacteria cultivated on RPA (by 4 standard grades: +, ++, +++, ++++) or absence of growth (−) was determined visually. In addition, we estimated character of growth, the average number of colonies per 1 cm^2 and average diameter of colonies (mm).

Growth of *Lactobacillus* was tested by changes in optical density of culture in liquid medium. Spectrophotometrical analysis at 440, 490 and 540 nm was used because difference between values of optical absorption of control and experimental samples measured at these wave lengths appeared to be the most informative.

3. DISCUSSION

It has been demonstrated that constant MF has no effect on growth of both strains of *Staphylococcus aureus* (202 and wild). Similarly, all above-mentioned microorganisms were not affected by alternating MF (**Tables 1** and **2**; **Figure 1**).

Cryoaction in all studied cases caused growth retardation of both *Staphylococci* and *Bacillus cereus* during the period of 1 - 2 days with the following growth restoration of selected cryostable subpopulation. Since cultural peculiarities of microorganisms survived at rapid freezing have to be interrogated, we preliminarily concluded that such an unexpected selection should be taken into consideration during cryomagnetic therapy of patients.

Intense growth stimulation of *Staphylococcus aureus* (202) induced by combined action of constant MF and freezing in comparison with intact control has been shown. On the other hand, combined influence of alternating MF and cryoaction had no effect on Staphylococcus aureus (wild) compared with freezing action alone (**Tables 1** and **2**; **Figure 1**).

Table 1. Influence of constant magnetic field (MF), deep freezing and their combination on growth of *Staphylococcus aureus* (202).

Type of influence	Intensity of growth over 1 day	A number of colonies per 1 cm^2 of RPA surface	Colony diameter, mm
Intact control	++	127 ± 20	0.21 ± 0.02
Post-MF	++	117 ± 20	0.20 ± 0.01
Cryo	+	28 ± 6	0.69 ± 0.07
Post-MF + Cryo	Conjoint growth	Can not be counted	Can not be counted

Table 2. Influence of alternating magnetic field (MF), deep freezing (Cryo) and their combined action (AltMF + Cryo) on growth of *Staphylococcus aureus* (wild).

Type of influence	Intensity of growth over			A number of colonies per 1 cm^2 of RPA surface on the 7-th day	Colony diameter on the 7-th day, mm
	1 Day	4 Days	7 Days		
Intact control	+++	+++	++++	74 ± 7	0.96 ± 0.05
AltMF	+++	+++	++++	77 ± 9	0.66 ± 0.07
Cryo	-	++	+++	27 ± 3	0.92 ± 0.09
AltMF + Cryo	-	++	+++	39 ± 3	0.84 ± 0.04

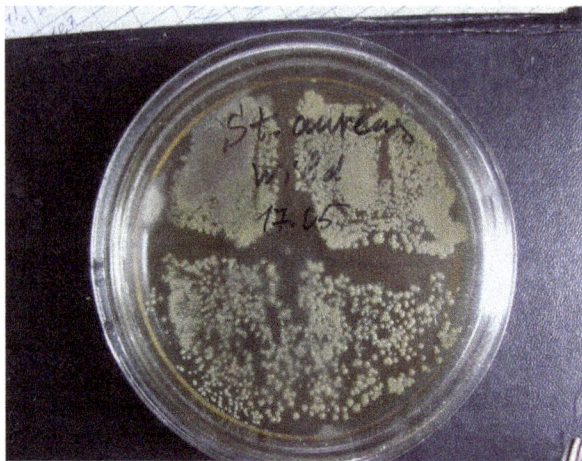

Figure 1. Influence of alternating magnetic field, deep freezing and their combined action on growth of *Staphylococcus aureus* (wild) culture: upper right quadrant—control; lower right quadrant—alternating MF; left lower quadrant—cryoaction; left upper quadrant—cryomagnetic action.

It has been shown that neither alternating MF nor constant MF have not induced any changes in further growth of both, *Micrococcus lysodeicticus* and *Pseudomonas aeruginosa* cultures. However deep and rapid freezing led to cryo-resistant clones appear ance. Since cultural features and degree of pathogenicity of novel sub-strains were not specifically examined, it should be taken into consideration that cryogenic therapy in clinics may attend appearance of microor microorganisms with unidentified antibiotic susceptibility.

It should be noted that, as numerous experiments testify on the cultures of blasts, complete death of all of cellular elements after one freezing cycle of does not take a place usually [3].

Authors reported [4] about research of subzero temperatures and MT combined action on some types of bacteria. In vitro experiments were performed on postoperative septic wound bacterial culture, and on humane large intestine bacteri—*Lactobacterium delbrukii*, and also—on *Bacillus cereus*. *Bacillacae* bacterium family (and *Bacillus* cereus as representative), is spore-forming bacteria which bear drying in a vacuum, extremely low

and high enough temperatures (spores perish only at 120°C). Appearing in favorable conditions, pathogenic microbes proceed in the viability and ill properties, becoming reason of such heavy diseases of animals and people, as anthrax, botulism, tetanus, emphysema but other. *Bacillus cereus* in our experiments is not pathogenic.

Our experiments proved that absorbancy of *Lactobacteriu delbuckii* suspension by comparison to control after Cryo does not change practically, after magnetic influence a bit grows (approximately on 8%). After combined action diminishes substantially (approximately on 25%) [4]. The amount of living microorganisms of *Staphylococcus aureus* (wild), by comparison to control, after Cryo (approximately in 3 times) decrease sharply. After magnetic influence remains practically unchanging, and after combined action (approximately twice) decrease substantially.

Besides, the most interesting result was obtained on the culture of *Bacillus cereus* [4].

When applied separately, cryoaction and magnetic field were on able to kill this microorganism, whereas combined action of these factors led to the complete devitalization of bacterial culture.

The regularity of freezing process and magnetic fields action on the biological systems is fundamental basis of these information study [5]. Staphylococcus *aureus*, *Pseudomonas aeruginosa* and some cultures of *Micrococcus* (luteus and roseus) is marked as a pathogen which accompanies the terminal stages of oncologic process and worsens his course [6].

On the whole, afore-mentioned experiments have very important character not only in the question of humane ecology, of health protection and improvement of his life quality. Development of industry conduces to appearance of waste, including such, which contain new anthropogenic components. In sense that authors succeeded to set some decline of survivability of some malignant bacteria, these experiments can be useful to the biotechnologies which disinfect these waste. Authors hope that as a result of deepening of such researches it will be succeeded to extend the list of malignant microorganisms which can

Comparative analysis of magnetic fields, low temperatures and their combined action on growth of some
conditionally pathogenic and normal humane microflora

39

be disinfect by described and other physical methods.

4. CONCLUSIONS

- It has been shown that neither alternating magnetic field (MF) (induction 52 mT) nor constant MF (induction 30 mT and frequency 50 Hz) applied separately from deep freezing have not induced any changes in further growth of both *Staphylococcus aureus* (202), *Staphylococcus aureus* (wild), *Micrococcus lysodeicticus* and *Pseudomonas aeruginosa* cultures.

- Deep freezing (up to –45˚C) achieved within 70 sec has suppressed intensity of growth of *Staphylococcus aureus* (202), *Staphylococcus aureus* (wild) *Micrococcus lysodeicticus* and *Pseudomonas aeruginosa* cultures. However, combined with alternating MF, this factor significantly stimulated (P < 0.05) growth of *Staphylococcus aureus* (202 and wild). Combined *Staphylococcus aureus* influence of cryoaction and alternating MF had no substantial effect on growth *Staphylococcus aureus* compared with the action of deep freezing applied alone.

- The amount of alive microorganism *Staphylococcus aureus* (wild) after cryoaction decrease sharply (approximately in 3 times). It has been demonstrated that both (202) and (wild) *Micrococcus lysodeicticus* and *Pseudomonas aeruginosa* respond to the impact of cryoaction by pronounced growth retardation followed by rapid increase in biomass increment of cryo-resistant clones selected during the experiment. Since cultural features and degree of pathogenicity of novel sub-strains were not specifically examined, it

should be taken into consideration that cryogenic therapy in clinics may attend appearance of microorganisms with unidentified characteristics.

The constant and alternating MFs affect bacteria in different ways. This fact can be explained by high resistance of procaryotic organisms towards influence of various surrounding factors. To summarize, our data are meant to stimulate further study in this area and perspectives of their practical application.

REFERENCES

[1] Buchanan, P.E. and Gibbons, K.E. (1974) Bergeu's manual of determinativ bacteriology. 8th Edition, Williams S. Wilkins, Baltimore.

[2] Smirnov, V.V., Reznik, S.R. and Sorokulova, I.B. (1980) Methodical recommendations for isolation and identification of Bacillus subtilis—Mesentericus bacteria group from human and animal organisms. Zdorov'e, Kiev, 26. (in Russian)

[3] Afanas'eva, N.I., Shevtsov, V.Ch. and Mysghychuk, A.V. (2002) Perspectives of application of cryo-surgery in oncology. *The International Medical Journal*, **1-2**, 156-159.

[4] Raichenko, O.I., Mosienko, V.S., Shlyakhovenko, V.O., Derev'yanko, O.V., Yanish, Y.V. and Karnaushenko, O.V. (2012) Combined action of low temperature and magnetic field of different intensities on growth of some bacterial species *in vitro*. *Health*, **4**, 249-252.

[5] Binhy, V.N. (2002) Magnetobiology: Underlying physical problems. Academic Press, San Diego.

[6] Clark, H.R. (1995) The cure for all diseases. New Century Press, Chulu Vista.

7

Broad bean cultivars increase extrafloral nectary numbers, but not extrafloral nectar, in response to leaf damage

Edward B. Mondor[*], **Carl N. Keiser, Dustin E. Pendarvis, Morgan N. Vaughn**

Department of Biology, Georgia Southern University, Statesboro, USA; [*]Corresponding Author

ABSTRACT

Phenotypic plasticity allows organisms to maximize fitness, by optimizing the expression of costly defensive traits. Broad bean, *Vicia faba* L. "Broad Windsor", produces increased numbers of extrafloral nectaries (EFNs) in response to leaf damage to attract mutualistic partners and reduce herbivory. It is currently unknown, however, whether EFN induction is cultivar-specific or is a more general phenomenon. It has also not been determined whether broad beans increase nectar secretion rates in conjunction with EFN induction. We hypothesized that: a) as all broad beans have conspicuous EFNs, all cultivars should produce additional EFNs in response to leaf damage, and b) overall nectar secretion rates should increase with EFN numbers, to attract additional mutualists. We tested our hypothesis by subjecting three broad bean cultivars, *Vicia faba* L. "Broad Windsor", "Stereo", and "Witkiem" to mechanical leaf damage. The degree of change in plant traits associated with growth, in addition to EFN induction, was assessed 1 week after leaf damage. Extrafloral nectar volumes were also assessed, every 24 hours, pre- and post-leaf damage. We confirmed our first, but rejected our second, hypothesis. All cultivars produced additional EFNs, but none increased extrafloral nectar volumes, when experiencing leaf damage. Further experimentation is required to determine if energetic tradeoffs limit multiple forms of defense (*i.e.*, EFN vs. nectar induction), or if this alternative strategy is adaptive for attracting and retaining mutualists. Understanding the costs and benefits of EFN vs. nectar induction will provide insight into the evolution of defensive mutualisms between plants and predatory arthropods.

Keywords: Broad Bean; Defense; Extrafloral Nectar; Extrafloral Nectary; Herbivory; Inducible Defense; Mutualism; Phenotypic Plasticity; *Vicia faba* L.

1. INTRODUCTION

When herbivory or predation risk increases, some plants and animals have the ability to alter their phenotype to decrease risk of further attack [1-4]. This plasticity enables organisms to express costly traits when required, and to reduce or eliminate these costs when such phenotypic expression is not necessary [5-7]. Inducible defenses, such as altered phenotypic expression, are adaptive when risk of attack is unpredictable and infrequent [2,8,9]. Conversely, if herbivory or predation is predictable and recurrent, constitutive defenses would be selected for, as the costs of developing and expressing defensive traits would be offset by the benefits of increased survival and (or) reproduction [2,8,9]. Irregardless of the modality, organisms optimize defensive traits to maximize fitness [10,11].

Defensive mutualisms, whereby one organism produces rewards to attract other organisms for protection, have been commonly observed [12-15]. In plants, over 93 families produce extrafloral nectaries (EFNs) [16], sugar producing structures outside of floral structures, which attract predatory arthropods, most commonly ants [14,17-20], but other insects as well [21,22]. While the evolution of EFNs has been debated [16,23,24], these structures frequently facilitate defensive mutualistic interactions between plants and natural enemies of the plant's herbivores [18,25-29]. Increased plant survival and (or) reproduction has been noted when predatory insects visit EFNs [19,20,27,28]. Increased plant fitness could maintain the expression of these sugar-producing organs, despite the initial selective pressures resulting in their evolution.

To facilitate the attraction and (or) retention of mutualists, many plants increase nectar secretion rates from existing EFNs in response to leaf damage [27,28,30,31].

More rarely, some species, such as broad bean *Vicia faba* L. [32-34] and sweet cherry *Prunus avium* L. [35], produce additional EFNs when damaged. Located on the leaf stipules at the base of the petioles, broad bean EFNs range in degree of purple coloration creating a conspicuous visual display (purple EFNs on a light green stipule—[32]), in contrast to an inconspicuous display (green EFNs on green leaf or stem tissue) found on many other plant species [29,36,37]. Herbivore-specific elicitors are not required for [32,33], and both abiotic and biotic factors influence [33,34,38], EFN induction in broad bean plants.

Thus far, EFN phenotypic plasticity has been examined in only two cultivated varieties (cultivar) of broad bean, *V. faba* "Broad Windsor" [32,33,38] and "Hangdown" [34]. It is not known whether other cultivars produce additional EFNs in response to herbivory (*i.e.*, if induction is a cultivar-specific response or a more general phenomenon in this species). To our knowledge, all plants of this species produce conspicuous EFNs [39,40], thus selection should promote similar induction responses in all varieties. It is also unknown whether broad bean plants simultaneously alter extrafloral nectar secretion rates. Though increased extrafloral nectar secretion rates are commonly observed in other plant species [27,28,36,41], other species have not been shown to increase EFN numbers. Thus, it is unknown if most broad bean cultivars increase EFN numbers, nectar secretion rates, or both.

We hypothesize that EFN induction in broad bean is not cultivar-specific, but rather, all cultivars will produce additional EFNs in response to leaf damage. As all broad bean cultivars have conspicuous EFNs [39,40], it would be adaptive to produce a larger visual display to attract additional defensive mutualistic partners (*i.e.*, predatory arthropods—[42]). We also hypothesize that plants will increase extrafloral nectar secretion rates, in response to leaf damage. If plants increase per-plant nectar production, they are more likely to attract and retain mutualistic partners [14,17-22]. Here, we test these two hypotheses by assessing the ability of three broad bean, *V. faba*, cultivars to produce additional EFNs and (or) extrafloral nectar over 1 week following mechanical leaf damage.

2. MATERIALS AND METHODS

2.1. Plants

Broad bean, *V. faba*, seeds were individually planted in Fafard 3B potting mix (Conrad Fafard Inc., Agawam, MA) in 1L round, black, plastic pots. Three cultivars were used for the experiments: Broad Windsor, Stereo, and Witkiem. Plants were watered daily, top-dressed with Osmocote 14-14-14 N-P-K slow-release fertilizer (Scotts-Sierra Horticultural Products, Marysville, OH) prior to seedling emergence, and grown under greenhouse condi-

tions (13°C - 6°C, 22% - 45% rh, natural lighting) in a computer-generated random order using JMP 8.0 [43], to control for any differential lighting effects.

2.2. Is EFN Induction, in Response to Leaf Damage, Cultivar Specific?

When plants were 15 cm tall with 4 true leaves (averaged across all experiments), we recorded initial plant traits: plant height, number of partially expanded leaves, number of fully expanded leaves, and number of EFNs. Presence of EFNs was determined using a magnifying glass (Merangue LG807BL; Merangue International Ltd., Markham, ON Canada), as nectaries can differ in degree of purple coloration [33]. After recording these traits, leaf damage treatments were administered. For replicates with leaf damage, the outer one-third of each fully expanded leaf pair was excised using floral scissors. To ensure that compounds were not transferred between plants, the scissors were cleaned with an alcohol swab after excising tissue from each plant.

We used mechanical leaf damage for our treatments because the degree of tissue removal, compared to real herbivory, can be carefully controlled [44]. Plants exhibit similar increases in nectar secretion rates in response to both natural and mechanical leaf damage [27,36]. It is believed that "damage self-recognition" cues resulting from plant tissue damage *per se*, and not a herbivore-specific elicitor, is the necessary stimulus for these induction responses [45]. Furthermore, in previous studies, mechanical damage clearly increased EFN numbers in *V. faba* [32,33] and *P. avium* [35].

After assessing initial plant traits, broad beans were allowed to grow for 1 week, at which time plant traits were assessed a second time. Pre-treatment values were subtracted from post-treatment values to quantify the degree of change in each character. Immediately after assessing traits, plant shoots were cut at soil level, roots washed, and both plant parts placed in separate paper bags for drying. Root and shoot portions of each plant were dried for 2 weeks, then weighed to the nearest 0.01 g (Ohaus GT4100 balance; Ohaus Corporation, Pine Brook, NJ, USA).

This experiment was replicated three times. In trials 1, 2, and 3, replicate numbers were: Broad Windsor, no leaf damage—12, 11, 15, leaf damage—13, 11, 15; Stereo, no leaf damage—10, 8, 8, leaf damage—13, 10, 10; Witkiem, no leaf damage—5, 10, 8, leaf damage—5, 5, 4 respectively. Witkiem replicate numbers were lower than the other cultivars due to low germination rates.

2.3. Do Nectar Volumes Increase, in Different Cultivars, in Response to Leaf Damage?

During our third trial, we also assessed extrafloral

nectar secretion rates. Prior to conducting the leaf damage treatments, we collected and measured total nectar volumes per plant using microcapillary pipets (Kimble Glass Inc, Vineland, NJ, USA). Nectar was removed from each plant 24 hours before treatments (to remove any "standing crop" of nectar), collected immediately prior to leaf damage treatments ("pre-treatment"), and then every 24 hours (±2 hours) for 4 days after damage ("post-treatment"). Every nectary on a plant was assessed for the presence of nectar, irrespective of whether a droplet was visible. Replicate numbers for this experiment were as previously stated.

2.4. Statistical Analyses

To determine if EFN induction was cultivar- and (or) damage-dependent, data were analyzed with a two-way Analysis of Covariance (ANCOVA). By using ANCOVA, a model combining ANOVA and linear regression, it is possible to determine the effects of key, nominal variables, while correcting for variability in and assessing the influence of continuous variables, on a dependent variable. Main factors in the analysis were: cultivar (Broad Windsor vs. Stereo vs. Witkiem) and leaf damage (no vs. yes). Covariates were: change in stem length, change in number of partially expanded leaf pairs, change in number of fully expanded leaf pairs (all 1 week post-leaf damage), dry shoot weight, and dry root weight. Experiment (1 - 3) was also included as a covariate, to control for differences between trials. The dependent variable was the change in the number of EFNs on each plant, over 1 week.

As shoot weight was highly significant in the previous analysis, and to further assess putative plant defense, an additional one-way ANCOVA was performed. The main factors and covariates were the same as in the EFN induction analysis. The dependent variable, however, was: EFNs per gram of dry shoot weight. This analysis allowed us to assess EFN induction per unit of plant biomass.

Nectar secretion rates, in response to leaf damage, were analyzed with Multivariate Analysis of Variance (MANOVA), "contrast" function. The "contrast" function allowed us to assess changes in nectar secretion across time, with respect to the pre-treatment amounts (i.e., Pre-treatment vs. 24 hours post-treatment, Pre-treatment vs. 48 hours post-treatment, etc.). The same independent variables were used as in the previous analysis. Covariates were: stem length, number of partially expanded leaves, number of fully expanded leaves, and number of EFNs (all immediately pre-treatment). Dry shoot and root weights were also included as covariates, to account for overall biomass effects. The dependent variable was total nectar volume (in microliters) per plant. Nectar volume was cube root transformed [x' = $^{3}\sqrt{x}$] [47], prior

to analysis to normalize the data. For all experimental analyses, statistical calculations were performed using JMP 8.0 [43].

3. RESULTS

3.1. Is EFN Induction, in Response to Leaf Damage, Cultivar Specific?

All broad bean cultivars tested produced additional EFNs in response to leaf damage. Different cultivars had different induction responses; Broad Windsor produced more EFNs than did Stereo or Witkiem ($F_{2,161} = 6.75$, $P = 0.0015$). Plants that suffered leaf damage produced more EFNs than undamaged plants ($F_{1,161} = 34.44$, $P < 0.0001$), and this effect was consistent across cultivars, as demonstrated by no significant interaction ($F_{2,161} = 2.43$, $P = 0.092$; **Figure 1**). There was no difference in plant responses between trials ($F_{1,161} = 1.01$, $P = 0.32$). Extrafloral nectary numbers increased in association with numbers of partially developed ($F_{1,161} = 3.91$, $P = 0.050$) and fully developed ($F_{1,161} = 30.70$, $P < 0.0001$) leaf pairs, and dry shoot weights ($F_{1,161} = 20.92$, $P = 0.0001$). Neither plant heights ($F_{1,161} = 1.73$, $P = 0.19$) nor dry root weights ($F_{1,161} = 0.0018$, $P = 0.97$) were associated with EFN induction responses.

As dry shoot weight was a significant covariate in EFN induction, and to provide a better estimate of plant phenotypic expression on tissue defense, EFN induction per gram of dry shoot weight was assessed. Using this standardized metric, Broad Windsor produced more EFNs than did Stereo, with Witkiem showing an intermediate response ($F_{2,163} = 3.75$, $P = 0.026$). There was still a large effect of leaf damage on induction responses ($F_{1,163} = 24.41$, $P < 0.0001$). Like in the previous analysis, there was no significant interaction between cultivar and leaf damage ($F_{2,163} = 1.17$, $P = 0.31$; **Figure 2**); damaged

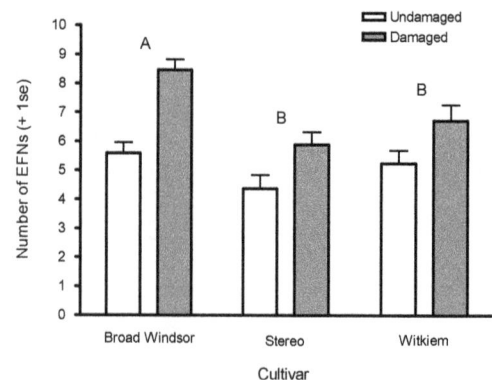

Figure 1. Mean extrafloral nectary (EFN) numbers produced by plants of three broad bean cultivars, *V. faba*, over one week in response to mechanical leaf damage. All cultivars produced more EFNs when damaged. Cultivars with different letters produced different numbers of EFNs overall, Tukey's HSD test ($P \leq 0.05$).

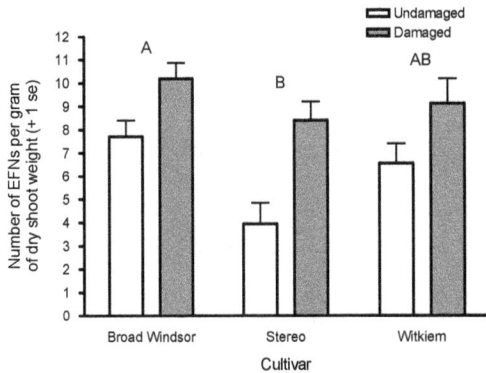

Figure 2. Mean EFN numbers produced by three broad bean cultivars, per gram of dry shoot weight. All cultivars produced more EFNs in response to mechanical leaf damage. Cultivars with different letters produced different numbers of EFNs overall, Tukey's HSD test ($P \leq 0.05$).

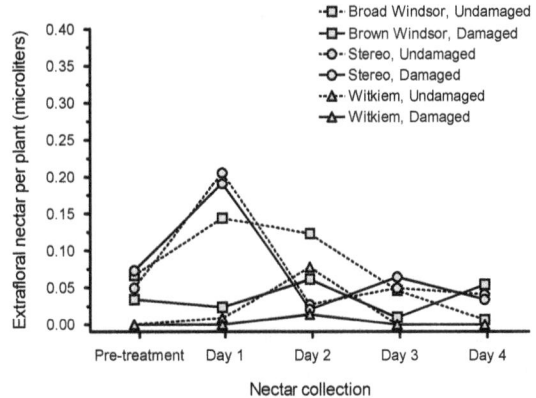

Figure 3. Mean volumes of extrafloral nectar produced, pre- and post-leaf damage, by three broad bean cultivars. There was no effect of leaf damage on extrafloral nectar secretion rates.

plants always produced more EFNs, irrespective of cultivar. Only two covariates were related to EFN induction responses per unit of biomass: trial ($F_{1.163} = 27.31$, $P < 0.0001$) and plant height ($F_{1.163} = 18.27$, $P < 0.0001$). Plants that grew less in height over 1 week, had larger induction responses. Numbers of partially developed ($F_{1.163} = 0.0001$, $P = 0.99$), and fully developed ($F_{1.163} = 1.38$, $P = 0.24$) leaf pairs were not reliable predictors of EFN induction responses.

3.2. Do Nectar Volumes Increase, in Different Cultivars, in Response to Leaf Damage?

We found, unequivocally, that nectar secretion rates in all three cultivars did not significantly increase in response to leaf damage. Compared to the pre-treatment nectar amounts, post-treatment nectar volumes did not differ among cultivars ($F_{8.90} = 1.40$, $P = 0.21$) nor did they increase in response to leaf damage ($F_{4.45} = 2.19$, $P = 0.086$).

In fact, the non-significant trend was for undamaged plants to have higher nectar secretion rates than damaged plants. There was no interaction between cultivar and leaf damage ($F_{8.90} = 1.13$, $P = 0.35$; **Figure 3**). Furthermore, none of the covariates: pre-treatment plant height ($F_{4.45} = 0.99$, $P = 0.42$), pre-treatment number of partially expanded leaves ($F_{4.45} = 0.28$, $P = 0.89$), pre-treatment number of fully expanded leaves ($F_{4.45} = 1.32$, $P = 0.28$), pre-treatment number of EFNs ($F_{4.45} = 0.25$, $P = 0.91$), dry shoot weight ($F_{4.45} = 1.52$, $P = 0.21$), and dry root weight ($F_{4.45} = 0.72$, $P = 0.59$) were significant.

4. DISCUSSION

Plants benefit by producing nutritious rewards to attract and retain mutualistic partners, to increase fitness [14,17-22,25,47]. As extrafloral nectaries (EFNs) and

nectar may be costly to produce [33,48-50], phenotypic plasticity allows plants to increase expression only when risk of herbivory increases. Previously, two broad bean cultivars, *Vicia faba* "Broad Windsor" [32,33] and "Hangdown" [34], were shown to alter EFN numbers in response to leaf damage (though see [38]). Here we tested the hypotheses that, in response to leaf damage: 1) all broad bean cultivars will produce additional EFNs, and 2) plants will also increase nectar secretion rates, to attract additional mutualists. Here, we present data that confirms our first, but rejects our second, hypothesis. Every cultivar produced additional EFNs, though there was no difference in nectar secretion rates, after experiencing leaf damage.

It is clear that EFN induction is a species-wide effect in broad bean plants. Broad Windsor has become the cultivar of choice for experiments on EFN induction [32,33,38], because it is easy to grow, is a large, robust plant, and has easily recognizable EFNs. It is interesting that the largest induction responses occurred in the Broad Windsor cultivar, of all the varieties tested. Some cultivars are more difficult to grow and are less robust (EB Mondor, Georgia Southern University, USA, unpubl. res.). Some researchers have suggested that a more appropriate measure of putative plant defense is looking at EFN induction per unit of leaf and stem tissue, *i.e.*, the actual plant area to be defended [28,51]. When looking at EFN induction per unit of biomass (*i.e.*, per gram of dry shoot weight), the smallest statured cultivar, Stereo, had the largest induction response. Thus, while plant vigor may be correlated with overall EFN numbers, the smallest plants may be the most heavily defended on the basis of EFNs per unit of above-ground biomass.

Most puzzling is why broad bean plants increase EFN numbers, but not nectar secretion rates in response to leaf damage. Many plant species increase extrafloral nectar secretions when damaged [27,28,36,41]. We propose at least three non-mutually exclusive reasons why broad bean plants might increase EFNs but not increase nectar secretions. First, plants may not have the resources to increase both EFN numbers and nectar secretion. Plants have finite resources available for growth, defense, and reproduction [52]. Though the overall costs of EFN induction have not yet been assessed, it is resource-dependent [33]. Nectar is also believed to be costly to produce [49,50]. Thus, this experiment may demonstrate a classic example of a defensive trade-off; an increase in one defensive trait decreases another [52]. These trade-offs may be especially prevalent in annual plants, like broad beans, as fitness (seed production) is solely determined in one year, as opposed to perennial plants where fitness is accrued over multiple years [53]. Second, it may be adaptive for plants to distribute the same level of resources over a more widely scattered area. By increasing EFNs, predatory arthropods would be required to search more plant surface area to obtain the same level of nutrients, as opposed to being directed to a specific area to defend [54]. It is postulated that herbivores would be more likely to be discovered and removed from the plant, due to the increased levels of search behavior over the plant surface, thereby further reducing herbivory [14,17, 18]. Third, it is possible that this effect has a physiological basis. Compensatory growth may occur in response to leaf damage [55,56], thereby accelerating stipule production prior to leaves unfolding, leading to more EFNs per leaf pair on a plant [33]. A transitory reallocation of resources from defense to growth, underlying this inducible defense, would explain how EFN numbers per leaf pair can change significantly in just a few days [34]. Further experimentation is required to better understand this phenomenon.

The adaptive significance of EFN vs. nectar induction needs to be addressed, in an array of plant species. Understanding the tradeoffs between these two defensive strategies may provide great insight into the selective pressures that mediate defensive mutualisms [31,45,57]. These two defensive strategies may function optimally for different mutualistic partners. For example, ants frequently visit EFNs, and their presence has been linked to increased plant survival and (or) reproduction [17,19,28, 47,58,59]. Many other predatory arthropods, however, are attracted to and feed on EFNs (e.g., predatory coccinellids, parasitic Hymenoptera—[21,22,60-62]). It should be investigated whether plants with conspicuous EFNs have different natural enemy guilds compared to those defending plants with inconspicuous EFNs.

Mutualistic interactions are dynamic associations that are highly dependent on the current ecological conditions [63,64] (though see [65]). Defensive mutualisms between plants and predatory insects are commonly mediated through EFNs and their nutrient-rich secretions [14,16,18]. Multiple broad bean cultivars produce additional EFNs in response to leaf damage, but none of these plants increase nectar secretion rates. A better understanding of the cost/benefit tradeoffs between EFNs and nectar [66], in both wild and cultivated plants, will provide great insight into the evolution of plant defense [1,2,53].

5. ACKNOWLEDGEMENTS

We thank M. Tremblay and M. Heil for comments on this manuscript. We also thank B. Cantwell, K. Duff, and A. Shepard for laboratory assistance. This work was supported by the Department of Biology and Georgia Southern University.

REFERENCES

[1]   Karban, R. and Baldwin, I.T. (1997) Induced responses to herbivory. University Press of Chicago, Chicago.

[2]   Tollrian, R. and Harvell, C.D. (1999) The ecology and evolution of inducible defenses. Princeton University Press, Princeton.

[3]   Sultan, S.E. (2000) Phenotypic plasticity for plant development, function, and life-history. *Trends in Plants Science*, **5**, 537-542.

[4]   Dewitt, T.J. and Scheiner, S.M. (2004) Phenotypic plasticity: functional and conceptual approaches. Oxford University Press, New York.

[5]   West-Eberhard, M.J. (1986) Alternative adaptations, speciation, and phylogeny. *Proceedings of the National Academy of Sciences of the USA*, **83**, 1388-1392.

[6]   West-Eberhard M.J. (1989). Phenotypic plasticity and the origins of diversity. *Annual Review of Ecology and Systematics*, **20**, 249-278.

[7]   Moran, N.A. (1992). The evolutionary maintenance of alternative phenotypes. *The American Naturalist*, **139**, 971-989.

[8]   Zangerl, A.R. and Rutledge, C.E. (1996) The probability of attack and patterns of constitutive and induced defense: A test of optimal defense theory. *American Naturalist*, **147**, 599-608.

[9]   Karban, R., Agrawal, A.A. and Mangel, M. (1997). The benefits of induced defenses against herbivores. *Ecology*, **78**, 1351-1355.

[10] Stearns, S.C. (1992). The evolution of life histories. Oxford University Press, Oxford.

[11] Steiner, U.K. and Pfeiffer, T. (2007) Optimizing time and resource allocation trade-offs for investment into morphological and behavioral defense. *The American Naturalist*, **169**, 118-129.

[12] Boucher, D.H., James, S. and Keeler, K.H. (1982) The ecology of mutualism. *Annual Review of Ecology and Systematics*, **13**, 315-347.

[13] Pierce, N.E., Kitching, R.L., Buckley, R.C., Taylor, M.F.J. and Benbow, K. (1987) Costs and benefits of cooperation between the Australian lycaenid butterfly *Jalmenus evagoras* and its attendant ants. *Behavioral Ecology and Sociobiology*, **21**, 237-248.

[14] Huxley, C.R and Cutler, D.F. (1991) Ant-plant interactions. Oxford University Press, New York.

[15] Morales, M.A. (2000) Mechanisms and density dependence of benefit in an ant-membracid mutualism. *Ecology*, **81**, 482-489.

[16] Koptur, S. (1992) Interactions between insects and plants mediated by extrafloral nectaries. In: Bernays, E., Ed. *CRC Series on Insect/Plant Interactions, Volume* 4, CRC Press, Boca Raton, 85-132.

[17] Janzen, D.H. (1966) Coevolution of mutualism between ants and acacias in Central America. *Evolution*, **20**, 249-275.

[18] Beattie, A.J. (1985) The evolutionary ecology of ant-plant mutualisms. Cambridge University Press, Cambridge.

[19] Ness, J.H. (2003) *Catalpa bignonioides* alters extrafloral nectar production after herbivory and attracts ant bodyguards. *Oecologia*, **134**, 210-218.

[20] Vesprini, J.L., Galetto, L. and Bernardello, G. (2003) The beneficial effects of ants on the reproductive success of *Dyckia floribunda* (Bromeliaceae), an extrafloral nectary plant. *Canadian Journal of Botany*, **81**, 24-27.

[21] Bugg, R.L. and Ellis, R.T. (1990) Insects associated with cover crops in Massachusetts. *Biology, Agriculture and Horticulture*, **7**, 47-68.

[22] Kost, C. and Heil, M. (2005) Increased availability of extrafloral nectar reduces herbivory in Lima bean plants (*Phaseolus lunatus*, Fabaceae). *Basic and Applied Ecology*, **6**, 237-248.

[23] Wagner, D. and Kay, A. (2002) Do extrafloral nectaries distract ants from visiting flowers? An experimental test of an overlooked hypothesis. *Evolutionary Ecology Research*, **4**, 293-305.

[24] Rosenweig, M.L. (2002). The distraction hypothesis depends on relatively cheap extrafloral nectaries. *Evolutionary Ecology Research*, **4**, 307-311.

[25] Bentley, B.L. (1977) Extrafloral nectaries and protection by pugnacious bodyguards. *Annual Review of Ecology and Systematics*, **8**, 407-427.

[26] Elias T.S. (1983). Extrafloral nectaries: Their structure and distribution. In: Bentley, B.L. and Elias, T.S., Eds. *The Biology of Nectaries*, Columbia University Press, New York, 174-203.

[27] Heil, M., Fiala, B., Baumann, B. and Linsenmair, K.E. (2000) Temporal, spatial and biotic variations in extrafloral nectar secretion by *Macaranga tanarius*. *Functional Ecology*, **14**, 749-757.

[28] Heil, M., Hilpert, A., Fiala, B. and Linsenmair, K.E. (2001) Nutrient availability and indirect (biotic) defence in a Malaysian ant-plant. *Oecologia*, **126**, 404-408.

[29] Doak, P., Wagner, D. and Watson, A. (2007) Variable extrafloral nectary expression and its consequences in quaking aspen. *Canadian Journal of Botany*, **85**, 1-9.

[30] Agrawal, A.A. and Rutter, M.T. (1998) Dynamic antiherbivore defense in ant-plants: The role of induced responses. *Oikos*, **83**, 227-236.

[31] Heil, M. and McKey, D. (2003) Protective ant-plant interactions as model systems in ecological and evolutionary research. *Annual Review of Ecology, Evolution, and Systematics*, **34**, 425-453.

[32] Mondor, E.B. and Addicott, J.F. (2003) Conspicuous extrafloral nectaries are inducible in *Vicia faba*. *Ecology Letters*, **6**, 495-497.

[33] Mondor, E.B., Tremblay, M.N. and Messing, R.H. (2006) Extrafloral nectary phenotypic plasticity is damage- and resource-dependent in *Vicia faba*. *Biology Letters*, **2**, 583-585.

[34] Jaber, L.R. and Vidal, S. (2009) Interactions between an endophytic fungus, aphids and extrafloral nectaries: Do endphytes induce extrafloral-mediated defenses in *Vicia faba*? *Functional Ecology*, **23**, 707-714.

[35] Pulice, C.E. and Packer, A.A. (2008) Simulated herbivory induces extrafloral nectary production in *Prunus avium*. *Functional Ecology*, **22**, 801-807.

[36] Wackers, F.L. and Wunderlin, R. (1999) Induction of cotton extrafloral nectar production in response to herbivory does not require a herbivore-specific elicitor. *Entomologia Experimentalis et Applicata*, **91**, 149-154.

[37] Paiva, E.A.S., Buono, R.A. and Delgado, M.N. (2007) Distribution and structural aspects of extrafloral nectaries in *Cedrela fissilis* (Meliaceae). *Flora—Morphology, Distribution, Functional Ecology of Plants*, **202**, 455-461.

[38] Laird, R.A. and Addicott, J.F. (2007). Arbuscular mycorrhizal fungi reduce the construction of extrafloral nectaries in *Vicia faba*. *Oecologia*, **152**, 541-551.

[39] Erith, A.G. (1930) The inheritance of colour, size, and form of seeds, and of flower colour in *Vicia faba* L. *Genetica*, **12**, 477-510.

[40] Ingels, C.A. (1998) Cover cropping in vineyards: A grower's handbook. University of California, Oakland.

[41] Koptur, S. (1989) Is extrafloral nectar production an inducible defense? In: Bock, J.H. and Linhart, Y.B., Eds., *The Evolutionary Ecology of Plants*, Westview, Boulder, 323-339.

[42] Katayama, N. and Suzuki, N. (2004) Role of extrafloral nectaries of *Vicia faba* in attraction of ants and herbivore exclusion by ants. *Entomological Science*, **7**, 119-124.

[43] SAS Institute Inc. (2008) JMP®, version 8.0. Cary.

[44] Tiffin, P. and Inouye, B. (2000) Measuring tolerance to herbivory: Accuracy and precision of estimates made using natural versus imposed damage. *Evolution*, **54**, 1024-1029.

[45] Heil, M. (2009) Damaged-self recognition in plant herbivore defence. *Trends in Plant Science*, **14**, 356-363.

[46] Quinn, G. and Keough, M. (2002) Experimental design and data analysis for biologists. Cambridge University Press, Cambridge.

[47] Koptur, S. Rico-Gray, V. and Palacios-Rios, M. (1998) Ant protection of the nectaried fern *Polypodium plebeium* in central Mexico. *American Journal of Botany*, **85**, 736-739.

[48] O'Dowd, D.J. (1979) Foliar nectar production and ant activity on a neotropical tree, *Ochroma pyramidale*. *Oecologia*, **43**, 233-248.

[49] Southwick, E.E. (1984) Photosynthate allocation to floral nectar a neglected energy investment. *Ecology*, **65**, 1775-1779.

[50] Pyke, G.H. (1991) What does it cost a plant to produce floral nectar? *Nature*, **350**, 58-59.

[51] Radhika, V., Kost, C., Mithofer, A. and Boland, W. (2010) Regulation of extrafloral nectar secretion by jasmonates in lima bean is light dependent. *Proceedings of the National Academy of Sciences of the USA*, **107**, 17228-17233.

[52] Zangerl, A.R. and Bazzaz, F.A. (1992) Theory and pattern in plant defense allocation. In: Fritz, R. and Simms, E.L., Eds., *Plant Resistance to Herbivores and Pathogens*, Uninversity of Chicago Press, Chicago, 363-392.

[53] Cohen, D. (1994) Modelling the coexistence of annual and perennial plants in temporally varying environments. *Plant Species Biology*, **9**, 1-10.

[54] Wackers, F.L., Zuber, D., Wunderlin, R. and Keller, F. (2001) The effect of herbivory on temporal and spatial dynamics of foliar nectar production in cotton and castor. *Annals of Botany*, **87**, 365-370.

[55] McNaughton, S.J. (1983) Compensatory plant growth as a response to herbivory. *Oikos*, **40**, 329-336.

[56] Trumble, J.T., Kolodny-Hirsch, D.M. and Ting, I.P. (1993) Plant compensation for arthropod herbivory. *Annual Review of Entomology*, **38**, 93-119.

[57] Bronstein, J.L., Alarcon, R. and Geber, M. (2006) The evolution of plant-insect mutualisms. *New Phytologist*, **172**, 412-428.

[58] Stephenson, A.G. (1982) The role of the extrafloral nectaries of *Catalpa speciosa* in limiting herbivory and increasing fruit production. *Ecology*, **63**, 663-669.

[59] De la Fuente, M.A.S. and Marquis, R.J. (1999) The role of ant-tended extrafloral nectaries in the protection and benefit of a Neotropical rainforest tree. *Oecologia*, **118**, 192-202.

[60] Bugg, R.L., Ellis, R.T. and Carlson, R.W. (1989) Ichneumonidae (Hymenoptera) using extrafloral nectar of faba bean (*Vicia faba* L., Fabaceae) in Massachusetts. *Biological Agriculture and Horticulture*, **6**, 107-114.

[61] Limburg, D.D. and Rosenheim, J.A. (2001) Extrafloral nectar consumption and its influence on survival and development of an omnivorous predator, larval *Chrysoperla plorabunda* (Neuroptera: Chrysopidae). *Environmental Entomology*, **30**, 595-604.

[62] Rose, U.S.R., Lewis, J. and Tumlinson, J.H. (2006) Extrafloral nectar from cotton (*Gossypium hirsutum*) as a food source for parasitic wasps. *Functional Ecology*, **20**, 67-74.

[63] Bronstein, J.L. (1994) Conditional outcomes in mutualistic interactions. *Trends in Ecology and Evolution*, **9**, 214-217.

[64] Thompson, J.N. (2005) The geographic mosaic of coevolution. University of Chicago Press, Chicago.

[65] Chamberlain, S.A. and Holland, J.N. (2009) Quantitative synthesis of context-dependency in ant-plant protection mutualisms. *Ecology*, **90**, 2384-2392.

[66] Rutter, M.T. and Rausher, M.D. (2004) Natural selection on extrafloral nectar production in *Chamaecrista fasciulata*: The costs and benefits of a mutualism trait. *Evolution*, **58**, 2657-2668.

8

Seasonal comparison of aquatic macroinvertebrate assemblages in a flooded coastal freshwater marsh

Sung-Ryong Kang[1], Sammy L. King[2]

[1]School of Renewable Natural Resources, Louisiana State University Agricultural Center, Baton Rouge, USA;
*Corresponding Author
[2]U.S. Geological Survey, Louisiana Cooperative Fish and Wildlife Research Unit, School of Renewable Natural Resources, Louisiana State University Agricultural Center, Baton Rouge, USA

ABSTRACT

Marsh flooding and drying may be important factors affecting aquatic macroinvertebrate density and distribution in coastal freshwater marshes. Limited availability of water as a result of drying in emergent marsh may decrease density, taxonomic diversity, and taxa richness. The principal objectives of this study are to characterize the seasonal aquatic macroinvertebrate assemblage in a freshwater emergent marsh and compare aquatic macroinvertebrate species composition, density, and taxonomic diversity to that of freshwater marsh ponds. We hypothesize that 1) freshwater emergent marsh has lower seasonal density and taxonomic diversity compared to that of freshwater marsh ponds; and 2) freshwater emergent marsh has lower taxa richness than freshwater marsh ponds. Seasonal aquatic macroinvertebrate density in freshwater emergent marsh ranged from 0 organisms/m^2 (summer 2009) to 91.1 ± 20.53 organisms/m^2 (mean ± SE; spring 2009). Density in spring was higher than in all other seasons. Taxonomic diversity did not differ and there were no unique species in the freshwater emergent marsh. Our data only partially support our first hypothesis as aquatic macroinvertebrate density and taxonomic diversity between freshwater emergent marsh and ponds did not differ in spring, fall, and winter but ponds supported higher macroinvertebrate densities than freshwater emergent marsh during summer. However, our data did not support our second hypothesis as taxa richness between freshwater emergent marsh and ponds did not statistically differ.

Keywords: Freshwater Emergent Marsh; Freshwater Pond; Aquatic Macroinvertebrate Assemblage; Hydrologic Connection

1. INTRODUCTION

Spatial and temporal variation in habitat conditions affects movements of aquatic macroinvertebrates [1,2] and contributes to the regulation of assemblages [3]. In riverine ecosystems, flow regime plays a major role in structuring patterns of biotic composition and diversity [1,2,4,5]. Similarly, marsh flooding and drying in freshwater marshes may be important factors affecting seasonal aquatic macroinvertebrate density and distribution. Lateral hydrologic connectivity between ponds and freshwater emergent marsh during flooded marsh conditions may decrease aquatic macroinvertebrate density in ponds while aquatic macroinvertebrate density in the freshwater emergent marsh may increase due to aquatic macroinvertebrate movement from ponds to the freshwater emergent marsh. Several studies suggest that ponds (or other habitats) that have a low degree of connection with adjacent waterways support relatively few organisms due to limited recruitment [6] and severe environmental conditions [7-9]. The effects of marsh flooding and drying between ponds and emergent marsh on aquatic macroinvertebrate assemblages in coastal marshes are relatively unknown and poorly studied.

Besides hydrologic connectivity, several studies suggest that hydroperiod affects the assemblages of wetlands macroinvertebrates [10-12]. [13] noted that desiccation stress and the physical environment (e.g., extreme temperatures, low dissolved oxygen) are expected to exert a dominant influence on aquatic macroinvertebrate assemblages in wetlands with short hydroperiods. [14] indicated that temporary ponds support relatively few aquatic macroinvertebrates when compared to more permanent sites. Also, [15] noted that ephemeral and temporary lakes tended to have fewer taxa than semi-permanent channel or terminal lake habitats in a central Australian arid-zone river. In this sense, the relatively long hydroperiod of ponds may allow for higher macroinvertebrate density than in the freshwater emergent marsh due to limited availability of macroinvertebrates as a result of drying in freshwater emergent marsh.

Coleopterans are known to possess physiological and behavioral mechanisms to survive desiccation during dry periods (e.g., Dytiscidae; [16]) and these traits may allow them to avoid the deep water habitat that commonly support relatively large and strong predators (e.g., fish, odonates; [17]).Thus, coleopterans in freshwater emergent marsh may be more abundant than in ponds. In contrast, odonates require a relatively long hydroperiod for the full development of nymphs even though they occur in shallow water [18,19]. Therefore, ondonates may avoid freshwater emergent marsh because of the risk of drying. In addition, macrophyte (e.g., SAV: submerged aquatic vegetation) cover appears to affect macroinvertebrate distribution by providing refuge from predators [20], and increasing the availability of food resources [21]. [17] characterized aquatic macroinvertebrate assemblages in freshwater marsh ponds, however, a paucity of information [22] exists on the aquatic macroinvertebrate assemblages in freshwater marshes and their similarity to assemblages in freshwater marsh ponds.

A clear understanding of the similarity and differences between freshwater emergent marsh and marsh ponds would enhance our understanding of aquatic macroinvertebrate habitat requirements in freshwater marshes because freshwater marsh and pond habitats represent very different physical and chemical conditions. The principal objectives of this study are to characterize the seasonal aquatic macroinvertebrate assemblage in a freshwater emergent marsh and compare aquatic macroinvertebrate species composition, density, and taxonomic diversity to that of freshwater marsh ponds. We hypothesize that 1) freshwater emergent marsh has lower seasonal density and taxonomic diversity compared to that of freshwater marsh ponds; and 2) freshwater emergent marsh has lower taxa richness than freshwater marsh ponds.

2. STUDY AREA AND METHODS

2.1. Study Area

This study was conducted in White Lake Wetlands Conservation Area (WLWCA, 29°52'N, 92°31'W) in the Chenier Plain of southwestern Louisiana. WLWCA is bounded on the south by White Lake (28.2 km north of the Gulf of Mexico) and is a 28,719 ha freshwater marsh.

Dominant vegetation is maidencane (*Panicum hemitomon* Schultes) and bulltongue arrowhead (*Sagittaria lancifolia* Linnaeus).

2.2. Data Collection

We randomly selected three sites for more intensive study in the edge of each emergent marsh site (*i.e.*, 100 m from channel or pond margin). To assess variation in environmental variables, monthly we measured salinity, dissolved oxygen (DO, mg/l), and water temperature (°C) with a YSI Model 85 Water Quality Meter. These variables were measured 2 - 3 cm above the sediment at each sampling point between 08:00 AM and 17:00 PM (one time per month). Percent cover of SAV in a 1×1-m frame was also randomly determined at three points around each sampling point (*i.e.*, sampling point + 2 random points) and the mean coverage of SAV in the 1 m^2 was calculated. We used a meter stick to check sampling point water depth (SPWD, water depth at sampling points).

To determine aquatic macroinvertebrate assemblage structure, we sampled each emergent marsh monthly from April 2009 to February 2010. For purposes of this study, seasons were defined as: 1) Spring 2009 (April-May); 2) Summer 2009 (June-August); 3) Fall 2009 (September-November); 4) Winter 2009 (December-February). We sampled water-column macroinvertebrates using a D-shaped sweep net with a 30-cm opening and 1-mm mesh size. Many previous studies [23-28] have found that a D-shaped sweep net is an effective sampling method for water-column macroinvertebrates in ponds. We conducted a total of 10 continuous sweeps of 2-m long each (surface covered 6 m^2; [23]) at a randomly selected point.

[17] provides detailed results of environmental variables that include salinity, DO, temperature, SAV, and SPWD and of aquatic macroinvertebrate species composition, density, and taxonomic diversity in two pond types (*i.e.*, permanently connected pond [PCP: permanently connected channel during all seasons]; temporarily connected pond [TCP: temporarily connected by surface water to the surrounding marsh but not permanently connected to a channel]]).

2.3. Statistical Analysis

Data are reported as mean ± SE, and significance level was set at $\alpha = 0.05$. Analyses of variance (ANOVA) and T-test (Proc Mixed, Version 9.3, SAS Institute, Cary, North Carolina) were used to test for statistical differences in environmental variables and aquatic macroinvertebrate density by season. We used a one-way ANOVA for each response variable. Models included the environmental variable, as well as density and taxonomic diversity. We conducted a one-way ANOVA with one fixed effect. Significant one-way ANOVA effects were tested using post-hoc comparisons of Tukey adjusted least squares means. For one-way ANOVA analyses, data were tested for normality with the Shapiro-Wilks test. In the event that the residuals were not normally distributed, the data were natural log-transformed.

3. RESULTS

Seasonal salinity was higher in spring and fall than in summer ($F_{3,8}$ = 12.0, p = 0.003). Comparison of DO and water depth indicated their highest values were in winter but temperature values were lowest in winter (DO: $F_{3,8}$ = 38.6, $p < 0.001$; temperature: $F_{3,8}$ = 132.1, $p < 0.001$; water depth: $F_{3,8}$ = 512.6, $p < 0.001$). Seasonal SAV cover did not differ. In freshwater ponds, salinity, temperature, and SAV were the lowest in winter but DO was the highest. The highest SPWD was in fall. The full results of water chemistry (i.e., salinity, DO, temperature), SAV cover, and SPWD in freshwater marsh ponds (i.e., PCPs and TCPs) are presented in **Table 1** and [17].

We collected 5,114 aquatic macroinvertebrates of 35 taxa from 33 samples. The dominant taxon during all flooded seasons was Chironomidae (spring 2009: 38.3%; fall 2009: 14.8%; winter 2009: 12.2%; **Table 2**). Seasonal aquatic macroinvertebrate density in freshwater emergent marsh ranged from 91.1 ± 20.53 organisms/m^2 (mean ± SE; spring 2009) to 0 organisms/m^2 (summer 2009; **Table 3**). Density in spring 2009 was higher than in all other seasons ($F_{3,7}$ = 31.3, $p < 0.001$). Taxonomic diversity did not differ. **Table 4** provides a detailed comparison of aquatic macroinvertebrate densities among freshwater emergent marsh and ponds. Aquatic macroinvertebrate density between freshwater emergent marsh and ponds did not differ in spring, fall, and winter but ponds supported higher macroinvertebrate densities than freshwater emergent marsh ($F_{2,6}$ = 13.2, p = 0.006) during summer. Taxa richness between freshwater emergent marsh and ponds did not statistically differ.

4. DISCUSSION

The present study considered the hypothesis that freshwater emergent marsh had lower aquatic macroinvertebrate density and taxonomic diversity than freshwater marsh ponds. When emergent marsh is flooded (i.e., lateral connectivity with pond or channel), aquatic macroinvertebrates can move from ponds into the freshwater emergent marsh, resulting in increased aquatic macroinvertebrate density in the freshwater emergent marsh. Our results indicated that aquatic macroinvertebrate density and taxonomic diversity among freshwater emergent marsh and ponds did not statistically differ in spring, fall, and winter (i.e., flooded marsh condition).

Table 1. Comparison of seasonal means (±SE) of water chemistry (n = 108), water depth (n = 108), and SAV cover (n = 108) in freshwater ponds and emergent marsh.

	FEM[*]	PCP[**]	TCP[***]
Spring			
Salinity (ppt)	0.4 (0.07)	1.2 (0.15)	0.4 (0.07)
Dissolved oxygen (mg/l)	3.1 (0.76)	2.8 (0.24)	3.3 (0.59)
Temperature (°C)	30.7 (1.64)	30.1 (2.05)	27.2 (1.95)
Water depth (cm)	23.2 (0.07)	34.4 (4.03)	36.4 (1.04)
SAV coverage (%)	40.0 (4.19)	34.4 (5.47)	32.2 (4.75)
Summer			
Salinity (ppt)	0.0 (0.00)	1.6 (0.08)	0.5 (0.07)
Dissolved oxygen (mg/l)	0.0 (0.00)	1.4 (0.33)	1.1 (0.46)
Temperature (°C)	0.0 (0.00)	31.4 (0.85)	30.8 (0.58)
Water depth (cm)	0.0 (0.00)	30.8 (3.15)	16.0 (0.80)
SAV coverage (%)	0.0 (0.00)	49.4 (20.69)	34.4 (12.03)
Fall			
Salinity (ppt)	0.5 (0.08)	0.5 (0.01)	0.3 (0.05)
Dissolved oxygen (mg/l)	2.7 (0.16)	2.4 (0.64)	1.2 (0.45)
Temperature (°C)	23.9 (0.76)	23.1 (3.33)	22.0 (2.21)
Water depth (cm)	27.9 (0.80)	40.9 (1.67)	47.9 (3.50)
SAV coverage (%)	46.7 (12.58)	37.2 (16.17)	36.7 (18.95)
Winter			
Salinity (ppt)	0.3 (0.06)	0.3 (0.02)	0.2 (0.00)
Dissolved oxygen (mg/l)	5.8 (0.10)	5.6 (0.14)	4.6 (1.03)
Temperature (°C)	12.3 (1.47)	11.9 (1.69)	11.8 (0.79)
Water depth (cm)	31.6 (0.97)	27.2 (0.76)	33.9 (3.78)
SAV coverage (%)	46.1 (17.33)	27.2 (2.00)	11.7 (2.55)

[*]FEM: freshwater emergent marsh [**]PCP: permanently connected pond-permanently connected channel during all seasons. [***]TCP: temporarily connected pond-temporarily connected by surface water to the surrounding marsh but not permanently connected to a channel

High variability in macroinvertebrate density within the emergent marsh and ponds suggests that macroinvertebrates in freshwater emergent marsh are patchily distributed. Furthermore, the lack of macroinvertebrates during the summer when the marsh dried also suggests that this resource is also temporally limited. In spite of the high variability and limited temporal availability, the freshwater emergent marsh is still an important and widely distributed habitat for aquatic macroinvertebrates along the Louisiana coast.

Table 2. Comparison of mean density (ind·m^{-2} (SE)) of aquatic macroinvertebrates in freshwater emergent marsh by season. No sample in summer (*i.e.*, June, July, August 2009) because of drying.

Order	Family	Genus	Spring	Fall	Winter
Odonata	Aeshnidae	*Coryphaeschna*		0.09 (0.07)	0.26 (0.13)
	Coenagrionidae	*Enallagma*	0.36 (0.08)	0.26 (0.13)	0.44 (0.33)
		Ischnura	0.06 (0.00)	0.02 (0.02)	0.06 (0.03)
	Libellulidae	*Erythemis*	0.14 (0.03)	0.04 (0.04)	0.11 (0.06)
		Pachydiplax	0.03 (0.03)	0.02 (0.02)	0.09 (0.04)
Coleoptera	Curculionidae	*Lissorhoptrus*	0.03 (0.03)	0.07 (0.04)	0.02 (0.02)
		Onychylis		0.02 (0.02)	
	Dytiscidae	*Celina*		0.02 (0.02)	0.04 (0.02)
		Cybister	0.19 (0.14)		
		Matus	0.06 (0.00)		
	Haliplidae	*Haliplus*		0.02 (0.02)	
	Hydrophilidae	*Berosus*	1.53 (0.97)	1.06 (0.86)	0.19 (0.04)
		Derallus	0.06 (0.00)	0.07 (0.07)	
		Enochruss	0.56 (0.22)	0.04 (0.04)	0.02 (0.02)
		Tropisternus	0.64 (0.42)	0.06 (0.06)	
	Noteridae	*Hydrocanthus*		0.02 (0.02)	0.02 (0.02)
Hemiptera	Belostomatidae	*Belostoma*	0.81 (0.25)	0.11 (0.03)	
	Corixidae	*Trichocorixa*	17.36 (10.31)	0.57 (0.49)	0.56 (0.13)
	Mesoveliidae	*Mesovelia*	0.14 (0.03)		
	Naucoridae	*Pelocoris*	0.11 (0.06)		0.02 (0.02)
	Nepidae	*Ranatra*	0.06 (0.06)		
	Notonectidae	*Notonecta*	0.03 (0.03)		
Ephemeroptera	Baetidae			0.02 (0.02)	0.15 (0.09)
	Caenidae	*Caenis*		0.22 (0.17)	1.43 (0.52)
Trichoptera	Leptoceridae	*Oecetis*	0.03 (0.03)		0.02 (0.02)
Diptera	Ceratopogonidae		0.03 (0.03)	0.06 (0.00)	
	Chironomidae		34.86 (0.97)	1.07 (0.58)	3.22 (1.33)
	Ephydridae		0.03 (0.03)		0.02 (0.02)
	Tabanidae			0.11 (0.08)	0.07 (0.02)
	Tipulidae			0.09 (0.09)	
Lepidoptera	Pyralidae				0.61 (0.50)
Amphipoda	Crangonyctidae	*Synurella*	1.47 (0.58)	0.91 (0.21)	3.63 (0.69)
	Hyalellidae	*Hyalella*	9.47 (2.25)	0.46 (0.23)	3.50 (1.43)
Isopoda	Asellidae	*Caecidotea*	22.89 (7.72)	1.72 (0.53)	8.17 (3.18)
		Lirceus	0.17 (0.00)	0.07 (0.02)	3.76 (2.65)

Table 3. Seasonal aquatic macroinvertebrate density (organisms/m^2 (±SE)) and taxonomic diversity (Index H') in freshwater emergent marsh from April 2009 to February 2010.

	FEM	PCP	TCP
Density			
Spring 2009	91.1 (20.53)	125.9 (34.6)	11.5 (2.22)
Summer 2009	0.0 (0.00)	24.7 (10.6)	9.2 (4.12)
Fall 2009	7.2 (0.54)	34.4 (9.6)	13.3 (1.27)
Winter 2009	26.4 (1.80)	45.1 (19.6)	6.9 (3.53)
Taxonomic diversity			
Spring 2009	1.0	3.1	3.1
Summer 2009	0.0	2.9	2.8
Fall 2009	2.1	3.0	2.8
Winter 2009	1.8	2.6	2.8

We hypothesized freshwater emergent marsh may have fewer aquatic macroinvertebrate taxa than freshwater marsh ponds due to the seasonal drying of the marsh. The dry condition in freshwater emergent marsh may decrease the taxa richness because of limited marsh accessibility for macroinvertebrates. However, our data did not support the hypothesis as taxa richness between freshwater emergent marsh and ponds did not statistically differ and common pond inhabitants were also common in the freshwater emergent marsh (**Table 4**). Similarly, [29] found that macroinvertebrate species in temporary communities are a nested subset of those in permanent communities.

Previous studies [14,30] emphasized the role of water permanence in determining the macroinvertebrate occur-

Table 4. Comparison of mean density (ind·m^{-2} (SE)) of aquatic macroinvertebrates in freshwater emergent marsh and ponds from April 2009-February 2010.

Order	Family	Genus	FEM	PCP	TCP
Odonata	Aeshnidae	*Coryphaeschna*	0.10 (0.05)	0.04 (0.01)	0.06 (0.02)
	Coenagrionidae	*Enallagma*	0.26 (0.10)	0.53 (0.13)	0.57 (0.22)
		Ischnura	0.03 (0.01)	0.15 (0.05)	0.08 (0.03)
	Libellulidae	*Erythemis*	0.07 (0.02)	0.59 (0.18)	0.24 (0.10)
		Pachydiplax	0.4 (0.02)	1.45 (0.53)	0.31 (0.10)
Coleoptera	Chrysomelidae	*Donacia*		0.02 (0.02)	0.01 (0.01)
	Curculionidae	*Lissorhoptrus*	0.03 (0.01)	0.05 (0.02)	0.10 (0.03)
		Onychylis	0.01 (0.01)	0.01 (0.01)	
		Stenopelmus		0.01 (0.01)	
	Dytiscidae	*Celina*	0.02 (0.01)	0.45 (0.18)	0.10 (0.05)
		Copelatus		0.08 (0.04)	0.05 (0.03)
		Cybister	0.04 (0.03)	0.04 (0.03)	
		Desmopachria			0.01 (0.01)
		Hydrovatus		0.28 (0.13)	0.08 (0.04)
		Laccophilus		0.01 (0.01)	0.01 (0.01)
		Matus	0.01 (0.01)	0.02 (0.02)	0.03 (0.03)
		Thermonectus		0.01 (0.01)	0.01 (0.01)
	Haliplidae	*Haliplus*	0.01 (0.01)	0.03 (0.01)	0.03 (0.01)
		Peltodytes		0.07 (0.04)	0.05 (0.05)
	Hydrophilidae	*Berosus*	0.62 (0.31)	0.23 (0.10)	0.11 (0.04)
		Derallus	0.03 (0.02)	0.01 (0.01)	0.03 (0.01)
		Enochruss	0.12 (0.07)	0.15 (0.09)	0.08 (0.02)
		Tropisternus	0.13 (0.10)	0.16 (0.11)	0.14 (0.05)
	Noteridae	*Hydrocanthus*	0.01 (0.01)	0.20 (0.08)	0.31 (0.16)
	Scirtidae	*Scirtes*		0.05 (0.04)	0.08 (0.05)
	Staphylindae	*Euaesthetus*			0.01 (0.01)
Hemiptera	Belostomatidae	*Belostoma*	0.18 (0.10)	0.22 (0.08)	0.04 (0.02)
	Corixidae	*Trichocorixa*	3.46 (2.50)	3.84 (0.84)	1.30 (0.64)
	Mesoveliidae	*Mesovelia*	0.03 (0.02)	0.15 (0.09)	0.09 (0.07)
	Naucoridae	*Pelocoris*	0.03 (0.02)	0.29 (0.15)	0.15 (0.05)
	Nepidae	*Ranatra*	0.01 (0.01)	0.05 (0.02)	0.04 (0.02)
	Notonectidae	*Notonecta*	0.01 (0.01)	0.12 (0.06)	0.12 (0.07)
Ephemeroptera	Baetidae		0.05 (0.03)	0.06 (0.02)	
	Caenidae	*Caenis*	0.45 (0.23)	0.97 (0.29)	0.10 (0.05)
Trichoptera	Hydroptilidae	*Oxyethira*		0.01 (0.01)	0.01 (0.01)
	Leptoceridae	*Oecetis*	0.01 (0.01)	0.02 (0.02)	0.01 (0.01)
Megaloptera	Corydalidae	*Chauliodes*			0.02 (0.01)

Continued

Order	Family	Genus	FEM	PCP	TCP
Diptera	Ceratopogonidae		0.02 (0.01)	0.37 (0.24)	0.03 (0.02)
	Chironomidae		7.51 (4.11)	7.20 (1.40)	5.98 (2.10)
	Culicidae				0.04 (0.04)
	Dolichopodidae			0.01 (0.01)	0.01 (0.01)
	Ephydridae		0.01 (0.01)		0.01 (0.01)
	Stratomyidae			0.01 (0.01)	0.03 (0.01)
	Tabanidae		0.05 (0.03)	0.17 (0.07)	
	Tipulidae		0.03 (0.03)	0.03 (0.01)	0.21 (0.08)
Lepidoptera	Pyralidae		0.17 (0.15)	0.27 (0.12)	
Amphipoda	Crangonyctidae	*Synurella*	5.88 (3.85)	4.78 (2.20)	3.01 (0.96)
	Hyalellidae	*Hyalella*	54.79 (21.40)	2.68 (0.67)	5.74 (2.33)
Isopoda	Asellidae	*Caecidotea*	13.78 (5.77)	4.28 (1.31)	5.19 (1.36)
		Lirceus	2.16 (1.61)	2.33 (1.04)	1.31 (0.28)

rence in freshwater wetlands. [26] noted that the hydroperiod gradient (*i.e.*, long, short) influenced the dominant macroinvertebrate genera. Our results, however, were not consistent with a hydroperiod effect on assemblage structure as we did not detect a difference in dominant species between ponds and freshwater emergent marsh. Based on [17], most aquatic macroinvertebrates in these coastal marshes appear to be poorly adapted to dry conditions. In freshwater emergent marsh, midges (Chironomidae) were relatively abundant species (*i.e.*, >10 %) during all flooded seasons. The family Chironomidae is frequently the most abundant group in freshwater communities [31] and their larvae are known to have a certain degree of resistance to desiccation although they live in water [32]. In this sense, Chironomidae in our study sites may be well adapted to desiccation stress, resulting in relatively high density in freshwater marsh ponds and freshwater emergent marsh.

Variation in life history traits of macroinvertebrates seems to be correlated with hydrologic condition (*i.e.*, flooding duration, water depth). The low fall densities in emergent marshes appear to be a life history strategy to avoid harsh conditions and maximize reproductive success by synchronizing egg hatch with more favorable winter conditions. It is also possible that low fall densities represent a residual effect of summer drying. In addition, individual taxa have different habitat requirements. According to previous studies [16,18,19], odonates require a relatively long hydroperiod for the full development of nymphs; thus, they may avoid the marshes due to the risk of dry conditions. However, coleopterans have relatively high desiccation tolerance;

therefore, they may avoid the ponds because they support an abundance of relatively large and strong predators (e.g., fish, dragonfly; [17]). Despite differences in physical and chemical conditions between marshes and ponds, our results indicated that the density of odonates and coleopterans between marshes and ponds did not statistically differ. These results could have been affected by our relatively small sample sizes. Additional research could provide important insights into aquatic macroinvertebrate use patterns in these habitats.

Flooding and drying conditions are common in wetlands and are an important part of the hydrological cycle. The relatively long inundation promotes higher densities and taxonomic diversity of aquatic macroinvertebrates. In our study, the results suggest that anthropogenic activities such as marsh management that increase or decrease the duration of lateral hydrologic connection between emergent marsh and adjacent waterbodies may potentially affect aquatic macroinvertebrate habitat value in freshwater marshes.

5. ACKNOWLEDGEMENTS

This project was supported by a Louisiana Department of Wildlife and Fisheries and U.S. Fish and Wildlife Service State Wildlife Grant with support also from the International Crane Foundation. We thank M. La Peyre, J. A. Nyman, R. Keim, A. Rutherford, and M. Ferro for their critical insights. The authors would like to acknowledge the field and laboratory contributions of J. Linscombe, R. Cormier, M. Huber, and A. Williamson. In addition, we extend gratitude to M. Kaller for helping with the aquatic macroinvertebrate identification and statistical analysis. Collections were made under Louisiana State University AgCenter

Animal Care and Use protocol (#AE2008-012). Any use of trade, firm, or product names is for descriptive purposes only and does not imply endorsement by the U.S. Government.

REFERENCES

[1] Paillex, A., Castella, E. and Carron, G. (2007) Aquatic macroinvertebrate response along a gradient of lateral connectivity in river floodplain channels. *Journal of the North American Benthological Society*, **26**, 779-796.

[2] Zilli, F.L. and Marchese, M.R. (2011) Patterns in macroinvertebrate assemblages at different spatial scales. Implications of hydrological connectivity in a large floodplain river. *Hydrobiologia*, **663**, 245-257.

[3] Leigh, C. and Sheldon, F. (2009) Hydrological connectivity drives patterns of macroinvertebrate biodiversity in floodplain rivers of the Australian wet/dry tropics. *Freshwater Biology*, **54**, 549-571.

[4] Poff, N.L., Allan, J.D., Bain, M.B., Karr, J.R., Prestegaard, K.L., Richter, B.D., Sparks, R.E. and Stromberg, J.C. (1997) The natural flow regime. A paradigm for river conservation and restoration. *BioScience*, **47**, 769-784.

[5] Puckridge, J.T., Sheldon, F., Walker, K.F. and Boulton, A.J. (1998) Flow variability and the ecology of large rivers. *Marine and Freshwater Research*, **49**, 55-72.

[6] Rozas, L.P. and Minello, T.J. (1999) Effects of structural marsh management on fishery species and other nekton before and during a spring drawdown. *Wetlands Ecology and Management*, 7, 121-139.

[7] Dunson, W.A., Friacano, P. and Sadinski, W.J. (1993) Variation in tolerance to abiotic stresses among sympatric salt marsh fish. *Wetlands*, **13**, 16-24.

[8] Rowe, C.L. and Dunson, W.A. (1995) Individual and interactive effects of salinity and initial fish density on a salt marsh assemblage. *Marine Ecology Progress Series*, **128**, 271-278.

[9] Gascon, S., Boix, D., Sala, J. and Quintana, X.D. (2008) Relation between macroinvertebrate life strategies and habitat traits in Mediterranean salt marsh ponds (Emporda wetlands, NE Iberian Peninsula). *Hydrobiologia*, **597**, 71-83.

[10] Wiggins, G.B., Mackay, R.J. and Smith, I.M. (1980) Evolutionary and ecological strategies of animals in annual temporary pools. *Archiv für Hydrobiologie*, **58**, 97-206.

[11] Spencer, M., Blaustein, L., Schwartz, S.S. and Cohen, J.E. (1999) Species richness and the proportion of predatory animal species in temporary freshwater pools; relationships with habitat size and permanence. *Ecology Letters*, **2**, 157-166.

[12] Zimmer, K.D., Hanson, M.A. and Butler, M.G. (2000) Factors influencing invertebrate communities in prairie wetlands: a multivariate approach. *Canadian Journal of Fisheries and Aquatic Sciences*, **57**, 76-85.

[13] Welborn, G.A., Skelly, D.K. and Werner, E.E. (1996) Mechanisms creating community structure across a freshwater habitat gradient. *Annual Review of Ecology and Systematics*, **27**, 337-363.

[14] Collinson, N.H., Biggs, J., Corfield, A., Hodson, M.J., Walker, D., Whitefield, M. and Williams, P.J. (1995) Temporary and permanent ponds: an assessment of the effects of drying out on the conservation value of aquatic macroinvertebrate communities. *Biological Conservation*, **74**, 125-133.

[15] Sheldon, F., Boulton, A.J. and Puckridge, J.T. (2002) Conservation value of variable connectivity: Aquatic invertebrate assemblages of channel and floodplain habitats of a central Australian arid-zone river, Cooper Creek. *Biological Conservation*, **103**, 13-31.

[16] Nilsson, A.N. (1986) Life cycles and habitats of the northern European Agabini (Coleoptera, Dytiscidae). *Entomologica Basiliensia*, **11**, 391-417.

[17] Kang, S.R. and King, S.L. (2013) Effects of hydrologic connectivity on aquatic macroinvertebrate assemblages in different marsh types. *Aquatic Biology*, **18**,149-160.

[18] Wissinger, S.A. (1988) Spatial distribution, life history, and estimates of survivorship in a fourteen-species assemblage of larval dragonflies (Odonata: Anisoptera). *Freshwater Biology*, **20**, 329-340.

[19] Zimmer, K.D., Hanson, M.A., Butler, M.G. and Duffy, W.G. (2001) Size distribution of aquatic invertebrates in two prairie wetlands, with and without fish, with implications for community production. *Freshwater Biology*, **46**, 1373-1386.

[20] Mittlebach, G.G. (1988) Competition among refuging sunfishes and effects of fish density on littoral zone invertebrates. *Ecology*, **69**, 614-623.

[21] Campeau, S., Murkin, H.R. and Titman, R.D. (1994) Relative importance of algae and emergent plant litter to freshwater marsh invertebrates. *Canadian Journal of Fisheries and Aquatic Sciences*, **51**, 681-692.

[22] Batzer, D.P., Pusateri, C.R. and Vetter R. (2000) Impacts of fish predation on marsh invertebrates: Direct and indirect effects. *Wetlands*, **20**, 307-312.

[23] Bolduc, F. and Afton, A.D. (2003) Effects of structural marsh management and salinity on invertebrate prey of waterbirds in marsh ponds during winter on the Gulf Coast Chenier Plain. *Wetlands*, **23**, 897-910.

[24] Batzer, D.P., Palik, B.J. and Buech, R. (2004) Relationships between environmental characteristics and macroinvertebrate communities in seasonal woodland ponds of

Minnesota. *Journal of the North American Benthological Society*, **23**, 50-68.

[25] Nicolet, P., Biggs, J., Fox, G., Hodson, M.J., Reynolds, C., Whitfield, M. and Williams, P. (2004) The wetland plant and macroinvertebrate assemblages to temporary ponds in England and Wales. *Biological Conservation*, **120**, 261-278.

[26] Tarr, T.L., Baber, M.J. and Babbitt, K.J. (2005) Macroinvertebrate community structure across a wetland hydroperiod gradient in southern New Hampshire, USA. *Wetlands Ecology and Management*, **13**, 321-334.

[27] Hornung, J.P. and Foote, A.L. (2006) Aquatic invertebrate responses to fish presence and vegetation complexity in western boreal wetlands, with implications for waterbird productivity. *Wetlands*, **26**, 1-12.

[28] Kratzer, E.B. and Batzer, D.P. (2007) Spatial and temporal variation in aquatic macroinvertebrates in the Okefenokee swamp, Georgia, USA. *Wetlands*, **27**, 127-140.

[29] Wissinger, S.A., Greig, H. and McIntosh, A. (2009) Absence of species replacements between permanent and temporary lentic communities in New Zealand. *Journal of the North American Benthological Society*, **28**, 12-23.

[30] Sanderson, R.A., Eyre, M.D. and Rushton, S.P. (2005) Distribution of selected macroinvertebrates in a mosaic of temporary and permanent freshwater ponds as explained by autologistic models. *Ecography*, **28**, 355-362.

[31] Pinder, L.C.V. (1986) Biology of freshwater Chironomidae. *Annual Review of Entomology*, **31**, 1-23.

[32] Suemoto, T., Kawai, K. and Imabayashi, H. (2004) A comparison of desiccation tolerance among 12 species of chironomid larvae. *Hydrobiologia*, **515**, 107-114.

9

Vegetation communities in estuarine tidal flats in the different river and basin environments of the four major rivers of Ise Bay (Suzuka, Tanaka, Kushida and Miya), Mie Prefecture, Japan

Korehisa Kaneko[1*], Seiich Nohara[2]

[1]Ecosystem Conservation Society-Japan, Tokyo, Japan;
[*]Corresponding Author

[2]Center for Environmental Biology and Ecosystem Studies, National Institute for Environmental Studies, Ibaraki, Japan

ABSTRACT

In this study, we compared and analysed vegetation communities in the estuarine tidal flats of the four major rivers of Ise Bay (Suzuka River, Tanaka River, Kushida River and Miya River) in Mie Prefecture, Japan. Along the Suzuka River, *Eragrostis curvula* of the exotic plant accounted for 60.0% or more of the entire surface area, and the plant volume was high. Along the Tanaka River, *Suaeda maritima* community occupied the sand-mud zone in the vicinity of the shoreline on gravel bars, while *Phragmites australis* community was distributed along a shallow lake upstream. In the Kushida River, a salt marsh plant community (a community type found in areas flooded at high tide) of *Suaeda maritima*, *Phragmites australis* and *Artemisia scoparia* was distributed on the sand-mud surface along the main river. A salt marsh plant community (a community type found in areas that do not flood at high tide) of *Phacelurus latifolius* accounted for least 50.0% of the entire surface area. Along the Miya River, the area covered by the annual salt marsh plant community type was larger than the area occupied by this community type along the other rivers. The flow volume of the Miya River was high in April, June and August-October of 2006, July and September of 2007 and April-June of 2008. The flow volume was especially high in July 2007, when it reached levels above 1500.0 m^3/s; change in flow volume was also large. We suggest that a large-scale disturbance occurred in the estuary, resulting in the formation of a gravelly sandy surface where an annual salt marsh plant community of *Suaeda maritima* and *Artemisia scoparia* has been established and grown as the annual precipitation and catchment volume of the basin have increased.

Keywords: Annual Salt Marsh Plant; Perennial Salt Marsh Plant; Flood Volume; Water Level; Disturbance

1. INTRODUCTION

In Ise Bay, Mie Prefecture, Japan, shoreline conservation for the purpose of national land conservation has been proposed in the coastal areas. Instead securing the area, naturally occurring estuarine tidal flats have decreased in the area and left the shoreline vulnerable. After securing the area, the recovery of ecosystem diversity was included in the "reproduction action plan of Ise Bay" that was established in 2007 [1].

In the estuarine tidal flats of the four major rivers (Suzuka, Tanaka, Kushida and Miya) of Ise Bay, there are distinct vegetation types including salt marsh plant communities and exotic-upland plant communities [2]. Factors that influence the formation of vegetation types in the estuarine tidal flats include the ground level [3], differences between tidal-level and superficial sediments [4,5] and ecosystem dynamics such as repeated disturbance and regeneration [6]. Additionally, the salt marsh plant communities found in low areas are influenced by the physical environment and sediment movement [7-10].

We aimed to determine whether the frequency of disturbance and the scale of change in flow volume and water level due to annual precipitation have greatly influenced the distribution of vegetation in the estuarine tidal flats. Given that there have been few previous studies addressing this problem, a detailed study would support the conservation and recovery of tidal flat environments.

In this study, we identified differences in vegetation communities and examined the relationship between vegetation communities and the environment (basin precipitation, water levels and flow volume) in the estuarine tidal flats of the four major rivers of Ise Bay.

2. STUDY SITES AND METHODS

2.1. Study Site

The study sites were located in the estuarine tidal flats of the four major rivers (Suzuka, Tanaka, Kushida and Miya) of Ise Bay, Mie Prefecture, Japan (**Figure 1**).

The study sites were selected from areas that experience the greatest influence from sediment deposition, both from upstream sediments and tide action. The river basin environments differ among the four rivers (**Table 1**). Suzuka River is a first-class river that is 38 km long with a basin area of 323 km^2, and the mean annual precipitation in the basin for 1986-2005 was approximately 1800 - 2000 mm in the plains and exceeded 2200 mm in mountainous areas [11].

Data collection for Tanaka River included a verbal survey from the Tsu City office in the Mie Prefecture. Tanaka River is a second-class river that is 4.9 km long with a basin area of 8.5 km^2, which includes Tsu City. The mean annual precipitation in the basin was approximately 1700 - 2300 mm.

Kushida River is a first-class river that is 85 km long with a basin area of 461 km^2. Mean annual precipitation in the basin for 1989-2008 was approximately 1600 - 2200 mm in the middle and downstream basin and exceeded 2500 mm in mountainous areas [12].

Miya River is a first-class river that is 91 km long with a basin area of approximately 920 km^2. Mean annual precipitation in the basin was greater than 2500 - 3000 mm in the upstream basin and approximately 2000 - 2500 mm in the middle and downstream basin [13].

2.2. Vegetation Survey

We used a survey of the study sites to map the distribution of plant communities (1:2500). Vegetation data were used to develop a physiognomic vegetation map with GIS (ArcView 3.1). For the plant community composition survey, we established transects to adequately sample all plant communities using the line transect method. Quadrants (1 m^2) were established along the lines in each community, and cover and plant height were recorded for all species in each quadrant. In total, 84 quadrants were sampled (21 at each study site). All quadrants were marked with a pole and a flag.

2.3. Statistical Methods

2.3.1. Classification of Vegetation Type

The vegetation types were classified into 1) salt marsh plants and 2) exotic and upland plants (herbs). Salt marsh plants were defined as plants growing in the areas surrounding marshes and bogs near the mouth of the lagoon and the river in the coastal zone, and upland plants were classified as plants growing in locations of strong anthropogenic influence such as reclaimed areas [14] and

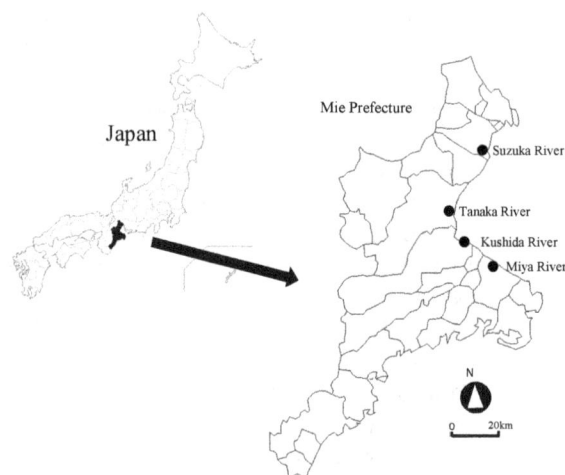

Figure 1. Location of the study sites.

Table 1. Total river length, basin area and mean annual precipitation for the four rivers.

	Suzuka River	Tanaka River	Kushida River	Miya River
Total length (km)	38	4.9	85	91
Basin Area (km^2)	323	8.5	461	920
Annual precipitation (mm)	Mauntains area: about 2,200 mm	1,700 - 2,300 mm	Mauntains area: about 2,500 mm	Mauntains area: 2,500 - 3,000 mm
	Plain area: 1,800 - 2,000 mm		Plain area: 1,600 - 2,200 mm	Plain area: 2,000 - 2,500 mm

※ Annual precipitation of Tanaka River is indicated the average annual precipitation (1889-2010) of Tushi city. ※ In Tanaka River, water level is used the observation data of Mie Prefecture in Japan.

Vegetation communities in estuarine tidal flats in the different river and basin environments of the four major rivers of
Ise Bay (Suzuka, Tanaka, Kushida and Miya), Mie Prefecture, Japan

57

in environments in which soil moisture is low [15]. Exotic plants were defined as plants brought in from foreign countries, and it is generally assumed that these plants were introduced after the Edo period in Japan [16].

2.3.2. Plant Volume

We calculated plant volume as follows:

Plant volume = mean cover value × plant height (m)

※ The mean cover values were determined using the cover classes 1, 2, 3, 4 and 5, which were converted into 0.1%, 2.5%, 15%, 37.5%, 62.5% and 87.5%, respectively, according to the Braun-Blanquet method [17].

3. RESULTS

3.1. The Area and Proportional Area of Plant Community

The area, proportional area and physiognomic vegetation map of each plant community are shown by studysite (**Table 2**, **Figure 2**). Along the Suzuka River, *Eragrostis curvula*, *Carex kobomugi*, *Imperata cylindrical* and *Solidago canadensis* communities were distributed on sand, while *Phragmites australis* and *Suaeda maritima* communities were distributed on sand and mud. However, the *Eragrostis curvula* community accounted for 60.0% or more of the entire surface area. The proportional area of salt marsh plant communities (a community type found in areas that flood at high tide) was low, with the *Suaeda maritima* and *Phragmites australis*

communities each occupying 8.8% of the entire surface area. The proportional area of coastal plant communities was low, with an area of 5.3% for *Carex kobomugi*, 6.1% for *Imperata cylindrical*, 0.9% for *Calystegia soldanella* and 0.6% for *Vitex rotundifolia*.

Along the Tanaka River, the coastal plant communities were distributed on sandy high ground that does not typically flood at high tide. *Suaeda maritima* community was distributed in a zone of sand and mud in the vicinityof the shoreline on gravel bars, and *Phragmites australis* community was distributed in a muddy lake in the upstream portion of the river.

Along the Kushida River, *Phacelurus latifolius* and *Hibiscus hamabo* communities were distributed on sandy high ground that does not typically flood at high tide, while *Suaeda maritima*, *Artemisia scoparia*, Zoysia sinica and *Phragmites australis* communities were distributed on the low ground of sandy mud in the vicinity of the shoreline.

Along the Miya River, *Suaeda maritime-Artemisia scoparia* community and *Phragmites australis* community were distributed on sandy mud and gravelly sand in the vicinity of the shoreline, while *Solidago canadensis* and *Eragrostis curvula* communities were distributed on the sandy high ground. The salt marsh plant (annual and perennial) communities (community types that are found in areas that flood at high tide) were dominated by *Suaeda maritime-Artemisia scoparia* (41.1%) and

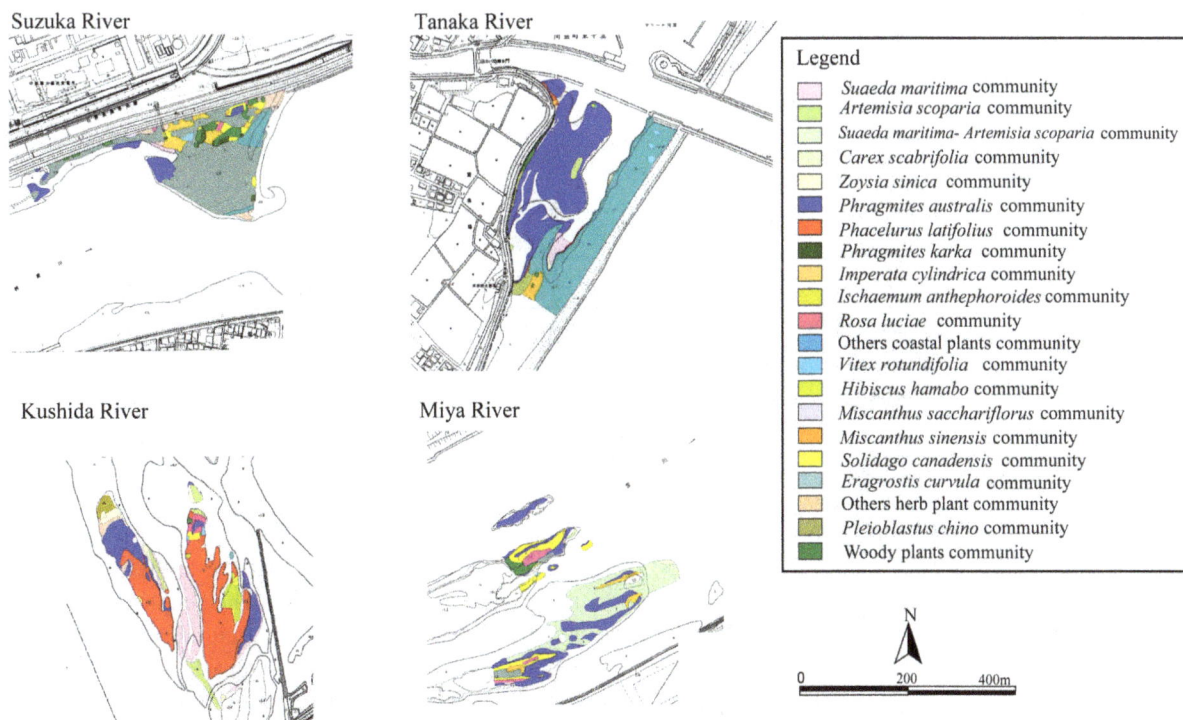

Figure 2. Physiognomic vegetation map of the estuarine tidal flats of 4 rivers in Ise Bay.

Table 2. Area and proportional area of plant community types at the study sites.

Vegetation types	Plant community	Suzuka River		Tanaka River		Kushida River		Miya River	
		Area (m²)	Proportional Area (%)	Area (m²)	Proportional Area (%)	Area (m²)	Proportional Area (%)	Area (m²)	Proportional Area (%)
Annual salt marsh plant community (flooding type at high tide)	*Suaeda maritima* community	297	0.8	948	1.9	3,906	10.6	92	0.3
	Suaeda maritima-Artemisia scoparia community	-	-	-	-	-	-	11,916	41.1
	Artemisia scoparia community	-	-	-	-	2,726	7.4	92	0.3
Perennial salt marsh plant community (flooding type at high tide)	*Carex scabrifolia* community	-	-	565	1.1	-	-	-	-
	Zoysia sinica community	-	-	325	0.7	986	2.7	176	0.6
	Phragmites australis community	2,903	8.0	22,486	45.2	6,113	16.6	9,091	31.4
Perennial salt marsh plant community (non-flooding type at high tide)	*Phragmites karka* community	678	1.9	-	-	-	-	-	-
	Phacelurus latifolius community	-	-	306	0.6	18,597	50.6	-	-
Coastal plant community	*Carex kobomugi* community	1,930	5.3	-	-	-	-	-	-
	Sonchus brachyotus community	-	-	95	0.2	-	-	-	-
	Ischaemum anthephoroides community	-	-	-	-	735	2.0	-	-
	Imperata cylindrica community	2,210	6.1	-	-	-	-	854	2.9
	Calystegia soldanella community	319	0.9	44	0.1	-	-	-	-
	Others costal plants community	-	-	21,481	43.2	78	0.2	-	-
	Hibiscus hamabo community	-	-	32	0.1	113	0.3	-	-
	Vitex rotundifolia community	223	0.6	262	0.5	1,605	4.4	31	0.1
	Rosa luciae community	-	-	-	-	630	1.7	-	-
Exotic-upland plant community (herb)	*Miscanthus sacchariflorus* community	114	0.3	50	0.1	-	-	-	-
	Miscanthus sinensis community	176	0.5	968	1.9	-	-	302	1.0
	Solidago canadensis Miscanthus sinensis community	-	-	-	-	-	-	529	1.8
	Solidago canadensis community	1,214	3.3	-	-	539	1.5	2,999	10.3
	Eragrostis curvula community	22,440	61.8	-	-	-	-	859	3.0
	Pueraria lobata community	-	-	49	0.1	-	-	-	-
	Lolium multiflorum community	342	0.9	-	-	-	-	-	-
	Sorghum halepense community	508	1.4	-	-	-	-	-	-
	Xanthium occidentale community	904	2.5	-	-	-	-	-	-
	Others herb plant community	-	-	827	1.7	649	1.8	24	0.1
Exotic-upland plant community (woody plant)	*Rosa multiflora* community	229	0.6	-	-	-	-	1,185	4.1
	Pleioblastus chino community	-	-	525	1.1	-	-	-	-
	Woody plants community	1,824	5.0	769	1.5	91	0.2	830	2.9
		36,311	100	49,732	100	36,768	100	28,980	100

※ The bold face is indicated the value that is the highest in the community.

Phragmites australis (31.4%) and accounted for 70.0% or more of the entire surface area.

3.2. Proportional Area and Distribution of Plant Community Types

The proportional area of exotic plant communities (surface area of exotic plant community) was highest along the Suzuka River; perennial salt marsh plant communities (found in areas that flood at high tide) and coastal plant communities had the highest proportional area along the Tanaka River. The proportional area for perennial salt marsh plant communities (found in areas that do not typically flood at high tide) was highest along the Kushida River, and the proportional area of annual salt marsh pl.ant communities was highest along the Miya River (**Figures 3** and **4**).

Vegetation communities in estuarine tidal flats in the different river and basin environments of the four major rivers of
Ise Bay (Suzuka, Tanaka, Kushida and Miya), Mie Prefecture, Japan

59

Suzuka River   Tanaka River

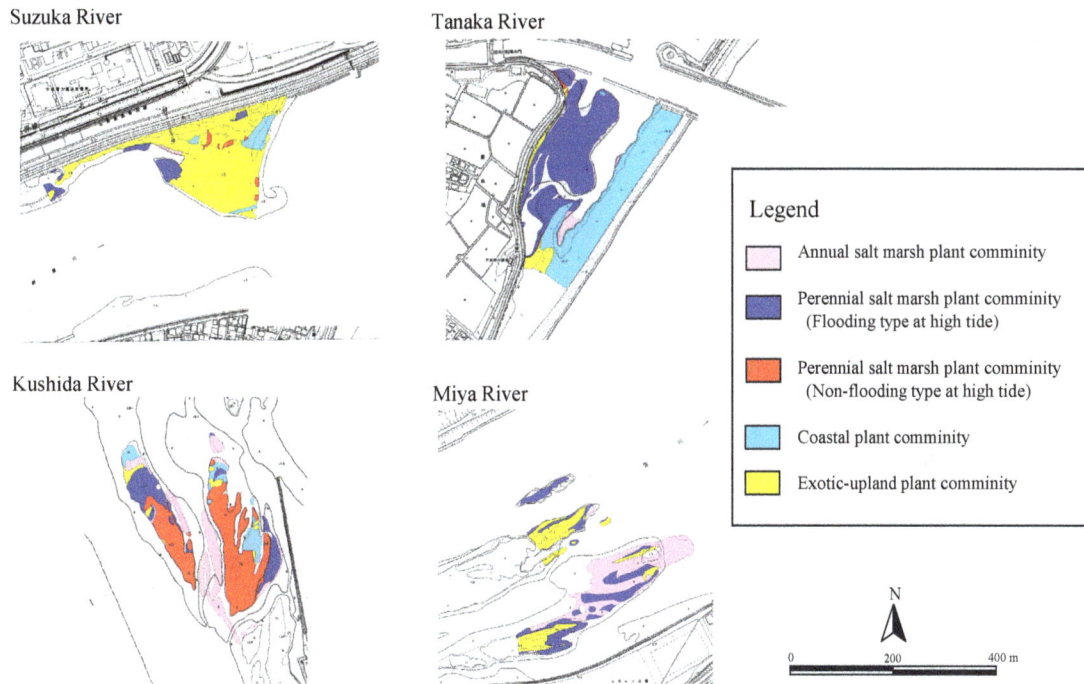

Kushida River   Miya River

Figure 3. Distribution map of plant community types in the estuarine tidal flats of 4 rivers in Ise Bay.

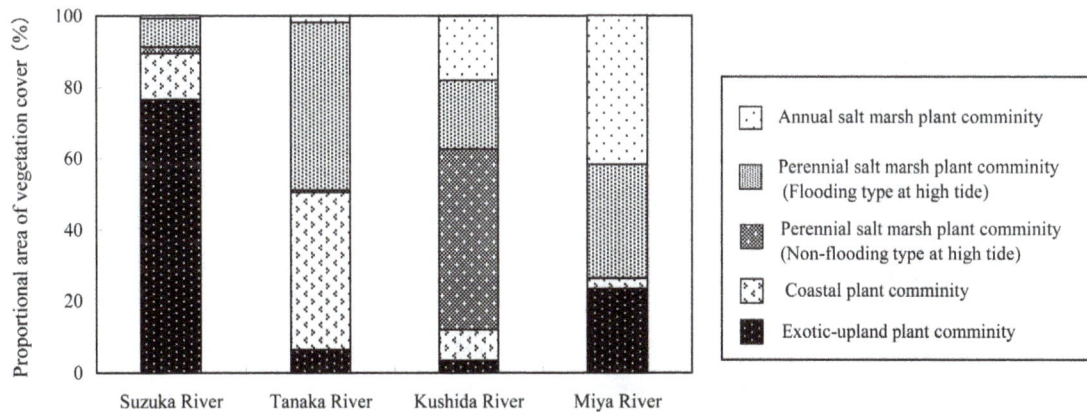

Figure 4. Proportional area of vegetation cover by plant community type at the study sites.

3.3. Species Diversity

The number of species per unit area for each plant type (flooded and non-flooded salt marsh plants, coastal plants and exotic-upland plants (herbs)) varied across the study sites. The number of species in the annual salt marsh plant (flooded) was relatively many along the Miya River and fewer along the Kushida, Tanaka and Suzuka rivers. Perennial salt marsh plant (flooded) had many species along the Miya River and fewer species along the Tanaka, Kushida and Suzuka rivers. Salt marsh plant (non-flooded) along the Kushida River had many species. Coastal plant had many species along the Tanaka River and fewer species along the Miya River. Exotic-upland plant had many species along the Suzuka River and a relatively fewer species along the Kushida,

Miya and Tanaka rivers (**Figure 5**).

3.4. Plant Volume

In comparing flooded or non-flooded salt marsh plants, coastal plants and exotic-upland plants at the study sites, the annual salt marsh plant (flooded) had the highest volume per area along the Miya River and lower volumes per area along the Kushida, Suzuka and Tanaka rivers. The perennial salt marsh plant (flooded) had the highest plant volume per area along the Miya River and a lower volume per area along the Tanaka, Kushida and Suzuka rivers. The salt marsh plant (non-flooded) had a higher volume per area along the Kushida River. The coastal plant had a high volume along the Kushida River and lower volumes along the Suzuka, Miya and Tanaka rivers. The exotic-upland had the highest volume per area along the Suzuka River and lower volumes along the Miya, Kushida and Tanaka rivers (**Figure 6**).

3.5. Hydrological Environment at Each Study Site

Flow volumes and water levels at each study site were examined using data for three-year period from January 2006 to June 2008 [18].

Regarding change in water level for each river, the water level in the Tanaka and Kushida rivers rose several times over the study period; however, these increases represented differences of less than 1.0 m from the minimum water levels for the rivers.

The water level in the Suzuka River rose to 1.0 m or more above the minimum water level during May-August of 2006, 2007 and May-June of 2008. Miya River showed frequently high levels except during winter and early spring. In particular, the water level in July 2007 reached a maximum that was approximately 4.0 m abovethe minimum water level (**Figure 7**). The change in water level was the largest in the Miya River (**Figure 8**).

Flow volume in the Miya River was highest in April, June and August-October of 2006, July and September of 2007 and April-June of 2008. Flow volumes in the Kushida and Suzuka Rivers were relatively low. In particular, the flow volume of the Miya River in July 2007 was extremely high, reaching over 1500.0 m^3/s. There are no data for the Tanaka River. Flow volume in the Miya River varied widely (**Figure 7**), and the largest change in flow volume occurred in the Miya River, relative to the other rivers (**Figure 8**).

4. DISCUSSION

4.1. Differences in Vegetation among the Study Sites

At the Suzuka River study sites, where sand was the major substrate, *Eragrostis curvula* of exotic plant was widely distributed and had highest plant volume (**Table 2**, **Figures 2** and **3**) *Eragrostis curvula* tolerates the movement of sand and water [19], and the accumulation of sand during high tides increases the amount of landsurface suitable for this species [20]. Additionally, *Eragrostis curvula* is used as a greening material in affore

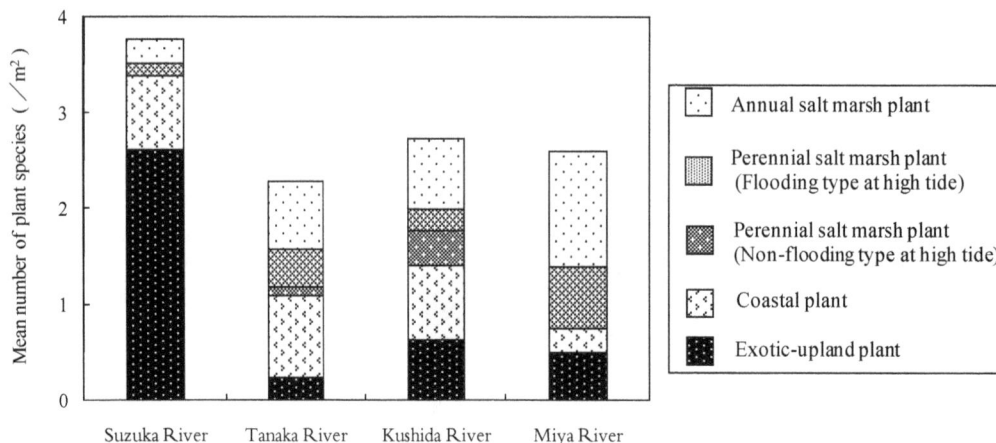

Figure 5. Mean number of plant species by plant community type at the study sites.

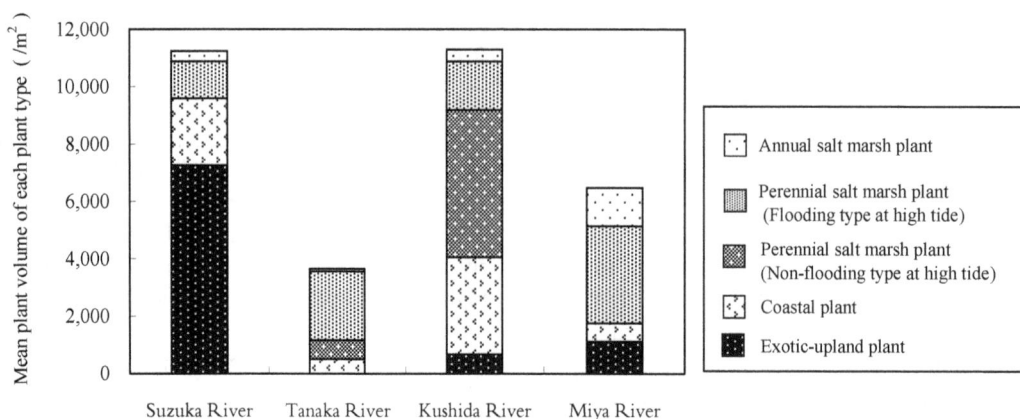

Figure 6. Mean plant volume by plant community type at the study sites.

Vegetation communities in estuarine tidal flats in the different river and basin environments of the four major rivers of Ise Bay (Suzuka, Tanaka, Kushida and Miya), Mie Prefecture, Japan

61

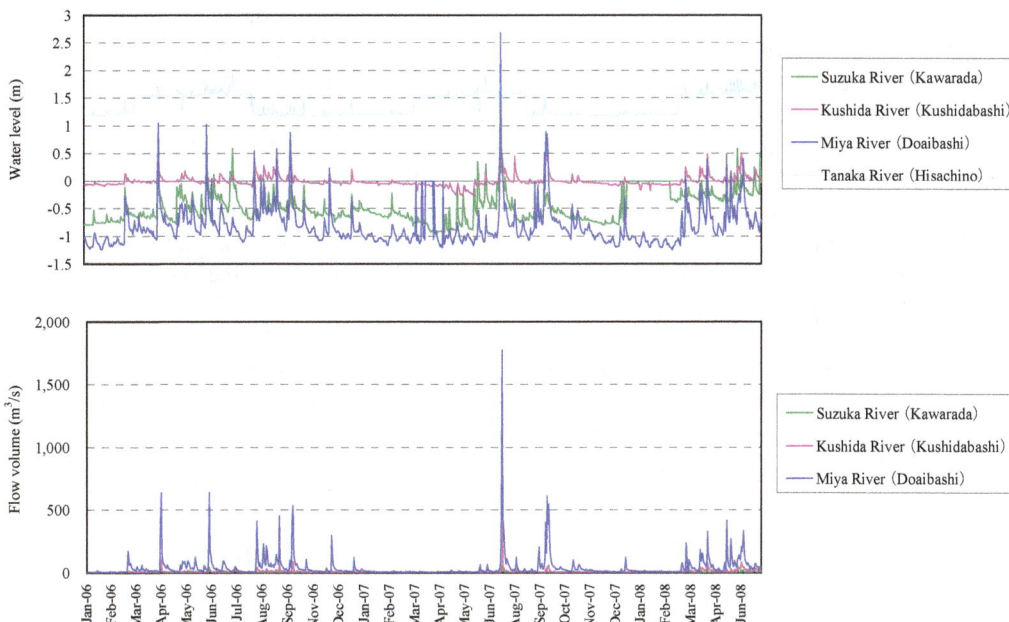

Figure 7. Changes in water level and flow volume for the four rivers (January 2006-June 2008). ※ For the Suzuka River, the Kushida River and the Miya River, data were obtained from the water information system of the Ministry of Land, Infrastructure, Transport and Tourism in Japan. ※ For the Tanaka River, water levels were determined using observation data from Mie Prefecture, Japan; however, there were no data for flow volume.

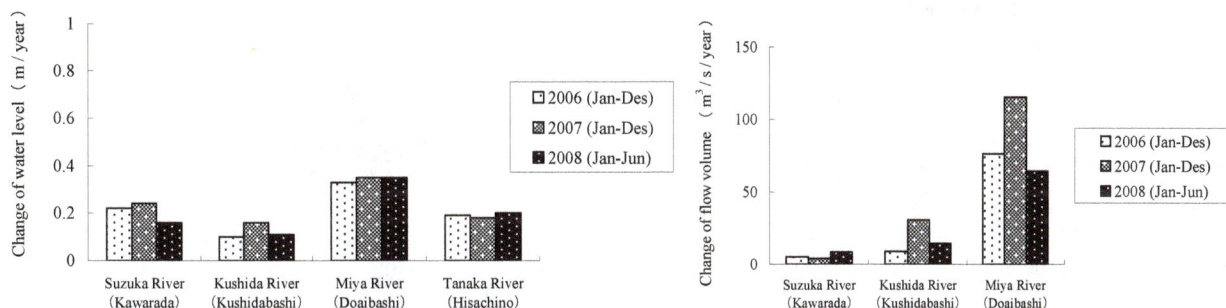

Figure 8. Changes in water level and flow volume per year (2006, 2007 and January-June 2008) for each river. ※ Change values indicate the standard deviation. ※ ( ) The location of data collection for the river. ※ For the Tanaka River, water level data were obtained from Mie Prefecture, Japan; however, there were no data for flow volume. For 2008, data for depth and flow volume are limited to the period from January-June because the study was conducted in July 2008.

station and soil erosion control projects [21].

We suggest that along the Suzuka River, there was a low frequency of disturbance and a comparatively stable alluvial environment because the flow volume was extremely low compared to the Kushida and Miya rivers. However, the water level rose in June 2008 and during June-August of 2006 and 2007. We suggest that sediment accumulated in the estuary when the river rose to high water levels, forming high ground that was not flooded at high tide. The seeds of *Eragrostis curvula* growing in the vicinity germinated and were established on the new surface through water dispersal and anemochory. [22] reported that this type of vegetation is stable when tidal levels remain constant. We suggest that once high ground

had formed in the estuarine tidal flats, the high surfaces were not easily flooded and the lack of flood disturbance allowed for the establishment of *Eragrostis curvula*.

In the estuarine tidal flats of the Tanaka River, *Phragmites australis* community, which is the perennial salt marsh (flooded) community type, occupied half of the surface area, while *Suaeda maritime* community which is the annual salt marsh plant community type was limited to sandy mud surfaces in the lagoon (**Figure 2**). The seeds of *Suaeda maritime* are dispersed by stream and other water movement, such as tides [23], and sometimes form seed banks [24]. The basin of the Tanaka River was extremely narrow compared to that of the other study sites, and the annual precipitation was relatively low

(**Table 1**).

We suggest that the catchment volume of the river and the disturbance frequency are low, limiting the growth and distribution of annual salt marsh plant community type such as *Suaeda maritime* community. In contrast to salt marsh plants, the seed of *Phragmites australis* is dispersed by the wind and can occupy new sandy surfaces once established. Dense vegetation cover forms through vegetative propagation by a vigorous underground stem system, excluding many other species [25]. Additionally, *Phragmites australis* is relatively tolerant of high soil salinity and grows in dense communities in the brackish waters of estuaries [26]. We suggest that sediments from upstream were deposited beyond the overflow embankment during flood events, forming surface deposits by sand sprays. *Phragmites australis* germinated and grew on these deposits, expanding its cover by spreading rhizomes.

In the estuarine tidal flats of the Kushida River, marsh plant communities of *Suaeda maritime*, *Phragmites australis* and *Artemisia scoparia* (flooded type) were distributed on sandy mud surfaces along the main river channel. Salt marsh plant (non-flooded) communities as *Phacelurus latifolius*, coastal plant communities and exotic-upland communities were distributed on sandy, slightly higher ground (**Figure 2**). The proportional area of *Phacelurus latifolius* was greater than 50% (**Table 2**). The flow volume of the Kushida River was the highest and reached a volume greater than 500.0 m^3/s in July 2007 (**Figure 7**). *Phacelurus latifolius* was distributed on surfaces that do not flood at high tide, and this species is an indicator species to the upland area. Due to the low level of disturbance, fine sand accumulated in non-flooded areas and the areas transformed into high ground [27].

We suggest that the basin area of Kushida River is smaller than that of Miya River, and 50.0% or more of the entire surface area in Kushida River is occupied by upland which is non-flooded areas at high tide. However, disturbances occurred when flow volume increased due to heavy rain, forming new gravelly surfaces on which the annual salt marsh plants of *Suaeda maritime* and *Artemisia scoparia* germinated and grew, forming a new community.

In the estuarine tidal flats of the Miya River, where gravelly sand surfaces are predominant, the proportional area of annual salt marsh plant communities was the highest among the study sites (**Figure 4**). The flow volume of the Miya River reached more than 500.0 m^3/s in April, July and October of 2006 and July and September of 2007. The flow volume of the Miya River was higher than that of the other study sites. The flow volume in July 2007 was extremely high, more than 1500.0 m^3/s (**Figure 7**), and the change in flow volume was large

(**Figure 8**). Because the basin area of the Miya River was the largest among the study areas, the amount of annual rainfall was also the highest. We suggest that large-scale disturbances occurred in the estuary and formed the gravelly sandy deposits when flow volume reached more than 500.0 m^3/s and water level rose to 1.0 m or more above the minimum water level. The annual salt marsh plants of *Suaeda maritime* and *Artemisia scoparia* germinated and grew as the annual precipitation and the catchment volume of the basin increased.

Based on our findings, we suggest that the scale and frequency of disturbance due to differences in basin area and annual precipitation greatly influence the vegetation in estuarine tidal flats.

REFERENCES

[1] Nishihara, T. (1997) As for recent movement of "Reproduction project of sea in Japan"—The action plan was settled on in the Ise Bay and the Hiroshima bay. *Japanese Journal of Coastal Zone Studies*, **19**, 3-5.

[2] Mie Prefecture, Environmental Conservation Agency (2005) Threatened wildlife of Mie 2005. Plants Mushroom, Mie Prefecture, Japan.

[3] Kaneko, K., Yabe, T. and Nohara, S. (2005) Vegetation changes and topographic feature in the delta of Obitsu River in Tokyo Bay. *Japanese Journal of Landscape Ecology*, **9**, 27-32.

[4] Kobayashi, S. (1996) The state environment of salt marsh plant community in the estuary-river course characteristics and salinity concentration environment. *Research Report of Ehime Prefectural Science Museum JAPAN*, **1**, 35-44.

[5] Kamata, M. and Ogura, Y. (2006) Habitat evaluation for plant communities as a salt marsh in the Naka River, Shikoku. *Ecology and Civil Engineering Society (Japan)*, **8**, 245- 261.

[6] Kikuchi, E. and Kurihara, Y. (1988) Tide river. In: Kurihara, Y., Ed., *Estuarine and Coastal Area of Ecology and Eco-Technology*, Tokai University Press, Tokyo, 150-160.

[7] Delaune, R., Patrick, W. and Buresh, R. (1978) Sedimentation rates determined by 137 Cs dating in a rapidly accreting salt marsh. *Nature*, **275**, 532-533.

[8] Deleeuw, J., Demunck, W., Olff, H. and Bakker, J. (1993) Does zonation reflect the succession of salt marsh vegetation? A comparison of an estuarine and a coastal island marsh in the Netherlands. *Acta Botanica Neerlandica*, **42**, 435-445.

[9] Dijkema, K.S. (1997) The influence of salt marsh vegetation on sedimentation. In: Eisma, D., Ed., *Intertidal Deposits River Mouths, Tidal Flats and Coastal Lagoons*, CRC Press, Boca Ration, 403-414.

[10] Hatton, R.S., Delaune, R.D. and Patrick, W.H.J. (1983) Sedimentation, accretion, and subsidence in marshes of Barataria Basin, Louisiana. *Limnology and Oceanogra-*

Vegetation communities in estuarine tidal flats in the different river and basin environments of the four major rivers of
Ise Bay (Suzuka, Tanaka, Kushida and Miya), Mie Prefecture, Japan

63

phy, **28**, 494-502.

[11] River Bureau of Ministry of Land, Infrastructure, Transport and Tourism. (2008) Outline of basin and river in Suzuka River water system (Plan). Ministry of Land, Infrastructure, Transport and Tourism, Tokyo.

[12] Mie Office of Rivers and National Highways of Ministry of Land, Chubu Region Maintenance Bureau, Infrastructure, Transport and Tourism (2010) Outline of basin in Kushida River. Ministry of Land, Infrastructure, Transport and Tourism, Tokyo.

[13] River Bureau of Ministry of Land, Infrastructure, Transport and Tourism (2007) Outline of basin and river in Miya River water system (Plan). Ministry of Land, Infrastructure, Transport and Tourism, Tokyo.

[14] Miyawaki, A., Okuda, S. and Suzuki, N. (1975) Vegetation in der umgebung der bucht von Tokyo. Institution for Transport Policy Studies, Tokyo.

[15] Kusanagi, T. (1986) Diagnosis of primary color weed. Rural Culture Association, Tokyo.

[16] Shimizu, T. (2003) Naturalized plants of Japan. HEIBONSHA, Tokyo.

[17] Braun-Blanquet, J. (1964) Pflanzensoziologie. Springer-Verlag, Wien.

[18] Water Information System in Ministry of Land, Infrastructure, Transport and Tourism, Japan.

[19] Nakatubo, T. (1997) Established and the influence of *Poaceae* exotic herbs in river flood field. *Japanese Journal of Conservation Ecology*, **2**, 179-187.

[20] Muranaka, T. and Washitani, I. (2001) Invasion of alien grass *Eragrostis curvula* on the gravelly floodplains of the Kinu River and decrease of river endemics: Necessity of urgent measures. *Japanese Journal of Conservation ecology*, **6**, 111-122.

[21] Japanese Society of Turfgrass Science. (1988) New revision lawn and greening. Soft Science, Inc., Tokyo.

[22] Scott, Warren. R. and William, A. Niering. (1993) Vegetation change on a northeast tidal marsh: Interaction of sea-level rise and marsh accretion, *Ecology*, 74, 96-103.

[23] Huiskes, A.H.L., Koutstaal, B.P., Herman, P.M.J., Beeftink, W.G., Markusse, M.M. and Munck, W.D.E. (1995) Seed dispersal of halophytes in tidal salt marshes. *Journal of Ecology*, **83**, 559-567.

[24] Hutchings, M.J. and Russel, P.J. (1989) The seed regeneration dynamics of an emergent salt marsh. *Journal of Ecology*, 77, 615-637.

[25] Kira, T. (1991) Ecology of reed memorandum future article explore the way of the conservation of the waterside. *Lake Biwa Research Center Report*, **9**, 29-37.

[26] Takada, H. (1974) Salt and biological: Basis of marine organism development. Sogensha, Osaka.

[27] Miyamoto, M. (2007) The effect of river channel-change on habitat and halophytes at Obitsu River estuary. Summary collection of master's thesis presentation in the 27th Annual Meeting of Kanto branch of the Ecological Society of Japan, 19.

10

Effectivity of arbuscular mycorrhizal fungi collected from reclaimed mine soil and tallgrass prairie

Mark Thorne[1*], Landon Rhodes[2], John Cardina[3]

[1]Environmental Science Graduate Program, The Ohio State University, Columbus, USA;
[*]Corresponding Author
[2]Department of Plant Pathology, The Ohio State University, Columbus, USA
[3]Department of Horticulture and Crop Science, Ohio Agricultural and Research Development Center,
The Ohio State University, Columbus, USA

ABSTRACT

We examined suitability of arbuscular mycorrhizal fungi (AMF) associated with cool-season non-native forages on reclaimed surface-mined land in southeast Ohio for establishment of native warm-season grasses. The goal of establishing these grasses is to diversify a post-reclamation landscape that is incapable of supporting native forest species. A 16-week glasshouse study compared AMF from a 30-year reclaimed mine soil (WL) with AMF from native Ohio tallgrass prairie soil (CL). Four native grasses were examined from seedling through 16 weeks of growth. Comparisons were made between CL and WL AMF on colonized (+AMF) and non-colonized plants (–AMF) at three levels of soil phosphorus (P). Leaves were counted at 4 week intervals. Shoot and root biomass and percent AMF root colonization were measured at termination. We found no difference between WL and CL AMF. Added soil P did not reduce AMF colonization, but did reduce AMF efficacy. Big bluestem (*Andropogon gerardii* Vitman), Indiangrass (*Sorghastrum nutans* (L.) Nash), and tall dropseed (*Sporobolus asper* (Michx.) Kunth) benefited from AMF only at low soil P while slender wheatgrass (*Elymus trachycaulus* (Link) Gould ex Shinners) exhibited no benefit. Establishment of tallgrass prairie dominants big blue-stem and Indiangrass would be supported by the mine soil AMF. It appears that the non-native forage species have supported AMF equally functional as AMF from a regionally native tallgrass prairie. Tall dropseed and slender wheatgrass were found to be less dependent on AMF than big bluestem or Indian-grass and thus would be useful in areas with little or no AMF inoculum.

Keywords: Arbuscular Mycorrhizal Fungi; Mycorrhizae; Ecosystem Restoration; Surface Mining; Calcareous Mine Soil; Prairie Grasses

1. INTRODUCTION

Surface coal mining negatively impacts landscapes by altering soil structure and chemistry, and negatively affects beneficial soil organisms such as AMF. Topsoil removal and stockpiling prior to mining destroys active AMF symbiosis and diminishes soil inoculum potential and AMF species composition [1,2]. This impact may inhibit establishment of AMF-dependent species during reclamation and restoration.

Arbuscular mycorrhizal fungi benefit establishment of many plant species on reclaimed mine soils [2-5]. The symbiotic function of these organisms is critical for supplying plants with minerals, primarily phosphorus, in exchange for organic energy compounds [6-10]. This relationship is critical to plant survival especially when soil phosphorus is low [11]. In addition, AMF may affect plant community composition and successional trajectories by differentially benefiting some plants over others [12-20].

While AMF symbiosis is common and occurs in nearly every terrestrial environment [9], differences in the effectiveness of AMF occur over the landscape and in response to management history [20,21]. Strains of AMF from infertile soil are more effective at phosphorus transfer to plants than AMF from fertile soil [22]. Greater effectivity has been found in AMF from zinc-contaminated soil as well other stressful habitats [23,24]. These studies suggest that in harsh, low-nutrient habitats, there is selection for superior AMF strains. Furthermore, a certain degree of host-plant-specificity occurs between AMF and host plants [25-28]. Therefore, more effective strains of AMF may benefit re-establishment of native vegetation in disturbed habitats, as long as host-specificity is not a barrier.

In southeast Ohio, surface coal mine reclamation practices since 1972 have converted nearly 80,000 hectares of native deciduous forestland to non-native forage grassland [29]. In 1972, reclamation laws required that overburden be contoured to approximate the original landscape form, and stockpiled topsoil be spread over the newly-constructed landscape causing severe compaction on the constructed landscape. Furthermore, revegetation did not require native species if preapproved plans stated otherwise [30]. In place of native forest species, non-native forage species such as tall fescue (*Festuca arundinaceae* Schreb.), Kentucky bluegrass (*Poa pratensis* L.), and birdsfoot trefoil (*Lotus corniculatus* L.) were planted because they established easily and tolerated soil compaction caused by the reclamation procedures. These cool-season forages produce a thick ground cover important for controlling erosion and have potentially maintained AMF across the landscape.

Replacing the non-native forage complex with regionally native prairie species is one alternative for increasing biodiversity and ecosystem function on reclaimed mine sites that are incapable of supporting native forest species. Tallgrass prairies are native to parts of Ohio and may represent a diverse set of species that could enhance the functional quality of the mined land [31-33]. However, it is unclear if AMF associated with the cool-season forage species currently growing on the reclaimed mined land would be effective in supporting tallgrass prairie vegetation. Warm-season tallgrass species are more dependent on AMF than cool-season grasses [11,34-36] and problems with host specificity or effectivity could delay or limit their establishment [37].

This research compares the infective and effective potential of AMF collected from a remnant central Ohio tallgrass prairie with AMF from reclaimed mine soil on growth of four native tallgrass prairie grasses. The grass species evaluated were big bluestem, Indiangrass, tall dropseed, and slender wheatgrass. Slender wheatgrass is a cool-season grass while big bluestem, Indiangrass and tall dropseed are warm-season species; all four occur throughout the central grassland region of North America, including tallgrass prairies [38]. The reclaimed mine soil examined in this study has supported a low-diversity non-native forage complex for 30 years. The tallgrass prairie remnant contains 177 plant species including signature tallgrass prairie species big bluestem, Indiangrass, tall dropseed, little bluestem (*Schizachyrium scoparium* (Michx.) Nash), and switchgrass (*Panicum virgatum* L.) [33].

The objectives of this research were 1) to determine if AMF associated with mine soil vegetation are as effective as native tallgrass prairie AMF in supporting native grass growth on reclaimed mine soil, and 2) to determine how these prairie grasses respond to each source of AMF when grown in soils with a range of soil phosphorus levels. The goal of this study was to identify growth responses of prairie grasses to AMF and phosphorus that would aid in developing strategies to increase biodiversity and ecosystem function on compacted reclaimed mine soil [39,40].

2. MATERIALS AND METHODS

2.1. AMF Sources and Pot Culture

Soil containing CL AMF was collected from the Claridon tallgrass prairie remnant near Marion, Ohio. Mine soil containing WL AMF was collected from a reclaimed surface mined area near Cumberland, Ohio. The CL site is a 2.2 ha linear remnant owned by the CSX Railroad and is overseen by the Marion County Historical Society [33]. The WL site is located on land that had been surface mined in the early 1980 s, and was once part of the Muskingum Mine, then owned and mined by Central Ohio Coal Company. In 1986, the land was donated to The International Center for the Preservation of Wild Animals, Inc. (*the* Wilds). The area is part of the Allegheny Plateau of southeast Ohio, which extends westward from the Allegheny Mountains and is a subdivision of the Appalachian Mountain Range.

Approximately 35 liters of soil were collected from each site during September, 2005. Soil from the surface 20 cm was collected from 15 to 20 randomly selected locations at each site. At the CL location, samples were collected alongside established prairie grasses big bluestem and Indiangrass, so that grass roots containing AMF would be included. At the WL location, soil was collected from an area supporting Kentucky bluegrass, tall fescue, and birdsfoot trefoil. These species were dominant throughout the reclaimed mined area. Pot cultures of each AMF source were prepared by mixing soil from each location, 1:1 by volume, with silica sand in a portable cement mixer, which was cleaned between mixes. The soil/sand mix was poured into 3.8-L plastic nursery containers and sown with white clover (*Trifolium repens* L.) as a host plant [41]. By using a legume inoculated with nitrogen fixing bacteria, instead of another grass, the pot cultures could be grown without having to add supplemental N. But, more importantly, this would reduce the chance of propagating pathogens specific to graminoids along with the AMF. The containers were placed on benches in a 20°C to 27°C glasshouse with artificial lighting 12 hr·day^{-1}. The pot cultures were watered daily without fertilizer for 10 months. Inoculum was prepared by chopping up soil and roots from each pot, discarding course roots and tops, then mixing all soil and fine roots together for each AMF source.

Sterile growing medium soil was prepared by mixing topsoil collected from the surface 20 cm at the WL site with silica sand, 1:1 by volume, in a portable cement mixer. The soil/sand mix was then steamed for 5 hr at 100°C. The sterile soil was stored in plastic bins for 21 days at 20°C prior to use in the experiment. Soil from each pot culture, the sterilized growing medium soil, and original WL topsoil were analyzed for pH, P, K, Ca, and Mg content by the Service Testing and Research Laboratory (STAR lab), The Ohio State University/Ohio Agricultural Research and Development Center, Wooster, OH (**Table 1**). Identification of AMF species was not attempted in this study; however, several *Glomus* species were noted in early examination of the source soils (personal observation). It is likely that WL AMF species were part of deciduous forest ecosystem present before mining; although much of the forested biome was deforested from agriculture in the late 1800s. The CL AMF was associated with a historical tallgrass prairie in Northwest Ohio.

2.2. Experiment Establishment and Design

Experimental units consisted of individual grass seedlings growing in 660-cm^3 pots (D40 Deepot®, Stuewe and Sons, Inc., Corvallis, OR) containing 500 cm^3 sterile growing medium soil plus one of four AMF inoculum treatments, and one of three P levels. Inoculum treatments included 100 cm^3 of CL or WL pot-culture soil, or 100 cm^3 of sterilized CL (CLS) or sterilized WL (WLS) pot culture soil. Sterilized inoculum soil was added to the AMF pots to control for possible fertilizer effects from the pot culture soil. The sterilized soil was prepared by autoclaving 8 L of each pot culture soil for 70 min at 130°C, and then resting the soil in plastic bags at 4°C for 96 h.

Three levels of P (P1, P2, P3) were established by mixing 0.0, 0.1, and 0.3 g triple super phosphate (0 - 45 - 0) (Bonide Products Inc., Oriskany, NY) per 500 cm^3 sterile soil plus the 100 cm^3 inoculum soil to reach target P levels of 5, 13, and 27 mg·kg^{-1}, respectively. Calculations were based on the recommendation that 10 mg·kg^{-1} P is required to increase available soil P 1 mg·kg^{-1} (Dr. Donald Eckert, The Ohio State University, personal communication). Each pot was standardized for bacteria by adding 100 ml of sievate corresponding to each particular AMF inoculum. The sievate for each inoculum was prepared by mixing 1000 cm^3 pot culture soil and 16 L water, allowing the slurry to briefly settle, and pouring the liquid and suspended matter through a 53-μm sieve.

The experiment was designed as a randomized complete block with a factorial arrangement of four levels of grass species, four levels of AMF source (CL, CLS, WL,

and WLS), and three levels of P (P1, P2, and P3). Each treatment was replicated six times. The four grass species (SPP) were "Bison" big bluestem (*Andropogon gerardii* Vitman), "Tomahawk" Indiangrass (*Sorghastrum nutans* (L.) Nash), "Revenue" slender wheatgrass (*Elymus trachycaulus* (Link) Gould ex Shinners), and tall dropseed (*Sporobolus asper* (Michx.) Kunth). Big bluestem, Indiangrass, and slender wheatgrass were purchased from Western Native Seeds, Coaldale, CO USA, and tall dropseed was purchased from Oak Prairie Farm, Pardeeville, WI USA. Seeds of each species were sown 10 - 20 per pot, and thinned to leave a single seedling in each pot.

Pots were placed in trays and arranged so that +AMF treatments were adjacent to −AMF control pots to allow for paired-pot comparisons of AMF sources. Trays were placed on a glasshouse bench in a randomized-block design such that blocks controlled for distance from the cooling/heating source on one end and exhaust fan at the other. Artificial lighting was set to maintain a minimum of 300 W·m^{-2} 16 h·day^{-1}, and temperature ranged between 19°C - 27°C.

2.3. Grass Leaf and Biomass Measurements

At 4, 8, 12, and 16 weeks following germination, living and dead leaves were counted on each plant. To reduce confusion in successive censuses, dead leaves were removed and stored for later biomass measurement. At the end of the 16-week experiment, plants were destructively harvested to assess aboveground and belowground biomass. Culms and leaves were clipped at the soil surface and put in paper bags along with dead leaves from earlier censuses. Roots were washed to remove the soil and then bagged separately from shoots. Biomass samples were dried at 55°C for a minimum of 96 h, and then weighed. Three small sub-samples were cut fresh from each root system to assess AMF colonization. The root sub-samples were approximately 10 × 25 mm each and cut from the top, middle, and bottom third of the root length. Root sub-samples were stored in a 48% ethanol solution until being processed for AMF evaluation.

2.4. AMF Colonization Assessment

Root samples were cleared and stained according to a modified Phillips and Hayman [42] procedure. During processing, root samples from each plant were contained in 28 × 5-mm tissue processing cassettes (Canemco Inc., Quebec, Canada). Roots were cleared in 10% KOH solution and autoclaved at 130°C for 10 min, and then acidified in a 1% HCL solution for 20 min at room temperature. Roots were stained in 0.05% Trypan blue staining solution containing 1:2:1 distilled water, lactic acid, and glycerin, and autoclaved for 7 min at 130°C. Following staining, roots were rinsed in tap water and stored in plastic Petri dishes covered with a 1:1 solution of distilled water and glycerin and kept in a 4°C cooler. Colonization was assessed using a gridline intersect method [43,44] in a plastic Petri dish with gridlines scored 13 mm apart on the bottom. For each sample, the first 50 roots bisecting gridlines were examined. Roots were designated colonized if the root segment contained hyphae, arbuscules, or vesicles; otherwise, they were designated non-colonized. Percent colonization was calculated by dividing the number of colonized roots by 50, then multiplying by 100. The root sample was then dried at 55°C and weighed, and the dry weight was added back to the total root biomass.

2.5. Statistical Analysis

Data were analyzed using PROC GLM in SAS/STAT® software [45] and significance was accepted at $\alpha = 0.05$. Main effects were SPP, AMF, and P. Post-hoc comparisons were made using protected Fisher's LSD test where differences were accepted only if the P-value calculated by PROC GLM was equal to or less than 0.05 [46]. Dependent variables were leaf number, shoot, root, and total biomass, root-to-shoot ratio (RSR), difference between +AMF and −AMF for shoot biomass difference (SDIFF), root biomass difference (RDIFF), total biomass difference (TDIFF), and AMF root colonization percent. Difference in biomass was calculated as a separate continuous random variable for each paired-pot comparison [47]. The null hypothesis for a paired-pot analysis is that the difference (D) between the two pairs is zero. Accepting the alternative hypothesis is based on the deviance from zero. Benefit from AMF inoculation was indicated by a positive outcome after subtracting the −AMF value from the +AMF value. Analysis of colonization percent only included the inoculated treatments in order to accurately reflect the level of infectivity of each inoculum.

3. RESULTS

3.1. AMF Colonization as Affected by Soil P, Inoculum, and Grass Species

The reclaimed mine soil used in this experiment initially averaged 12 mg·kg⁻¹ P, 3768 mg·kg⁻¹ calcium (Ca), and pH of 7.3 (**Table 1**). These values indicate a calcareous soil with limited available P. Mixing the soil with silica sand reduced the available P to 5 mg·kg⁻¹ creating critically low soil P for plant growth. Low P is conducive for testing and comparing the efficacy of the AMF strains. Furthermore, the addition of P acts as a control for the activity of AMF because it tests the efficacy of the AMF. Other nutrients were likely limiting, *i.e.* N, but P is the primary nutrient associated with AMF

Table 1. Soil properties of reclaimed mine and AMF inoculum soil used to compare growth of prairie grasses with different concentrations of phosphorus (P) and different sources of arbuscular mycorrhizal fungi (AMF).

Soil Parameter[a]	Mine[b] topsoil (0 - 20 cm)	Sterile mine soil/sand mix (1:1)	AMF source[c]	
			CL	WL
pH	7.3	7.3	7.7	7.9
P (mg·kg⁻¹)	12	5	<1	7
K (mg·kg⁻¹)	161	80	77	41
Ca (mg·kg⁻¹)	3768	1722	1345	1262
Mg (mg·kg⁻¹)	321	198	235	198

[a]Soil analyzed by STAR lab, Wooster, OH. P analyzed with Bray P1 method; K, Ca, and Mg analyzed with ammonium acetate extract method. [b]Soil collected from *the* Wilds 30-yr reclaimed surface-mined land near Cumberland, OH. [c]CL collected from Claridon tallgrass prairie remnant near Marion, OH; WL collected from *the* Wilds mine soil supporting non-native forage grasses. AMF inoculum soil prepared as pot-cultures containing a 1:1 mix of soil and silica sand.

function. Adding N to the pots would have increased growth, but then the activity of the AMF may have been confounded by amount of N taken up by each species, and in turn, how each species converted the added N into photosynthates to fuel the AMF. In our study, we found that neither AMF source nor P concentration had any effect on colonization percent when averaged over all other factors (**Table 2**). Grasses inoculated with either CL or WL averaged slightly greater than 50% AMF colonization, indicating that both AMF cultures were equally accepted by the host grasses. The AMF colonization trended lower from 56% to 49% as P increased, but those differences were not significant (**Table 2**). Colonization differed among species, as tall dropseed and slender wheatgrass had the highest percentages with 70% and 55%, respectively (**Table 2**). Big bluestem and Indiangrass had lowest colonization with 51% and 36%, respectively.

3.2. AMF and Soil P Effect on Plant Growth

Response to soil P concentration was predictable, as an increase in P resulted in an increase in biomass production, and was most consistent for shoot biomass (**Table 3**). Total biomass also followed the same pattern, increasing with each increased level of P. However, root biomass did not increase from P2 to P3, thus P2 had the highest RSR, as the increased shoot growth at the higher P was not matched by a corresponding increase in root growth. This is likely a result of space limitation in the pots and not a lack of response to increased P, as pots with the highest P level were densely packed with roots when harvested.

Slender wheatgrass produced the greatest total biomass, which was similarly split between root and shoot

Table 2. Percent colonization by arbuscular mycorrhizal fungi (AMF) as affected by AMF sources, soil phosphorus level (P), and grass species (SPP) in a 16-week glasshouse experiment.

Parameter	Colonization[a]
AMF	(%)
CL	52.3 a
WL	53.1 a
	(P = 0.7917)
P[b]	
P1	55.6 a
P2	53.7 a
P3	48.9 a
	(P = 0.2271)
SPP	
Big bluestem	51.0 bc
Indiangrass	35.5 c
Tall dropseed	69.9 a
Slender wheatgrass	54.7 ab
	(P = 0.0226)

[a]Colonization percents reflect only plants inoculated with AMF. Non-inoculated plants had 0% AMF colonization. Differences within each variable are determined using protected Fisher's LSD ($\alpha = 0.05$). Means followed by the same lower-case letter (a, b, c) are not different. [b]P target levels were P1 = 5 mg·kg^{-1} P; P2 = 13 mg·kg^{-1} P; P3 = 27 mg·kg^{-1} P.

Table 3. Biomass production as affected by grass species (SPP) and three levels of soil phosphorus (P) in a 16-week glasshouse experiment.

Main effects	Dependent variables[a]			
	Shoot	Root	Total	RSR[b]
SPP	(g dry weight)			(g/g)
Big bluestem	0.7 c	1.5 a	2.2 b	2.4 a
Indiangrass	0.7 c	1.1 b	1.8 c	1.8 b
Tall dropseed	1.1 b	0.7 c	1.8 c	0.7 c
Slender wheatgrass	1.5 a	1.3 ab	2.8 a	0.9 c
	(P < 0.0001)	(P < 0.0001)	(P < 0.0001)	(P < 0.0001)
P[c]				
P1	0.6 c	0.7 b	1.4 c	1.5 ab
P2	1.0 b	1.4 a	2.4 b	1.6 a
P3	1.3 a	1.4 a	2.7 a	1.3 b
	(P < 0.0001)	(P < 0.0001)	(P < 0.0001)	(P = 0.0231)

[a]P-values represent the probability of differences within each dependent variable for each main-effect. Differences within each variable are determined using protected Fisher's LSD ($\alpha = 0.05$). Means followed by the same lower-case letter (a,b,c) are not different. [b]Root to shoot ratio (RSR) calculated by dividing root weight by shoot weight. [c]See **Table 2** for target P levels.

biomass (**Table 3**). Big bluestem had similar root biomass compared with slender wheatgrass, but only produced half the shoot biomass. The difference in allocation of resources between these species was reflected in the RSR, as slender wheatgrass averaged 0.9 while big

bluestem averaged 2.4 (**Table 3**). Indiangrass and tall dropseed produced the least total biomass; however, Indiangrass allocated more resources to root biomass, while tall dropseed allocated more resources to shoot production. This indicates that big bluestem and Indiangrass appear to direct more resources, proportionately, to root growth during seedling establishment, compared with slender wheatgrass and tall dropseed.

3.3. AMF Effectivity in Paired-Pot Comparison

To compare the effectiveness of the AMF cultures, a paired-pot arrangement was used to examine the difference in biomass accumulation between colonized and non-colonized plants. By subtracting the biomass of a −AMF plant from an adjacent +AMF plant for each component (shoot/root/total), new variables were created that, if positive, indicated AMF benefit, and if negative, indicated AMF detriment. The GLM analysis indicated that SPP and P had the greatest influence on shoot difference (SDIFF), root difference (RDIFF), and total difference (TDIFF), with P-values < 0.0001 (**Table 4**). In contrast, AMF had a significant impact only on SDIFF (**Table 4**).

There was a significant interaction between AMF and P for RDIFF and TDIFF, influenced mainly by RDIFF (**Table 4**). For SDIFF, AMF benefit was positive for both CL and WL at P1, but decreased for CL with each increase in P, whereas there was no decreased benefit for WL at P3 (**Figure 1(a)**). There was no difference in AMF benefit for RDIFF between P1 and P2 for CL AMF, which was negative at all three P levels (**Figure 1(b)**); however, WL AMF was positive at P1 but negative at both P2 and P3, with the least benefit occurring at P2 (**Figure 1(b)**). For TDIFF, the benefit of both AMF sources was positive at P1, but negative at P2 and P3 (**Figure 1(c)**). The interactions between AMF and P, for all three difference variables, occurred as the response to increasing P differed between AMF sources.

Significant interactions were also found between SPP and P for all three difference variables. Overall, the only AMF benefit occurred with big bluestem, Indiangrass only at the lowest P level (**Figure 2**). For SDIFF, benefit decreased as P increased for big bluestem, Indiangrass, and slender wheatgrass, but tall dropseed was not affected (**Figure 2(a)**). In addition, slender wheatgrass experienced negative benefit at all three P levels. For RDIFF, both tall dropseed and slender wheatgrass had negative benefit at all three P levels and was not affected by an increase in P (**Figure 2(b)**). Big bluestem RDIFF declined with each increase in P while Indiangrass declined only from P1 to P2. The interactions with TDIFF were very similar to RDIFF (**Figure 2(c)**).

The interaction between SPP and P was also evident in the number of leaves produced during the 16-week expe-

Table 4. Analysis of variance table (PROC GLM) for the full model with a factorial arrangement testing the difference in biomass production for grass species (SPP) colonized with arbuscular mycorrhizal fungi (+AMF) and non-AMF (–AMF) plants. Dependent variables shoot difference (SDIFF), root difference (RDIFF), and total difference (TDIFF) were produced by subtracting biomass of –AMF plants from +AMF plants in a paired-pot glasshouse experiment examining the effects of AMF source and P on growth of prairie SPP grown in sterilized mine soil.

		SDIFF		RDIFF		TDIFF	
Model	DF	F value	P > F	F value	P > F	F value	P > F
BLOCK	5	0.4	0.8753	2.2	0.0601	1.2	0.3225
SPP	3	15.5	<0.0001	19.2	<0.0001	25.5	<0.0001
AMF	1	6.0	0.0162	0.6	0.4294	0.2	0.6244
P	2	38.9	<0.0001	29.4	<0.0001	51.2	<0.0001
SPPxAMF	3	0.7	0.5354	2.8	0.0461	2.8	0.0428
SPPxP	6	3.9	0.0013	3.3	0.0053	5.2	<0.0001
AMFxP	2	2.8	0.0629	15.6	<0.0001	13.1	<0.0001
SPPxAMFxP	6	0.5	0.8391	0.7	0.6449	0.9	0.5276
TOTAL	141						

riment (**Figure 3**). At P1, leaf production at each census was greater for mycorrhizal big bluestem, Indiangrass, and tall dropseed compared with non-mycorrhizal plants. Indiangrass appeared to have the greatest AMF benefit, whereas +AMF slender wheatgrass produced slightly more leaves only at 4 weeks (**Figure 3**).

At P2 and P3, AMF effect was less evident for all grass species except slender wheatgrass where –AMF plants produced the greatest number of leaves (**Figure 3**). The interaction between SPP and P at these levels of P would indicate that the cool-season slender wheatgrass is negatively affected by AMF when P is abundantly available, whereas the warm-season grasses (big bluestem, Indiangrass, and tall dropseed) are less affected (**Figure 3**). Slender wheatgrass appears to gain little, if any, benefit from AMF, which would suggest this species would be useful in restoration plantings where AMF is not initially present.

4. DISCUSSION

A number of studies have shown that AMF colonization is reduced by higher soil P [48-51], but that was not evident in this study. If plants can obtain P on their own then symbiosis would be less beneficial. However, there is often no clear relationship between colonization percent and P uptake or plant growth response [52-54], meaning that efficacy is not necessarily related to the magnitude of colonization. It is known that warm-season grasses tend to be more dependent on AMF than cool-season grasses, especially when P is limited [35]. Cool-season grasses tend to have finer root systems that are better suited for P uptake, while warm-season grasses tend to have more coarse root systems. The high abundance of AMF in slender wheatgrass roots was unexpected. Big bluestem is known to be very dependent on AMF [55] and is a dominant species in tallgrass prairies

across North America. The mycorrhizal status of tall dropseed has not been reported, but a related species, *Sporobolus heterolepis*, is mycorrhizal [56].

Big bluestem and Indiangrass responded to AMF and P as expected according to previous research [55]. Both species benefited from AMF when soil P was low, and showed less benefit as P increased. Both of these grasses allocated more resources to roots than aboveground tissue, which is important for access to nutrients and water during periods of stress. Harris [57] determined that competitive success of non-native downy brome (*Bromus tectorum* L.) was due to its ability to establish a deep root system during autumn and winter when native bluebunch wheatgrass (*Pseudoroegneria spicata* (Pursh) A. Löve) was dormant. During spring, a downy brome infestation depleted soil moisture before bluebunch wheatgrass was able to complete its reproductive cycle. The dominance of big bluestem and Indiangrass in tallgrass prairies is likely due to their ability to establish deep root systems over time, as well as their association with AMF when soil P is limited.

Tall dropseed and slender wheatgrass both appear facultative in their response to AMF. Tall dropseed is a warmseason prairie grass, but seems to respond to AMF and P similarly to facultative cool-season grasses. Greater production of aboveground biomass compared with root biomass and low dependence on AMF would suggest that tall dropseed can quickly establish following disturbance in habitats where P and AMF may be limiting. Slender wheatgrass forms association with AMF, but is clearly not dependent on AMF. It is able to access P when soil levels are low, and can be very productive when soil P is higher.

5. CONCLUSIONS

Results of this research indicate that AMF associated with reclaimed mine soil are not likely a barrier for es-

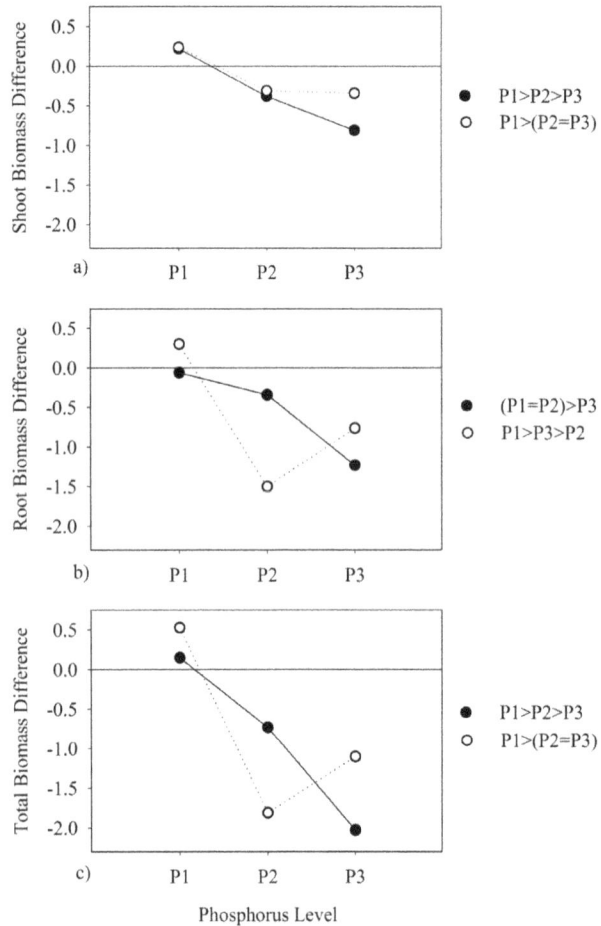

Figure 1. Interaction of arbuscular mycorrhizal fungi (AMF) from the Claridon tallgrass prairie remnant (CL) and *the* Wilds reclaimed mine soil (WL) in Ohio and soil phosphorus level (P1, P2, P3). Values represent the difference in biomass between AMF-colonized and non-colonized grasses [a]in a 16-week glasshouse experiment measuring shoot biomass difference (SDIFF), root biomass difference (RDIFF), and total biomass difference (TDIFF). Solid lines (—) represent CL-AMF comparisons; dotted lines (····) represent WL-AMF comparisons. Differences at each phosphorus level within each graph ($\alpha = 0.05$) are shown to the right.

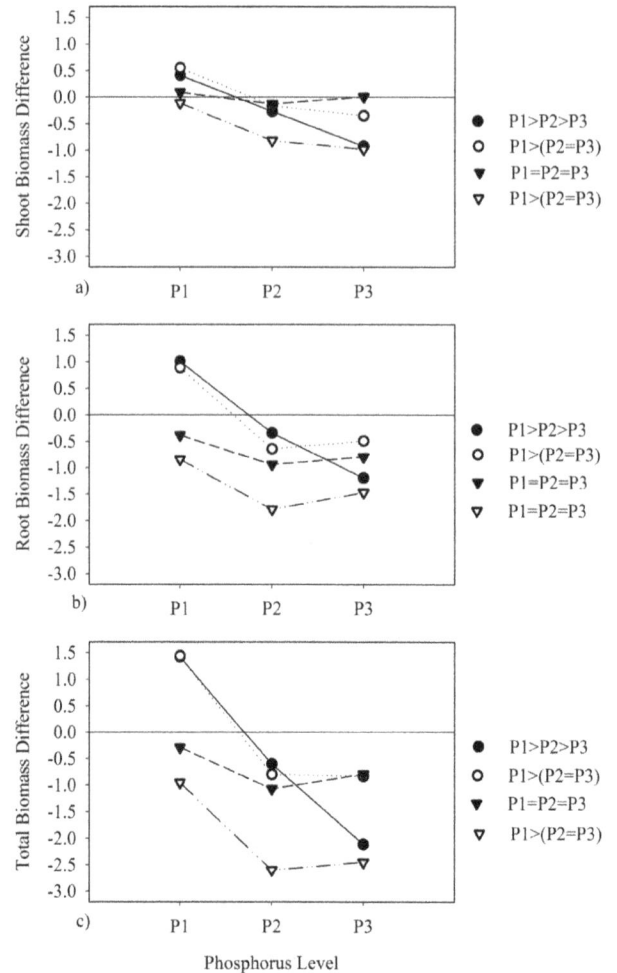

Figure 2. Interaction of grass species (SPP) and soil phosphorous concentration (P) on the difference in biomass between grasses colonized with arbuscular mycorrhizal fungi (AMF) and non-colonized grasses in a 16-week glasshouse experiment measuring shoot (SDIFF), root (RDIFF), and total biomass measurements (TDIFF). Solid lines (—) represent big bluestem; dotted lines (····) represent Indiangrass; dashed lines (– –) represent tall dropseed; dashed and dotted lines (– ·· –) represent slender wheatgrass. Differences at each phosphorus level within each graph are shown to the right and determined using a protected Fisher's LSD ($\alpha = 0.05$).

tablishing tallgrass prairie species. Colonization levels were similar between the two AMF inoculums. This would also suggest that host specificity is not a deterrent for native grass establishment even though the mine soil AMF have been associated with non-native cool-season forage species for 30 years. It appears that poor soil conditions of the mine soil, *i.e.* compacted calcareous soil with low available phosphorus, may have selected an effective AMF community, which could benefit native tallgrass prairie grasses.

Tall dropseed and slender wheatgrass both appear to establish well when P is low, with or without AMF, and would be useful in early establishment of a prairie community on reclaimed mine soil. Big bluestem and Indiangrass are more dependent on AMF, but did benefit from

the mine soil AMF in this study. This study suggests that years of growth by non-native cool-season forage species on reclaimed compacted mine soil in southeast Ohio have propagated AMF that would aid in the establishment of native AMF-dependent warm season prairie grasses. The ecological significance of these findings is that in highly disturbed landscapes there are many potential ways for ecosystems to self organize. The non-native species planted during reclamation were a valuable nurse crop for the indigenous AMF, but did not yield a diverse landscape. By adding more species that can utilize the AMF, the low diversity issue may be addressed and a new era of self organization could lead to more function, structure, and

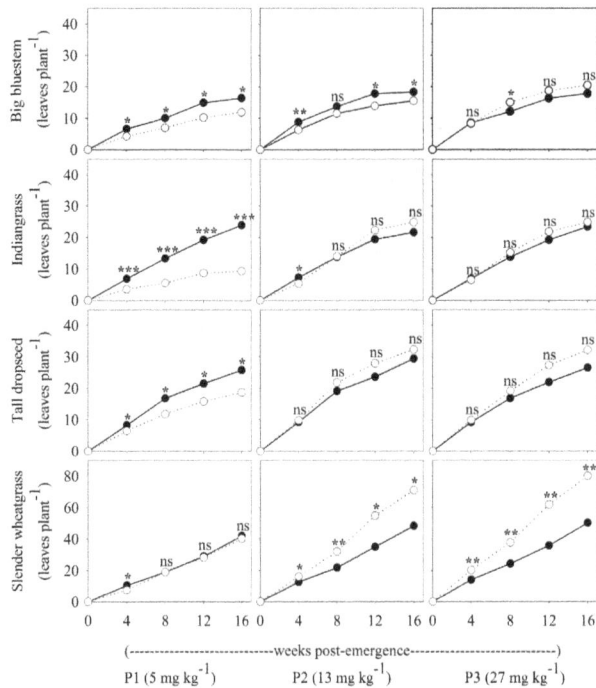

Figure 3. Arbuscular mycorrhizal fungi (AMF) effect on number of leaves produced by prairie grasses at three concentrations of soil phosphorus (P) level during a 16-week glasshouse experiment. Solid lines (—) represent AMF colonized; dotted lines (····) represent non-AMF colonized plants. P-values are signified as follows: * = 0.05 < P < 0.001, ** = 0.001 < P < 0.0001, *** = P < 0.0001, and ns = P > 0.05 between AMF and non-AMF plants at each sampling.

productivity from the landscape.

6. ACKNOWLEDGEMENTS

We thank Dr. Nicole Cavender for access to soil at the Wilds; Dr. Robert Klips for access to soil at the Claridon Prairie; Dr. Marc Evans for statistical consultation; and Drs. Craig Davis and James Metzger for comments on the manuscript. We also thank David Snodgrass and Jim Vent of the Howlett Greenhouse facility at The Ohio State University for assistance with this project. Salaries and research support were provided by state and federal funds appropriated to the Ohio Agriculture Research and Development Center, The Ohio State University. Manuscript No. HCS13-04.

REFERENCES

[1] Gould, A.B. and Liberta, A.E. (1981) Effects of topsoil storage during surface mining on the viability of vesicular-arbuscular mycorrhiza. *Mycologia*, **73**, 914-921.

[2] Waaland, M.E. and Allen, E.B. (1987) Relationship between VA mycorrhizal fungi and plant cover following surface mining in Wyoming. *Journal of Range Management*, **40**, 271-276.

[3] Lambert, D.H. and Cole Jr., H. (1980) Effects of my-

corrhizae on establishment and performance of forage species in mine spoil. *Agronomy Journal*, **72**, 257-260.

[4] Allen, E.B. (1989) The restoration of disturbed arid landscapes with special reference to mycorrhizal fungi. *Journal of Arid Environments*, **17**, 279-286.

[5] Hetrick, B.A.D., Wilson, G.W.T. and Figge, D.A.H. (1994) The influence of mycorrhizal symbiosis and fertilizer amendments on establishment of vegetation in heavy metal mine spoil. *Environmental Pollution*, **86**, 171-179.

[6] Gerdemann, J.W. (1968) Vesicular-arbuscular mycorrhiza and plant growth. *Annual Review of Phytopathology*, **6**, 397-418.

[7] Rhodes, L.H. and Gerdemann, J.W. (1975) Phosphate uptake zones of mycorrhizal and non-mycorrhizal onions. *New Phytologist*, **75**, 555-561.

[8] Barrow, N.J., Malajczuk, N. and Shaw, T.C. (1977) A direct test of the ability of vesicular-arbuscular mycorrhizae to help plants take up fixed soil phosphate. *New Phytologist*, **78**, 269-276.

[9] Smith, F.A. and Smith, S.E. (1997) Structural diversity in (vesicular)-arbuscular mycorrhizal symbiosis. *New Phytologist*, **137**, 373-388.

[10] Cavagnaro, T.R., Smith, F.A., Smith, S.E. and Jakobsen, I. (2005) Functional diversity in arbuscular mycorrhizas: exploitation of soil patches with different phosphate enrichment differs among fungal species. *Plant Cell and Environment*, **28**, 642-650.

[11] Brejda, J.J., Yocom, D.H., Moser, L.E. and Waller, S.S. (1993) Dependence of 3 Nebraska Sandhills warm-season grasses on vesicular-arbuscular mycorrhizae. *Journal of Range Management*, **46**, 14-20.

[12] Janos, D.P. (1980) Mycorrhizae influence tropical succession. *Biotropica*, **12**, 56-64.

[13] Allen, E.B. and Allen, M.F. (1984) Competition between plants of different successional stages: Mycorrhizae as regulators. *Canadian Journal of Botany*, **62**, 2625-2629.

[14] Crowell, H.F. and Boerner, R.E.J. (1988) Influences of mycorrhizae and phosphorus on belowground competition between two old-field annuals. *Environmental and Experimental Botany*, **28**, 381-392.

[15] Gange, A.C., Brown, V.K. and Farmer, L.M. (1990) A test of mycorrhizal benefit in an early successional plant community. *New Phytologist*, **115**, 85-91.

[16] Allen, M.F. and Allen, E.B. (1992) Mycorrhizae and plant community development: Mechanisms and patterns. In: Carrol, G.C. and Wicklow, D.T., Eds., *The Fungal Community: Its Organization and Role in the Ecosystem. Mycology Series* 9, Marcel Dekker, Inc., New York.

[17] Gange, A.C., Brown, V.K. and Sinclair, G.S. (1993) Vesi-

cular-arbuscular mycorrhizal fungi: A determinant of plant community structure in early succession. *Functional Ecology*, **7**, 616-622.

[18] Hartnett, D.C., Samenus, R.J., Fischer, L.E. Hetrick, B.A.D. (1994) Plant demographic response to mycorrhizal symbiosis in tallgrass prairie. *Oecologia*, **99**, 21-26.

[19] Koske, R.E. and Gemma, J.N. (1997) Mycorrhizae and succession in planting of beachgrass in sand dunes. *American Journal of Botany*, **84**, 118-130.

[20] Gillespie, I.G. and Allen, E.B. (2006) Effects of soil and mycorrhizae from native and invaded vegetation on a rare California forb. *Applied Soil Ecology*, **32**, 6-12.

[21] Scullion, J., Eason, W.R. and Scott, E.P. (1998) The effectivity of arbuscular mycorrhizal fungi from high input conventional and organic grassland and grass-arable rotations. *Plant and Soil*, **204**, 243-254.

[22] Henkel, T.W., Smith, W.K. and Christensen, M. (1989) Infectivity and effectivity of indigenous vesicular-arbuscular mycorrhizal fungi from contiguous soils in southwestern Wyoming, USA. *New Phytologist*, **112**, 205-214.

[23] Shetty, K.G., Hetrick, B.A.D. and Schwab, A.P. (1995) Effects of mycorrhizae and fertilizer amendments on zinc tolerance of plants. *Environmental Pollution*, **88**, 307-314.

[24] Thorne, M.E., Zamora, B.A. and Kennedy, A.C. (1998) Sewage sludge and mycorrhizal effects on Secar bluebunch wheatgrass in mine spoil. *Journal of Environmental Quality*, **27**, 1228-1233.

[25] Zhu, Y.G., Laidlaw, A.S., Christie, P. and Hammond, M.E.R. (2000) The specificity of arbuscular mycorrhizal fungi in perennial ryegrass-white clover pasture. *Agriculture Ecosystems and Environment*, **77**, 211-218.

[26] Ronsheim, M.L. and Anderson, S.E. (2001) Population-level specificity in the plant-mycorrhizae associations alters intraspecific interactions among neighboring plants. *Oecologia*, **128**, 77-84.

[27] Bevor, J.D. (2002) Host-specificity of AM fungal population growth rates can generate feedback on plant growth. *Plant and Soil*, **244**, 281-290.

[28] Sanders, I.R. (2003) Preference, specificity and cheating in the arbuscular mycorrhizae symbiosis. *Trends in Plant Science*, **8**, 143-145.

[29] Kaster, G. and Vimmerstedt, J.P. (1996) Tree planting on strip-mined land. In: Norland, E.R. and Ervin, M.S., Eds., *Forest Resource Issues in Ohio* 1996, *Legislator's Handbook*, 2nd Edition. Ohio Society of American Foresters, Columbus.

[30] SMCRA (1977) (Surface Mining Control and Reclamation Act) Office of Surface Mining Reclamation and Enforcement, US Department of Interior, Washington DC.

[31] Transeau, E.N. (1935) The prairie peninsula. *Ecology*, **3**, 423-437.

[32] Sala, O.E., Patron, W.J., Joyce, L.A. and Lauenroth, W.K. (1988) Primary production of the central grassland region of the United States. *Ecology*, **69**, 40-45.

[33] Klips, R.A. (2003) Vegetation of Claridon railroad prairie, a remnant of the Sandusky Plains of central Ohio. *Castanea*, **68**, 135-142.

[34] Loree, M.A.J. and Williams, S.E. (1987) Colonization of western wheatgrass (*Agropyron smithii* Rydb.) by vesicular-arbuscular mycorrhizal fungi during the revegetation of a surface mine. *New Phytologist*, **106**, 735-744.

[35] Hetrick, B.A.D., Wilson, G.W.T. and Leslie, J.F. (1991) Root architecture of warm- and cool-season grasses: relationship to mycorrhizal dependence. *Canadian Journal of Botany*, **69**, 112-118.

[36] Noyd, R.K., Pfleger, F.L. and Russelle, M.P. (1995) Interactions between native prairie grasses and indigenous arbuscular mycorrhizal fungi: Implications for reclamation of taconite iron ore tailing. *New Phytologist*, **129**, 651-660.

[37] Cavender, N. and Knee, M. (2006) Relationship of seed source and arbuscular mycorrhizal fungi inoculum type to growth and colonization of big bluestem (*Andropogon gerardii*). *Plant and Soil*, **285**, 57-65.

[38] Hitchcock, A.S. (1971) Manual of the grasses of the United States. Dover Public, New York.

[39] Taheri, W.I. and Bevor, J.D. (2010) Adaptation of plants and arbuscular mycorrhizal fungi to coal tailings in Indiana. *Applied Soil Ecology*, **45**, 138-143.

[40] Simmons, J.A., Currie, W.S., Eshleman, K.N., Kuers, K., Monteleone, S., Negley, T.L., Pohlad, B.R. and Thomas, C.L. (2008) Forest to reclaimed mine land use change leads to altered ecosystem structure and function. *Ecological Applications*, **18**, 104-118.

[41] Liu, R. and Wang, F. (2003) Selection of appropriate host plants used in trap culture of arbuscular mycorrhizal fungi. *Mycorrhiza*, **13**, 123-127.

[42] Phillips, J.M. and Hayman, D.S. (1970) Improved procedures for clearing roots and staining parasitic and vesicular-arbuscular mycorrhizal fungi for rapid assessment of infection. *Transactions of the British Mycological Society*, **5**, 158-161.

[43] Newman, E.I. (1966) A method of estimating the total length of root in a sample. *Journal of Applied Ecology*, **3**, 139-145.

[44] Giovannetti, M. and Mosse, B. (1980) An evaluation of techniques for measuring vesicular arbuscular mycorrhizal infection in roots. *New Phytologist*, **84**, 489-500.

[45] SAS Institute Inc. (2002) SAS OnlineDoc, Version 9.1.3.

[46] Milliken, G.A. and Johnson, D.E. (1984) Analysis of messy data, Volume I: Designed experiments. Wadsworth, Inc., Belmont.

[47] Zar, J.H. (1999) Biostatistical analysis. 4th Edition, Prentice-Hall, Inc., Upper Saddle River.

[48] Mosse, B. (1973) Plant growth responses to vesicular-arbuscular mycorrhiza IV. In soil given additional phosphate. *New Phytologist*, **72**, 127-136.

[49] Schubert, A. and Hayman, D.S. (1986) Plant growth responses to vesicular-arbuscular mycorrhiza. XVI. Effectiveness of different endophytes at different levels of soil phosphate. *New Phytologist*, **103**, 79-90.

[50] Sainz, M.J. and Arines, J. (1988) Effects of native vesicular-arbuscular mycorrhizal fungi and phosphate fertilizer on red clover growth in acid soils. *Journal of Agricultural Science Cambridge*, **111**, 67-73.

[51] Al-Karaki, G.N. and Al-Omoush, M. (2002) Wheat response to phosphogypsum and mycorrhizal fungi in alkaline soil. *Journal of Plant Nutrition*, **25**, 873-883.

[52] Lioi, L. and Giovannetti, M. (1987) Variable effectivity of three vesicular-arbuscular mycorrhizal endophytes in *Hedysarum coronarium* and *Medicago sativa*. *Biology and Fertility of Soils*, **4**, 193-197.

[53] Sanders, I.R. and Fitter, A.H. (1992) The ecology and functioning of vesicular-arbuscular mycorrhizas in co-existing grassland species. *New Phytologist*, **120**, 525-533.

[54] Mohammad, M.J., Pan, W.L. and Kennedy, A.C. (1995) Wheat response to vesicular-arbuscular mycorrhizal fungal inoculation of soils from eroded toposequence. *Soil Science Society of America Journal*, **59**, 1086-1090.

[55] Hetrick, B.A.D., Kitt, D.G. and Wilson, G.T. (1986) The influence of phosphorus fertilization, drought, fungal species, and nonsterile soil on mycorrhizal growth response in tall grass prairie plants. *Canadian Journal of Botany*, **64**, 1199-1203.

[56] Dhillion, S.S. (1992) Evidence for host-mycorrhizal preference in native grassland species. *Mycological Research*, **96**, 359-362.

[57] Harris, G.A. (1967) Some competitive relationships between *Agropyron spicatum* and *Bromus tectorum*. *Ecological Monographs*, **37**, 89-111.

11

Retrospective analysis of two Northern California wild-land fires via Landsat five satellite imagery and Normalized Difference Vegetation Index (NDVI)

Bennett Sall[1], Michael W. Jenkins[2*], James Pushnik[1,3]

[1]Department of Biological Sciences, California State University, Chico, USA
[2]Department of Ecology and Evolutionary Biology, University of California, Santa Cruz, USA;
[*]Corresponding Author
[3]Institute for Sustainable Development, California State University, Chico, USA

ABSTRACT

Wild-land fires are a dynamic and destructive force in natural ecosystems. In recent decades, fire disturbances have increased concerns and awareness over significant economic loss and landscape change. The focus of this research was to study two northern California wild-land fires: Butte Humboldt Complex and Butte Lightning Complex of 2008 and assessment of vegetation recovery after the fires via ground based measurements and utilization of Landsat 5 imagery and analysis software to assess landscape change. Multi-temporal and burn severity dynamics and assessment through satellite imagery were used to visually ascertain levels of landscape change, under two temporal scales. Visual interpretation indicated noticeable levels of landscape change and relevant insight into the magnitude and impact of both wild-land fires. Normalized Burn Ratio (NBR) and delta NBR (ΔNBR) data allowed for quantitative analysis of burn severity levels. ΔNBR results indicate low severity and low re-growth for Butte Humboldt Complex "burned center" subplots. In contrast, ΔNBR values for Butte Lightning Complex "burned center" subplots indicated low-moderate burn severity levels.

Keywords: Wild-Land Fire; Burn Severity; Vegetation Recovery; Normalized Difference Vegetative Index (NDVI); Normalized Burn Ratio (NBR)

1. INTRODUCTION

In June 2008, two wild-land fires consumed large areas in rural Butte County, California, USA. On June 11, the Butte Humboldt Complex (BHC) fire broke out and spread rapidly to over 23,344 acres causing the destruction of 87 homes, ten injuries, and 20.5 million dollars in damages [1]. On June 21, the Butte Lightning Complex (BLC) ignited and quickly spread to over 59,440 acres, causing the destruction of 106 homes, 71 injuries, and 85.3 million dollars in damages [1]. Fire regime measurements of severity, frequency and vegetation recovery, are all directly related to fire impact, in both environmental and economic terms. These are complex and dynamic systems with factors including but not limited to climate, local weather, fuel loading and encompass scales from regional to global [2-4]. However, complicated fire regimes are to study and quantify there does seem to be one universal and coherent view; fires are and will continue to become more severe and frequent under future global climate projections [2-7].

Ground based vegetation reflectance measurements and associated reflectance indices have been effective in looking at several noteworthy vegetation reflectance trends residing in the blue and red spectral regions. The blue and red spectral regions typically denote increased absorbance, due to photosynthetic activities, and more specifically absorption by vegetation pigments [8]. Reflectance in the infrared spectral region can also be used to determine water content detection in canopies [9]. Quick and reliable measurements of plant water concentration via ground based indices can lead to important information for irrigation practices, drought assessment of natural communities, and the definition of wildfire risk [10-12].

The Normalized Difference Vegetation Index (NDVI) is well established and utilized by researchers around the globe, with a multitude of ecological applications including pre- and post-wild-land fire assessments, multi-time series landscape change analysis, and landscape dynamics and drought influences on wild-land fires [13,14]. NDVI has the ability to accurately utilize specific band regions of the wavelength spectrum involved with vegetation physiological characteristics [15]. More centrally, relying on links between increased absorption in the red visible region and higher chlorophyll content [16], as well as increased absorption in the near infrared region. NDVI illustrates a decreased ability of vegetation to reflect heat, denoting vegetation stress [17]. NDVI has been shown to have vegetation values spanning above zero to one [15], and can be broadly viewed with the equation (Near infrared-Red visible region/Near infrared + Red visible region) [18], with red visible region ranging from 0.63 to 0.69 μm and near infrared region ranging from 0.75 to 0.80 μm [18]. NDVI has been shown to be effective in shrub-land communities for detection of vegetation responses related to drought influences as well as monitoring vegetation development [13].

Technological advancements in recent decades in high resolution spectral imaging sensors and analysis software have permitted remote sensing to emerge as a valuable tool to comprehend, measure, and investigate a variety of different environments in a biological, physical, and chemical nature with high levels of accuracy and have been instrumental in gaining a better understanding of trends involved with vegetation physiology and plant stress [19-22]. This approach may provide a better alternative to traditional field methods because it is less costly and destructive allowing for results to be obtained in a timelier manner for analysis [20,23,24].

Research has also revealed by way of satellite derived indices, utilization of the near IR and mid infrared regions and more specifically, water absorption bands ranging from 0.950 to 0.970 μm, 1.150 to 1.260 μm, and 1.520 to 1.540 μm, to be highly effective in assessment of water content in canopies in relation to visible and near infrared regions [9]. These wavelengths are able to be absorbed further into the canopy structure, providing a larger more accurate representation of water content in comparison to the visible region, due to increased reflectance [9].

Entire areas affected by the Butte Humboldt and Lightning Complexes were examined via satellite remote sensing Landsat 5 imagery and ENVI 4.8 analysis software [25,26]. In an effort to gain a unique qualitative sense of magnitude and impact of these fires on vegetation and associated landscape change, as well as to properly designate unburned, burned edge, burned center, and reference subplot locations.

Burn severity can be described as the amount of change inflicted by fire disturbance on a particular area [27]. Fire disturbance has been shown to affect vegetation in numerous ways including the decreased water

content and health, changes in soil properties, density, species types, and arrangements [28]. In the near infrared and mid infrared regions of wavelength spectrum, changes linked to wild-land fire can be observed using Landsat Enhanced Thematic Mapper Plus (ETM+) and previous counterpart Landsat Thematic Mapper (TM) [28]. In post-wild-land fire landscapes, reflectance in the near infrared region and more specifically band 4, has been shown to be lower, while in contrast, mid infrared regions associated with band 7, has been shown to have the largest reflectance escalations [31]. ΔNBR results from subtraction of pre- and post-wild-land fire NBR values, then multiplied by a 1000 [29,30], with known values spanning from −1 to 1 [32]. Positive ΔNBR values indicate decreased amounts of vegetation with negative NBR values indicating increased vegetation growth between image acquisition dates [32].

Collection of pre-, post-NBR, and ΔNBR data from original BHC and BLC subplots allowed for quantitative and applicable assessment of burn severity levels [31]. Low fire severity for savannah grassland type ecosystems have been reported, with observed fire behavior to be rapidly sweeping in nature, however, differences still exist among severity levels between savannah and chaparral communities, both of which encompass BHC affected areas [33]. Nevertheless, low burn severity levels are expected for BHC, with increasing burn severity levels expected for BLC.

Based on the close proximity between BHC and BLC, both surrounding landscapes share similar vegetation types and compositions [34]. However, BHC encompasses lower elevation with savannah type grasslands and a mosaic of sparsely spatially oriented large oaks (*P. santrons*), foothill pine (*Pinus sabiniana*), and California Buckeye *(Aesculus californica)* [34]. An understory of herbaceous shrubs and herbs are present and include Christmas berry (*Heteromeles arbutifolia*), California coffeeberry (*Rhamnus californica*), fescue bunchgrasses (*F. occidentalis, Festuca californica*), and hedge nettle (*Stachys rigida*) [34]. As elevation increases, the landscape transitions into chaparral type vegetation that include Manzanita (*Arctostaphylos sp.*), Scrub oak (*Quercus dumosa*), and chamise (*Adenostoma fasciculatum*) [34]. In contrast, areas affected by the BLC were at an overall higher elevation and located in the Sierra Nevada Mountain Range in Plumas National Forest [34]. This mixed conifer forest has dense softwood vegetation that include Jeffrey pine (*Pinus jeffreyi*), Ponderosa pine (*Pinus ponderosa*), and Incense cedar (*Calocedrus decurrens*) along with an understory of shrubs consisting of Greenleaf Manzanita (*Arctostaphylos patula*), Squaw Carpet (*Ceanothus prostratus*), and Rabittbush Goldenweed (*Haplopappus bloomeri*) [34].

In an effort to quantify wildfire vegetation recovery

and visually interpret landscape change, ground based reflectance indices, remote sensing techniques including multi-temporal imagery, NDVI grey-scale imagery, and burn severity indices were incorporated into this study to gain a more diverse perspective of pre- and post-wild-land fire landscapes.

2. MATERIALS AND METHODS

2.1. Ground Based Remote Sensing

With the use of original satellite subplot coordinates, a total of thirty subplots were identified using a hand held GPS unit [35], on April 19, 21, and 23, 2010. A total of ten individual reflectance scans per subplot were collected using a Unispec SC portable spectrometer [36]. The objective of this field sampling process was to stay within the 30 by 30 meter barriers, while also striving for an even and broad distribution of individual scan locations within each subplot, in an effort to gain a uniform composite of the landscape. At each subplot locations reflectance measurements were taken at a distance of approximately one meter from the ground, with the optic sensor held slightly offset and angularly away, in a vertical out-stretched arm, while standing to insure the least possible interference due to human appendages, such as feet, and also to gain fair representation vegetation diversity. At this height, the Unispec SC portable spectrometer scans the landscape in a helical cone shape becoming larger as height is increased, with a measured circular pattern of 2 m^2 [36]. Each individual scan provided an individual complete spectral signature ranging from 310 to 1100 nm, in 10nm increments [36]. Following field spectrometer instructions and protocol, a white reflectance board was used to minimize differences in light versus cloudy conditions between scan which enabled calibration by means of reference [36,37]. Applying dark scans prior to data collection to achieve increased accuracy reduced device noise. Reflectance values for each wavelength increment across the spectrum were calculated for each subplot by dividing the subplot data by reference white board data, with averages for each subplot used to derive NDVI values [36].

2.2. Satellite Remote Sensing Imagery

In an effort to visually determine varying levels of landscape change, magnitude and direction of affected areas, multi-temporal false color RGB images were constructed using Landsat 5 images, WRS path 44 row 32 and ENVI 4.8 computer software [25,26]. Two time comparisons featuring seven-year pre wild-land fire and one year pre- and post-wild-land fire were selected for both Butte Humboldt and Lightning Complexes. Starting with the longer time period of seven year pre wild-land fire, multi-temporal false color images for both com-

plexes were constructed by shifting the red monitor pixels to become the most recent entire Landsat 5 image taken on 04/02/2008, then shifting both green and blue monitor pixels further back in time to the entire Landsat 5 image taken on 04/28/2000 [25,26]. Next, the same methods were repeated for one-year pre- and post-wild-land fires with Landsat 5 images taken on 04/05/2009 and 04/02/2008, respectively. Burned center two subplot coordinates for each complex were selected to give an overall uniform orientation of images for visual analysis. A zoom magnification factor of two was utilized, with an estimated coverage distance of over twenty miles between vertical and horizontal borders.

Selection of subplot locations involved the use of post-fire images for both Butte Humboldt and Lightning Complexes. Based on the observable burn scars, subplot locations were determined. Each subplot was given labels that correspond to its location in relation to the fire disturbances. For both Complexes, three subplots were designated as burned center, three subplots were designated as unburned, three subplots were designated as burned-edge, and one subplot was designated as reference, totaling ten subplots per fire.

Time-based data imagery was acquired by downloading eighty-five available Landsat 5 images for spring and summer months dating back twelve years from USGS Landsat data archives WRS path 44, Row 32 [25]. Spectral analysis was performed using ENVI 4.8 computer software for each of the 30 total designated subplots, with NDVI values recorded for the months of April, May, June, July, and August, starting with April 2010 and spanning back 12 years to April 1998. April thru August months were selected for the NDVI temporal based data analysis in an effort to minimize phenological vegetation differences occurring throughout yearly summer-winter vegetation cycles, as well as achieving the optimal time frame for studying Butte Humboldt and Lightning Complexes pre- and post-wild-land fire vegetation responses and recovery.

The time intervals associated with these wild-land fires made for ideal spectral measurement properties associated with peak vegetation growth. Other benefits included reduced cloud cover and rainfall amounts in relation to winter months and ideal crowning vegetation growth. The onsets of both wild-land fires occurring in early summer in conjunction with related seasonality contributed to the improved accuracy of noted results. The NDVI values were recorded at each subplot location with an individual pixel size of 30 by 30 meters. Using Landsat 5 imagery and ENVI software analysis with a 30 by 30 meter pixel size was ideal for overall spatial resolution because of the vast areas associated with wild-land fire.

2.3. Burn Severity Assessment

Burn severity, pre-NBR (Normalized Burn Ratio), post-NBR, and ΔNBR values were calculated for both Butte Humboldt and Lightning Complex original sub-plots. The burn severity index NBR is composed of bands (B4-B7)/(B4 + B7) [29,30], and has been shown to be connected to moisture levels of vegetation [14]. Based on previous research illustrating the ecological relevance of post-wild-land fire data collection within a year time frame [28], Landsat 5 images dated 04/02/2008, 04/05/2009, and 08/14/2010 were selected to perform band-math calculations to collect pre- and post-NBR values, using ENVI 4.8 computer software. ΔNBR was calculated by subtracting Pre-NBR from post-NBR values, then multiplied by 1000 [29,30].

3. RESULTS

3.1. Study Sites

Located in the eastern foothill regions of Chico, CA spanning towards Paradise, CA, ten 30 by 30 meter sub-plots were designated in Butte Humboldt Complex fire affected areas (**Table 1**, **Figure 1**). Additionally, located north east of Concow, California, ten 30 by 30 meter subplots were designated in Butte Lightning Complex fire affected areas (**Table 1**, **Figure 1**). All subplots were selected and categorized as unburned, burned edge, burned center, and control (**Table 1**, **Figure 1**).

3.2. Ground Based Reflectance and NDVI

BHC and BLC post-wild-land fire ground based NDVI values revealed vegetation recovery for both BHC and BLC in burned edge and burned center subplots (Figure 2). BHC and BLC control showed higher NDVI values in relation to other subplot categories, with a couple execptions (**Figure 2**). Both BHC and BLC unburned subplots illustrated a range of NDVI values similar but slightly enhanced to burned edge and burned center sub-plot categories (**Figure 2**).

3.3. Satellite Remote Sensing Imagery

Multi-temporal Landsat 5 images of the areas affected by BHC were visually assessed using band five (1.55 - 1.75 μm): R (04/02/2008), G (04/28/2000), B (04/28/2000) spanning over a seven-year period prior to the wild-land fire and band five (1.55 - 1.75 μm): R (04/05/2009), G (04/02/2008), B (04/02/2008) spanning over a one-year period pre- and post-wild-land fire (**Figure 3**). Multi-temporal image spanning over seven years prior to the wild-land fire showed a light red hue over the entire region (**Figure 3**). In contrast, the multi-temporal image spanning over a year pre- and post-wild-land fire

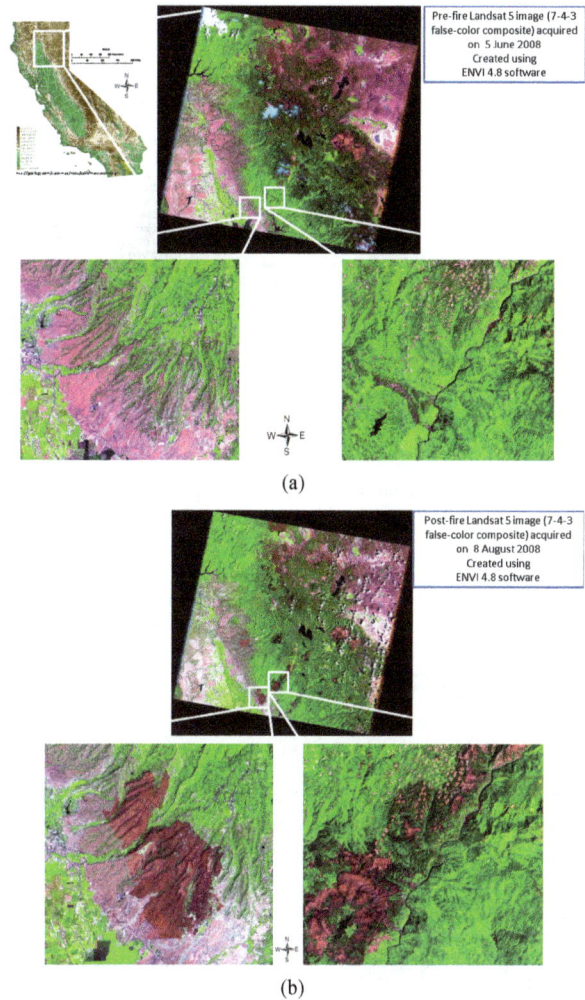

(a)

(b)

Figure 1. Pre-fire Landsat 5 image (7-4-3 false-color composite) acquired on 5 June 2008 and Post-fire Landsat 5 image (7-4-3 false-color composite) acquired on 8 August 2008. Both created using ENVI 4.8 software.

showed a light blue hue with regions of slightly lighter shades of blue present.

Multi-temporal Landsat 5 images of the areas affected by BLC were also visually assessed using band five (1.55 - 1.75 μm): R (04/02/2008), G (04/28/2000), B (04/28/2000) spanning over a seven-year period prior to the wild-land fire (**Figure 2** left, and band five (1.55 - 1.75 μm): R (04/05/2009), G (04/02/2008), B (04/02/ 2008) spanning over a one-year period pre- and post-wild-land fire (**Figure 3**). Multi-temporal image spanning over seven years prior to the wild-land fire showed both dark red and deep bright blue regions, while in contrast, the multi-temporal image spanning over a year pre- and post-wild-land fire showed affected areas in bright intense red (**Figure 3**).

Landsat 5 NDVI grey-scale images of pre-(04/02/2008) and post-(04/05/2009) BHC were visually assessed (**Figure 3**). In contrast, the post-wild-land fire image clearly

Butte Humboldt Complex, June 2008

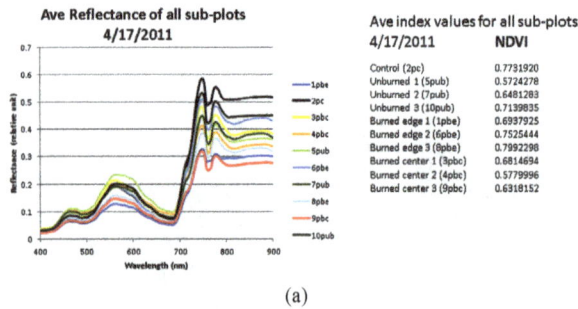

(a)

Butte Lightning Complex, June 2008

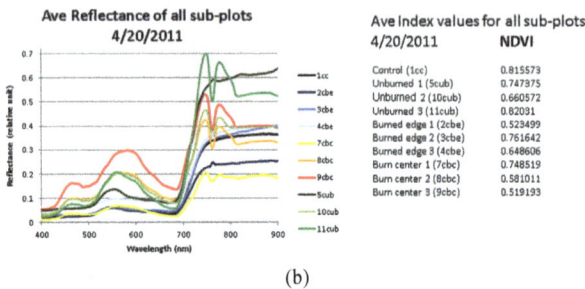

(b)

Figure 2. Ground based reflectance measurements of Butte Humboldt Complex and Butte Lightning Complex subplots.

Table 1. Butte Humboldt Complex and Butte Lightning Complex subplot locations and elevations.

Butte Hunboldt Complex	Latitude	Longitude	Elevation (ft)
Reference	39°42.708	−121°46.774	242
Unburned 1	39°44.744	−121°39.657	1387
Unburned 2	39°40.680	−121°43.720	416
Unburned 3	39°43.728	−121°39.656	1207
Burned Edge 1	39°44.736	−121°46.770	309
Burned Edge 2	39°42.705	−121°44.736	511
Burned Edge 3	39°41.690	−121°42.702	587
Burned Center 1	39°43.724	−121°41.692	1023
Burned Center 2	39°43.731	−121°40.681	1125
Burned Center 3	39°42.708	−121°40.677	908
Reference	39°40.175	−121°31.563	1049
Butte Lightning Complex	Latitude	Longitude	Elevation (ft)
Unburned 1	39°43.726	−121°30.508	1876
Unburned 2	39°42.716	−121°31.522	1991
Unburned 3	39°42.705	−121°31.530	1853
Burned Edge 1	39°45.761	−121°28.477	2080
Burned Edge 2	39°47.797	−121°26.449	1522
Burned Edge 3	39°53.888	−121°21.363	1666
Burned Center 1	39°47.798	−121°29.488	2755
Burned Center 2	39°47.799	−121°29.491	2667
Burned Center 3	39°47.797	−121°29.493	2543

showed visible signs of destruction illustrated by the magnitude and direction of the subsequent burn scar (**Figure 3**). Landsat 5 NDVI grey-scale images of pre-(04/02/2008) and post-(04/05/2009) BLC were visually assessed (**Figure 3**). The pre-wild-land fire NDVI grey-scale image showed the presence of healthy vegetation with a slight semi-circular ring of destruction in the upper center area (**Figure 3**). In contrast, the post-wild-land fire image clearly showed visible signs of sizable destruction illustrated by the magnitude and direction of the subsequent burn scar (**Figure 3**).

Analysis of satellite derived NDVI for BHC reference and unburned subplots, April thru August, over a twelve-year span, showed temporal oscillation patterns with unburned subplot number three, consisting of the highest overall NDVI values followed by unburned one subplot, reference subplot, and unburned sub-plot two (**Figure 4**). Unburned sub-plot three NDVI values decreased to no data from 7/7/2008 to 8/8/2008, followed by a sharp increase, with the highest NDVI value for this increase occurring on 5/7/2009, and a slightly lower than the previous NDVI oscillation pattern observed thereafter (**Figure 4**). Unburned sub-plot two followed a similar trend with the lowest NDVI values overall recorded on 7/7/2008 and 7/23/2008 (**Figure 4**). Unburned sub-plot two there was a sharp increase in NDVI values on 6/5/2002 followed by a decrease to no data from 6/21/2002 to 8/24/2002 (**Figure 4**). A decrease in NDVI values was also observed for both unburned sub-plot two and reference sub-plots on 8/22/2007. Unburned sub-plot one maintained temporal oscillation patterns throughout the entire time duration (**Figure 4**).

Analysis of NDVI for BHC reference and burned edge subplots, April thru August, over a twelve-year span, showed temporal oscillation patterns, with burned edge one and two subplots showing the highest overall NDVI values followed by reference and burned edge three subplots (**Figure 4**). Burned edge three subplot decreased to no data from 7/7/2008 to 7/23/2008 followed by a sharp increase in NDVI values with the highest NDVI value for this increase occurring on 4/5/2009 (**Figure 4**). Burned edge sub-plot three showed a decrease in NDVI values relative to the oscillation trends on 7/7/2002 (**Figure 4**). Both burned edge two and reference subplots maintained NDVI temporal oscillation patterns throughout the time duration, except for a decrease to near zero for reference subplot occurring on 8/22/2007 (**Figure 4**).

Analysis of NDVI for BHC reference and burned center subplots, April thru August, over a twelve-year span, showed temporal oscillation patterns, with burned center two and three subplots showing the highest overall NDVI values followed by burned center one and reference subplots (**Figure 4**). Burned center one and three subplots showed initial decreases in NDVI prior to the

Retrospective analysis of two Northern California wild-land fires via Landsat five satellite imagery and Normalized Difference Vegetation Index (NDVI)

79

Butte Humboldt Complex   Butte Lightning Complex

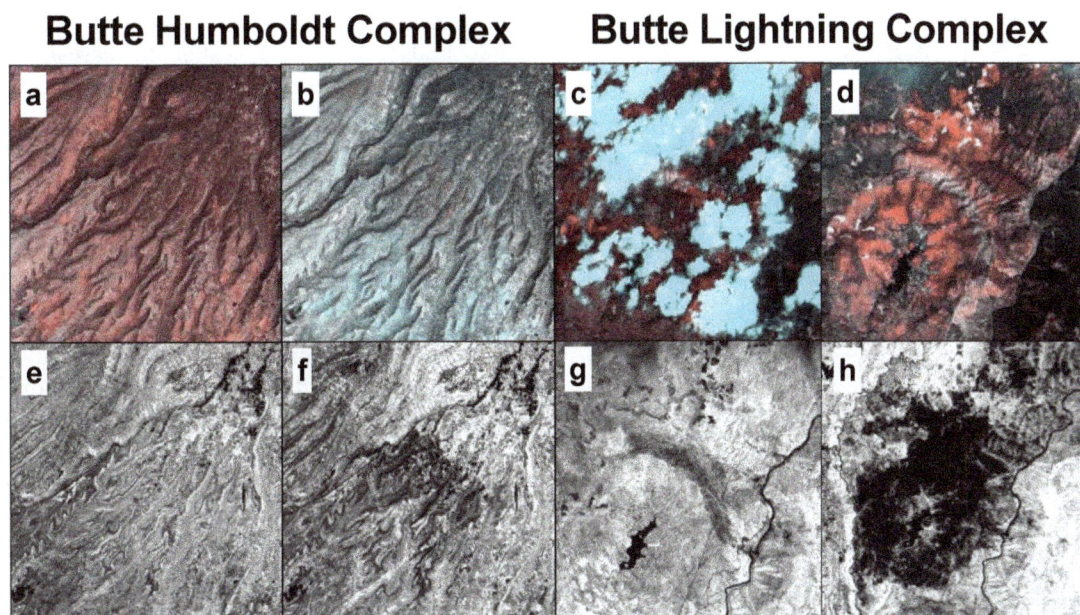

Figure 3. Landsat TM and ETM+ derived false color composites multi-temporal images of (a) (b) Butte Humboldt Complex, band five: (a) Red (04/02/2008), Green (04/28/2000), Blue (04/28/2000); (b) Red (04/05/2009), Green (04/02/2008), Blue (04/02/2008)/(c) (d) Butte Lightning Complex, band five: (c) Red (04/02/ 2008), Green (04/28/2000), Blue (04/28/2000); (d) Red (04/05/2009), Green (04/02/2008), Blue (04/02/2008)/NDVI pre- and post-wild-land fire grey-scale images of (e) (f) Butte Humboldt Complex: (e) (04/02/2008) and (f) (04/05/2009)/(g) (h) Butte Lightning Complex: (g) (04/02/2008) and (h) (04/05/2009) (Landsat 5 images, THOR atmospherically corrected, WRS path 44, row 32) (USGS 2010; ENVI 4.8 2010).

BHC, then both subplots decreased further to no data from 7/7/2008 to 7/23/2008, followed by a sharp increase in NDVI with the highest NDVI values for these increases occurring on 4/5/2009 (**Figure 4**). Burned center two subplot decreased to near zero from 7/7/2008 to 7/23/2008, followed by an increase in NDVI with the highest NDVI value for this increase occurring on 4/5/ 2009 (**Figure 4**). Reference subplot maintained NDVI temporal oscillation patterns throughout the entire time duration, except for a decrease to near zero occurring on 8/22/2007 (**Figure 4**).

Analysis of NDVI for BLC reference and unburned subplots, April thru August, over a twelve-year span, showed temporal oscillation patterns, with the reference subplot showing the highest overall NDVI values, followed by unburned two, unburned three, and unburned one subplots (**Figure 4**). Unburned one subplot started the time duration with most NDVI values resulting in above 0.4, and then decreased significantly to below 0.05 from 5/4/2002 to 8/24/2002, followed by an increase and oscillation pattern not reaching NDVI values above 0.35 (**Figure 4**). Unburned two, three, and reference subplots maintained NDVI temporal oscillation patterns throughout the entire time duration except for the reference subplot showing three noticeable decreases in NDVI values on 5/1/2001, 8/11/2003, and 7/23/2008 (**Figure 4**).

Analysis of NDVI for BLC reference and unburned subplots, April thru August, over a twelve-year span,

showed temporal oscillation patterns, with burned edge one and two subplots showing the highest overall NDVI values. However, burned edge one subplot showed no data on 4/28/2000 (**Figure 4**). Reference and burned edge three subplots both showed lower NDVI values overall with three noticeable decreases (**Figure 4**). Decreased NDVI values were observed on 5/1/2001, 8/11/2003, and 7/23/2008 for reference subplot and 4/28/ 2000, 4/26/2005 and 4/2/2008 for burned edge three subplot (**Figure 4**).

Analysis of NDVI for BLC reference and burned center subplots, April thru August, over a twelve-year span, showed temporal oscillation patterns, with the reference and burned center three subplots showing the highest overall NDVI values, followed by burned center two and one subplots (**Figure 4**). Burned center one, two, and three subplots showed oscillation patterns with noticeable decreases on 4/28/2000 and an increases on 7/23/ 2002 for burned center two subplot and on 5/18/2007 for burned center one subplot (**Figure 4**).

Reference subplot showed temporal oscillation patterns throughout the entire time duration except for three noticeable decreases in NDVI values on 5/1/2001, 8/11/2003, and 7/23/2008 (**Figure 4**). Burned center one subplot showed an initial slight decrease in NDVI values from 5/4/2008 to 6/5/2008, then decreasing further to no data from 7/7/2008 to 7/23/2008, with a zero value observed on 8/8/2008 (**Figure 4**). Burned center one sub-

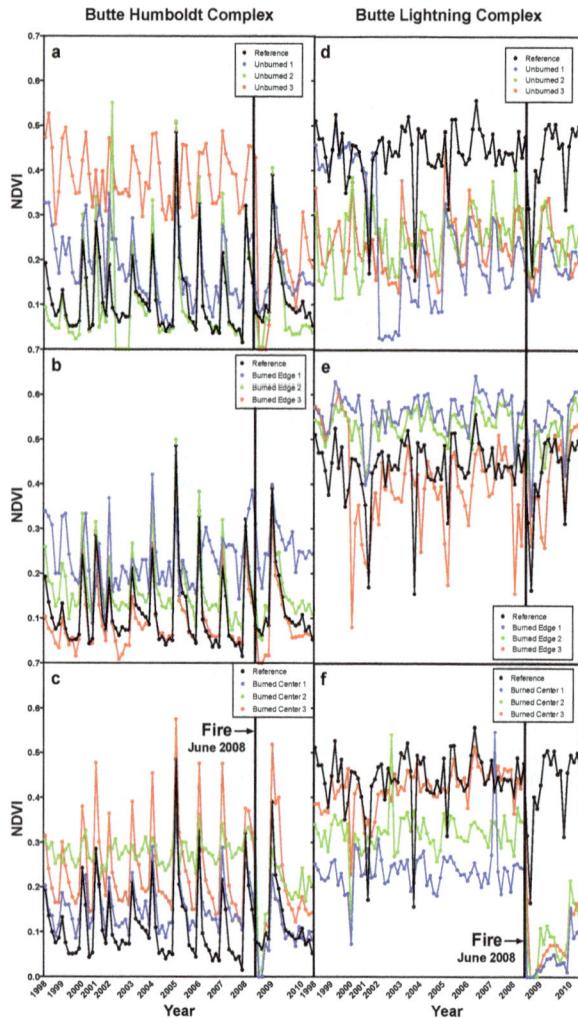

Figure 4. Butte Humboldt Complex (a-c) and Butte Lightning Complex (d-f) 12-year (1998-2010) April through August, Landsat derived NDVI time based data. (a, d) reference and unburned subplots, (b, e) reference and burned edge subplots, (c, f) reference and burned center subplots.

plot then showed a slight increase and decrease in NDVI values staying below .05 from 8/24/2008 to 8/27/2009, followed by an increase on 6/27/2010 (**Figure 4**). Burned center two subplot showed an initial slight decrease in NDVI values from 5/4/2008 to 6/5/2008, and then decreased to below an NDVI value of 0.1 on 7/7/2008, followed by a further decrease to no data from 7/23/2008 to 8/24/2008 (**Figure 4**). Burned center two subplot then observed a low NDVI value on 4/5/2009, trailed by a small oscillation and decreasing NDVI values incrementally from 4/21/2009 to 7/26/2009, with a subsequent increase in NDVI values recorded from 8/11/2009 to 6/27/2010 (**Figure 4**). Burned center three subplot showed an initial slight increase in NDVI values from 5/4/2008 to 6/5/2008, then decreased to below 0.2 on 7/7/2008, followed by a further decrease to no data on 7/23/2008, with a zero value observed on 8/8/2008 (**Figure 4**). Burn-

ed center three subplot then observed a near zero value on 8/24/2008, followed by a slight increase and decrease in NDVI values staying below 0.08 from 4/5/2009 to 8/27/2009, followed by an increase on 6/27/2010 (**Figure 4**).

3.4. Re-NBR, Post-NBR, and ΔNBR Values for Subplots

The majority of pre- and post-NBR values were within normal ranges and not above or below a value of 1 (**Table 2**) [32]. Exceptions were observed for BLC burned center one subplot with a value of zero and burned center two and three subplots with large post-NBR values (**Table 2**). Delta NBR values for BHC unburned three subplot and BLC unburned two and burned center one subplots indicated low to moderate levels of burn severity (**Table 2**).

ΔNBR values for BHC unburned three subplot and BLC unburned two and burned center one subplots indicated low to moderate levels of burn severity (**Table 2**). In contrast, BHC unburned one and two, and burned edge two subplots indicated a burn severity of unburned (**Table 2**). Also, BLC reference, unburned one and three subplots, and all burned edge subplots observed ΔNBR values indicating unburned levels of burn severity (**Table 2**). ΔNBR values for BHC reference, burned edge one and three, and burned center three subplots indicated low re-growth (**Table 2**). ΔNBR values for BHC burned center one and two subplots indicated low levels of burn severity (**Table 2**). Unexpectedly, ΔNBR values for burned center subplots two and three were not identifiable. However, more recent post-NBR imagery of BLC burned center two and three subplots, yielded values within normal ranges, with ΔNBR values indicating low and mid to high burn severity (**Table 2**).

4. DISCUSSION

4.1. Ground-Based Reflectance and NDVI

Ground based reflectance spectra for the BHC fire in the control subplot shows the maximum reflected PAR in the UV wavelength region (**Figure 2**). BLC reflectance spectra for unburned subplots 1 and 3 shows the maximum reflected PAR in the UV wavelength region (**Figure 2**). Increased reflectance in the UV wavelength region is generally indicative of vegetations ability to reflect photons of non-photosynthetically active wavelengths under stressful conditions, via secondary pigments such as xanthophylls and caronenoids [38,39]. Expected trends in shifts of the red-edge region and enhanced absortion of PAR for control and unburned subplots vs. burned edge and burn center subplots were not observed (**Figure 2**). However, a few of the burn center and burned edge subplots in both fires do show

Table 2. Pre-NBR, Post-NBR, ΔNBR and corresponding burn severity level for sub-plots of Butte Humboldt Complex and Butte Lightning Complex.

Fire	Sub-Plot Location	Pre-NBR	Post-NBR	dNBR	Severity Level
		(band4-band7)/ (band4+band7)	(band4-band7)/ (band4+band7)	(Pre-NBR-Post-NBR)	
Butte Hunboldt Complex	Reference	0.2203	0.3273	−0.1069	Low Regrowth
	Unburned 1	0.3509	0.4312	−0.0803	Unburned
	Unburned 2	0.2475	0.3028	−0.0552	Unburned
	Unburned 3	0.4118	0.0448	0.367	Low-Moderate
	Burned Edge 1	0.264	0.3857	−0.1217	Low Regrowth
	Burned Edge 2	0.2035	0.2941	−0.0906	Unburned
	Burned Edge 3	0.1698	0.2952	−0.1254	Low Regrowth
	Burned Center 1	0.2039	0.102	0.1018	Low
	Burned Center 2	0.3684	0.2235	0.1448	Low
	Burned Center 3	0.3393	0.4407	−0.1014	Low Regrowth
Butte Lightning Complex	Reference	0.4949	0.5385	−0.0435	Unburned
	Unburned 1	0.1485	0.0811	0.0674	Unburned
	Unburned 2	0.4	0.0204	0.3797	Low-Moderate
	Unburned 3	0.3125	0.2642	0.0484	Unburned
	Burned Edge 1	0.6129	0.6452	−0.0323	Unburned
	Burned Edge 2	0.6176	0.6066	0.0111	Unburned
	Burned Edge 3	0.4217	0.4783	−0.0566	Unburned
	Burned Center 1	0.3581	0	0.358	Low-Moderate
	Burned Center 2	0.3556	2.7021	−2.3466	N/A
	Burned Center 3	0.4898	2.1296	−1.6398	N/A

enhanced reflectance in the visable spectrum, which has been previously described as chlrophyll degradation [38,39]. General expected trends of NDVI values were observed in most subplots in both fires, with higher NDVI observations found in control and unburned sub-plots with a few exceptions (**Figure 2**). Caution must be taken when examining this studies ground based reflectance results and discussion must include the well documented varaiable shifts in reflectance and NDVI values when measurements include temoral differences, vegetation types and post-fire recovery.

4.2. Multi-Temporal Imagery

Multi-temporal imagery has revealed relationships associated with the colors red and blue and the shade and intensity of the red and blue colors and the corresponding visual representation of various levels of landscape change [22]. The color red and its shade and intensity

(**Figures 3(a)-(d)**) may represent negative ecosystem impacts, such as deforestation from logging as well as other possible disturbances, while in contrast, blue and it's shade and intensity (**Figures 3(a)-(d)**) may denote new vegetation growth [22].

The seven-year pre-fire multi-temporal image of BHC fire showed the area of study encompassed in a low intensity red hue (**Figure 3(a)**), which represents negative landscape change and may be attributed to previously documented fire disturbances, water stress and/or a combination of environmental vegetative stressors [1,22]. In contrast, one-year pre-and post-wild-land fire multi-temporal image of BHC showed the entire area encompassed in a light blue hue (**Figure 3(b)**), which indicates some form of vegetation re-growth [22]. This positive re-growth may possibly be linked to increased recovery due to the elasticity of species typical of savannah grass-land type vegetation in relation to fire disturbance

[22,40,41]. Both multi-temporal images of BHC displayed noticeable levels of landscape change.

Multi-temporal images in areas affected by the BLC revealed extensive levels of landscape change (**Figures 3(c)** and **(d)**). The seven-year pre-wild-land fire multi-temporal image showed patchy areas of deep blue and deep red (**Figure 3(c)**), which may indicate noticeable levels of both positive and negative landscape change. However, the bright blue patches in the seven-year pre-wild-land fire are likely representative of atmospheric interference, more specifically cloud formations (**Figure 3(c)**). One year pre- and post-wild-land fire multi-temporal image showed the entire area encompassed in varied red color intensity (**Figure 3(d)**), indicating various possible levels of fire induced negative landscape change. Comparison of multi-temporal images illuminated significant visual differences between both time periods (**Figures 3(c)** and **(d)**). While the seven year pre-wild-land fire multi-temporal image consisted of both positive and negative differences in growth potentials, the one year pre- and post-wild-land fire image revealed a uniquely comprised array of various shades and intensity of red (**Figures 3(c)** and **(d)**).

4.3. NDVI Grayscale Imagery

One-year pre- and post-fire NDVI images for BHC and BLC fires were incorporated into this study as another efficient and relatively quick tool to visually interpret fire impact on vegetation and landscape change. Comparison of NDVI grey scale images of BHC and BLC fires yielded visually apparent differences in pre- and post-fire vegetation (**Figures 3(e)** and **(h)**). BHC pre-fire areas displayed the presence of light pixel regions representative of healthy vegetation (**Figure 3(e)**) [22]. The BHC post-fire affected areas displayed very low vegetation values mainly occurring in the fire scar, which is easily evident in the post-fire image (**Figure 3(f)**). The pre-fire NDVI image of BLC affected area revealed a semi-circular region of dark pixels indicative of decreased NDVI values and stressed vegetation, which could possibly be linked to prior fire disturbances (**Figure 3(g)**) [22,40,42]. Comparison of both complexes post-fire NDVI images gives a unique sense of scale and magnitude in relation to each other. A significantly larger and darker post-fire burn scar was observed for areas affected by the BLC (**Figure 3**), which indicates greater vegetation damage and potential landscape change, and may be attributed to the prior abundance of woody vegetation, typically observed in this mixed conifer area [22,34]. Vegetation types and post-fire vegetation re-growth dynamics are believed to have been contributing factors leading to less noticeable BHC burn scar and increased NDVI recovery (**Figures 3(e)** and **(h)**) [22,40, 41]. Results of the NDVI post-fire image comparisons

between complexes, however, may have been influenced by differences in NDVI recovery time, because in efforts aimed at consistency, collection of post-fire NDVI images for both complexes were derived from a further dated singular post-fire image, then transformed into NDVI grey-scale images. This inconsistency resulted in the BHC experiencing a longer recovery time due to image acquisition date and prior containment in relation to the BLC. Consistency among multi-temporal imagery is difficult to maintain, due to a combination of potential variations including but not limited to sensor characteristics, atmospheric conditions, solar angle, and sensor view angle [22].

4.4. Satellite Derived NDVI

The time based NDVI data leading up to the fire complexes of 2008 yielded interesting and informative results (**Figure 4**). During this twelve year pre-fire span, all subplots for both complexes observed NDVI temporal oscillations patterns indicative of the vegetation seasonality [17,45], which has been shown to be linked to climate fluctuations in precipitation and temperature (**Figure 4**) [46]. Peak NDVI values were observed occurring at the height of vegetation growth around August, proceeded by minimum NDVI values occurring in the spring months of April and May (**Figure 4**). Timing of these NDVI temporal oscillation patterns corresponded with precipitation and maximum and minimum temperatures (**Figure 4**). More specifically, when precipitation was high (climate data not shown), while minimum and maximum temperatures were low [47], decreased NDVI values were observed for the same time period, with an opposite trend observed for drier and hotter months (**Figure 4**).

These results while adding further evidence to the relationship between NDVI and meteorological data was not an unexpected outcome because of previously documented relationships between oscillation trends in NDVI values and meteorological variability [42,46]. There were however, several anomalies including decreases and increases of NDVI values as well as no data observed being inconsistent with associated seasonal oscillation patterns (**Figure 4**) [46]. Closer examination revealed several trends that are consistent with previous studies illustrating correlations between NDVI decreases and wild-land fire disturbance (**Figure 4**) [40,42]. For example, BHC unburned subplot two displayed an increase in NDVI values on 6/5/2002, followed by a decrease to no data from 6/21/2002 to 08/24/2002 (**Figure 4**). In addition, BLC unburned one showed a significant decrease in NDVI values starting on 5/4/2002 and continuing through 8/24/2002 (**Figure 4**). During this time period the presence of fire disturbance in BHC unburned two subplot area was verified through local fire records [1]. Other

Retrospective analysis of two Northern California wild-land fires via Landsat five satellite imagery and Normalized Difference Vegetation Index (NDVI)

83

decreased NDVI values were recorded for several subplots prior to the onset of BHC and BLC and may be contributed to one or many factors including fire disturbance, vegetation stress, decreased temperatures, increased precipitation, and cloud cover [15,40,48].

Prior studies have established relationships between lack of precipitation and drought to decreased NDVI values [40,42,46,49]. Comparison of local meteorological data to NDVI values of burned center subplots leading up to of the onset of both BHC and BLC illustrated similar trends (**Figure 4**). Paradise weather stations noted low precipitation associated with drought-like conditions occurred for an extended time period prior to both wild-land fires [47]. BHC and BLC burned center subplots experienced normal NDVI values in the spring months of April and May prior to both fire disturbances (**Figure 4**) indicating gradual vegetation growth, due to relationships between NDVI and above ground net primary production [45]. However, decreased NDVI values were observed directly before both onsets for BHC and BLC, burn center subplots one and three and one and two respectively (**Figure 4**). A likely explanation could be related to a combination of drought-like conditions and elevated temperatures associated with mid-summer months, coupled with previous buildup of vegetation growth becoming drier and moisture stressed, led to decreased NDVI values prior to fire onset [40,42,46,49], which may have added significant influence on both ignition and nature of both fire disturbances [15]. Other fire studies have found similar results [50]. For example, the Lookout fire occurring in 1999, in north east areas of Plumas National Forest, caused extensive damage of over a thousand (ha) hectares with the onset attributed to lack of precipitation coupled with increased temperatures associated with early, mid, and late summer months [50]. During the temporal period when both Butte Complexes were active, NDVI values for BHC burned center subplots one and three and BLC burned center subplots one, two, and three decreased to no data (**Figure 4**). NDVI decreases of this nature indicate vegetation loss and exposure of bare rock and soil [22].

NDVI post-fire recovery trends were clearly observed for both BHC and BLC burned center subplots, though spatially different in nature. BHC burned center subplots showed a steady linear post-wild-land fire recovery with NDVI values peaking on 04/05/2009 (**Figure 4**), while on the same date, BLC burned center two subplot had just started NDVI recovery with an initial recorded low value after experiencing a no data period previously (**Figure 4**). In addition, BLC burned center one and three subplots exhibited small gradual bell shaped curves, followed by all three BLC burned center subplots exhibiting steady increases in NDVI values (**Figure 4**). Previous research has noted the

possibility of full regeneration of forest type vegetation to pre-fire conditions in a time period of thirteen years following a major wild-land fire [40]. Changes in species composition and initial increases in species abundance have been observed in the first stages of forest recovery [51], on temporal scale of one to three years, followed by lower levels in subsequent years [41,52]. However, increased levels of wild-land fire severity have been shown to decrease overall species abundance [53]. It is believed this initial post-fire temporal period which illustrated NDVI recovery for BLC (**Figure 4**), represents along with existing burned and partially burned vegetation, an understory re-growth consisting of herbs, forbs, and annuals, which have been shown to advantageously colonize previously unexposed areas [51]. Following this initial period, species abundance has been observed to decrease over time with increased competition from softwood and shrub species shaping the landscape [51].

This 12-year retrospective analysis was able to illustrate post-wild-land fire NDVI recovery of vegetation to pre-wild-land fire NDVI levels. More specifically, BLC burned center three subplot exhibited a full recovery to pre-wild-land fire NDVI levels, with burned center one and two subplots exhibiting slightly less enhanced values (**Figure 4**). In contrast, even though the onset of BLC was documented approximately ten days later, BLC post-wild-land fire NDVI values for all three burned center subplots were not able to achieve full recoveries during the course of this study (**Figure 4**). It is important to note, however, that while NDVI can temporally display overall vegetation responses in designated areas, limitations including lack of ability to individually discriminate plant types and sensitivities involved with mountainous terrain, slope, and aspect can affect spectrometer sensors and cause associated errors [22,40]. Also, diverse elevations and plant community structure at subplot locations may have added further complications to NDVI results [46]. Previous research has reported variability in NDVI results among different plant community types, related to different ranging elevations and weather conditions involving, cloud cover, temperature, and precipitation fluctuations occurring at individual locations [46,48] There is a possibility that due to the short time span of twelve years pertaining to this time based study, trends associated with longer durations may not be distinguishable and or present [42].

4.5. Interpretation of NBR and ΔNBR Results

The majority of pre- and post-NBR values for BHC and BLC unburned, burned edge subplots, were above zero indicating vegetation growth, resulting in ΔNBR values corresponding to burn severities ranging from unburned to low re-growth (**Table 1**) [32]. This illustrates that these areas most likely either remained relatively unchanged or experienced slight increased vegetation growth [31,32]. Negative pre- and post-NBR values represent locations with no vegetation growth [32]. More specifically result in band 4 having a smaller reflectance in relation to band 7 [32]. BHC unburned three and BLC unburned two subplots observed ΔNBR values indicating low-moderate burn severity (**Table 2**). BHC burned center three subplot observed ΔNBR values indicating low re-growth, while in contrast, BLC burned center subplots two and three revealed high negative ΔNBR values and thus could not be identified (**Table 2**). Noted discrepancies, however, resided in high positive post-wild-land fire NBR values, particular to BLC burned center two and three subplots, thus effecting ΔNBR values (**Table 2**). However, through the use of a later post-NBR acquisition date, NBR values for BLC burned center two and three subplots revealed normal ranges of post-NBR values (data not shown), and were consistent with low and mid to high burn severity levels respectively [32].

Previous studies also have shown the NBR index can be sensitive to differences in types of vegetation classes [27]. NBR values have been shown to be more accurate for forest types in relation to savannah-grasslands areas, while post-fire forest areas with an abundance of burned woody material remnants may cause spectral shifts [27]. An additional study found recovery in riparian areas in the mixed coniferous forest of Plumas National Forest to consist of focal competitors such as white fir sugar pine, mountain alder, Douglas fir, and red-osier dogwood [50]. These softwood vegetation types have been shown to prevail in post-fire re-growth and recovery and can be attributed to fire resistance capabilities, seed bank, and dispersal dynamics [50]. Post-wild-land fire vegetation re-growth in Savannah type grassland areas in contrast have been observed to be less predictable and inconsistent with broad areas ranging from bare ground and sparse grasses and herbs to densely situated fire adapted shrubs and forbs which may have affected accuracy and discrepancies in results [27,53].

BLC burn center one subplot showed a zero post-NBR value, which in turn resulted in a ΔNBR value indicating low to moderate severity burn (**Table 2**). Previous research has established a wide range of possibilities for the zero post-NBR values found in BLC burned center one subplot, such as sunlight, thickness and dimensions of pre-wild-land fire vegetation, angles of terrain, and elevation [40]. Previous fire severity assessments illustrated low severity burn levels for Sierra Nevada foothill and mountain regions [55].

ΔNBR indicating low severity burn for BHC burned center one and two subplots, may be generally described by low-land surface type fires, contrasted with elevated mortality of undersized trees [33]. In contrast, ΔNBR

indications of low to moderate fire severity for BLC burned center one subplot may also be related and consistent with burned bark on surviving trees, along with significant destruction of undersized and crown tree species [33]. BHC and BLC vegetation types may be dissimilar in species and composition. However, they both occupy Mediterranean type ecosystems and are comprised of numerous species of vegetation resistant and well fitted to fire disturbances [40].

5. CONCLUSION

This study's practical approach allowed for real world application and integration of biological principles with a variety of remote sensing techniques and applications. NBR and ΔNBR values yielded relevant insight into both BHC and BLC, fire affected ecosystems. The majority of BHC and BLC ΔNBR results quantified burn severity dynamics; while satellite imagery illustrated differences between BHC and BLC pre- and post-wild-land fire impacts on vegetation and possible landscape change.

REFERENCES

[1] Calfire (2010)

[2] Conedera, M., Tinner, W., Neff, C., Meurer, M., Dickens, A. and Krebs, P. (2008) Reconstructing past fire regimes: Methods, applications, and relevance to fire management and conservation. *Quaternary Science Reviews*, **28**, 555-576.

[3] Taylor, A. and Beaty, R. (2005) Climate influences of fire regimes in northern Sierra Nevada Mountains, Lake Tahoe Basin, Nevada, USA. *Journal of Biogeography*, **32**, 425-438.

[4] Fried, J., Torn, M. and Mills, E. (2004) The impact of climate change on wildfire severity: A regional forecast for Northern California. *Climatic Change*, **64**, 161-191.

[5] Flannigan, M.D., Stocks, B.J. and Wotton, B.M. (2000) Climate change and forest fires. *Science of the Total Environment*, **262**, 221-229.

[6] Dale, V.H., Joyce, L.A., McNulty, S., Neilson, R.P., Ayres, M.P., Flannigan, M.D., Hanson, P.J., Irland, L.C., Lugo, A.E., Peterson, C.J., Simberloff, D., Swanson, F.J., Stocks, B.J. and Wotton, B.M. (2001) Climate change and forest disturbances. *BioScience*, **51**, 723.

[7] Whitlock, C., Shafer, S.L. and Marlon, J. (2003) The role of climate and vegetation change in shaping past and future fire regimes in the northwestern US and the implications for ecosystem management. *Forest Ecology and Management*, **178**, 5-21.

[8] Zomer, R.J., Trabucco, A. and Ustin, S.L. (2009) Building spectral libraries for wetlands land cover classification

and hyperspectral remote sensing. *Journal of Environmental Management*, **90**, 2170-2177.

[9] Sims, D.A. and Gamon, J.A. (2003) Estimation of vegetation water content and photosynthetic tissue area from spectral reflectance: A comparison of indices based on liquid water and chlorophyll absorption features. *Remote Sensing of Environment*, **84**, 526-537.

[10] Chandler, C., Cheney, P., Thomas, P., Trabaud, L. and William, D. (1983) Fire in forestry. *Forest Fire Behavior and Effects*, Vol. I. John Wiley & Sons Ltd., New York.

[11] Peñuelas, J., Filella, I., Biel, C., Serrano, L. and Savé, R. (1993) The reflectance at the 950 - 970 nm region as an indicator of plant water status. *International Journal of Remote Sensing*, **14**, 1887-1905.

[12] Peñuelas, J., Piñol, J., Ogaya, R. and Filella, I. (1997) Estimation of plant water concentration by the reflective water index WI (R900/R970). *International Journal of Remote Sensing*, **18**, 2869-2875.

[13] Filella, I., Peñuelas, J., Llorens, L. and Estiarte, M. (2004) Reflectance assessment of seasonal and annual changes in biomass and CO_2 uptake of Mediterranean shrubland submitted to experimental warming and drought. *Remote Sensing of Environment*, **90**, 308-318.

[14] Veraverbeke, S., Verstraeten, W.W., Lhermitte, S. and Goossens, R. (2010) Evaluating Landsat Thematic Mapper spectral indices for estimating burn severity of the 2007 Peloponnese wildfires in Greece. *International Journal of Wildland Fire*, **19**, 558-569.

[15] Pettorelli, N., Olav Vik, J., Mysterud, A., Gaillard, J.M., Tucker, C.J. and Stenseth, N.C. (2005) Using the satellite-derived NDVI to assess ecological responses to environmental change. *Trends in Ecology and Evolution*, **20**, 503-510.

[16] Lichtenhaler, H.K., Wenzel, O., Buschmann, C. and Gitelson, A. (1998) Plant stress detection by reflectance and fluorescence. *Annals of the New York Academy of Sciences*, **851**, 271-285.

[17] Guerschman, J.P., Paruelo, J.M. and Burke, I.C. (2003) Land use impacts on the Normalized Vegetation Index in temperate Argentina. *Ecological Applications*, **13**, 616-628.

[18] Tucker, C.J. (1979) Red and photographic infrared linear combinations for monitoring vegetation. *Remote Sensing of Environment*, **8**, 127-150.

[19] Salazar. L., Kogan, F. and Roytman, L. (2007) Use of remote sensing data for estimation of winter wheat yield in the United States. *International Journal of Remote Sensing*, **28**, 3795-3811.

[20] Rosso, P.H., Pushnik, J.C., Lay, M. and Ustin, S.L. (2005) Reflectance properties and physiological responses of *Salicornia virginica* to heavy metal and petroleum contamination. *Environmental Pollution*, **137**, 241-252.

[21] Ustin, S.L. and Gamon, J.A. (2010) Remote sensing of plant functional types. *New Phytologist*, **186**, 795-816.

[22] Horning, N., Robinson, J.A., Sterling, E.J., Turner, W. and Spector, S. (2010) Remote sensing for ecology and conservation: A handbook of techniques. Oxford University Press, New York, 4-371.

[23] Inoue, Y. (2003) Synergy of remote sensing and modeling for estimating ecophysiological processes in plant production. *Plant Production Science*, **6**, 3-16.

[24] Black, S.C. and Guo, X. (2008) Estimation of grassland CO_2 exchange rates using hyperspectral remote sensing techniques. *International Journal of Remote Sensing*, **29**, 145-155.

[25] US Geological Survey (2011)

[26] ENVI 4.8 (2010) Exelis visual information solutions. Boulder, CO.

[27] Epting, J., Verbyla, D. and Sorbel, B. (2005) Evaluation of remotely sensed indices for assessing burn severity in interior Alaska using Landsat TM and ETM+. *Remote Sensing of Environment*, **96**, 328-339.

[28] Soverel, N.O., Perrakis, D.D.B. and Coops, N.C. (2010) Estimating burn severity for Landsat dNBR and RdNBR indices across western Canada. *Remote Sensing of Environment*, **114**, 1896-1909.

[29] Smith, A.M.S., Lentile, L.B., Hudak, A.T. and Morgan, P. (2007) Evaluation of linear spectral unmixing and ΔNBR for predicting post-fire recovery in a North American ponderosa pine forest. *International Journal of Remote Sensing*, **28**, 5159-5166.

[30] Key, C.H. and Benson, N.C. (2006) Landscape assessment: Sampling and analysis methods: Firemon: Fire effects monitoring and inventory system. General Technical Report. USDA Forest Service, Rocky Mountain Research Station, Fort Collins CO., RMRS-GTR-164-CD.

[31] Van Wagtendonk, J.W., Root, R.R. and Key, C.C. (2004) Comparison of AVIRIS and Landsat ETM+ detection capabilities for burn severity. *Remote Sensing of Environment*, **92**, 397-408.

[32] Miller, J.D. and Thode, A.E. (2007) Quantifying burn severity in a heterogeneous landscape with a relative version of the data Normalized Burn Ratio (dNBR). *Remote Sensing of Environment*, **109**, 66-80.

[33] Skinner, C. and Chang, C. (1996) Fire regimes, past and present. Sierra Nevada Ecosystem Project: final report to Congress.

[34] Barbour, M.G. and Major, J. (1977) Terrestrial vegetation of California. University of California Davis, Davis, 18-

[35] Garmin (2008) Garmin International Inc. KS 66062, Olathe.

[36] Unispec, S.C. (2008) PP Systems International Inc., Amesbury.

[37] Letts, M.G., Phelan, C.A., Johnson, D.R.E. and Rood, S.B. (2008) Seasonal photosynthetic gas exchange and leaf reflectance characteristics of male and female cottonwoods in a riparian woodland. *Tree Physiology*, **28**, 1037-1048.

[38] Jenkins, M., Krofcheck, J.D., Pushnik, J., Teasdale, R. and Houpis, J. (2012) Exploring the edge of a natural disaster. *Open Journal of Ecology*, **2**, 222-232.

[39] Furuuchi, H., Jenkins, M., Houpis, J., Senock, R. and Pushnik, J. (2013) Estimating plant crown transpiration and water use efficiency by vegetative reflectance indices associated with chlorophyll fluorescence. *Open Journal of Ecology*, **3**, 122-132.

[40] Vicente-Serrano, S.M., Perez-Cabello, F. and Lasanta, T. (2011) *Pinus halepensis* regeneration after a wildfire in a semiarid environment: Assessment using multitemporal Landsat images. *International Journal of Wildland Fire*, **20**, 195-208.

[41] Hernandez-Clemente R., Cerrillo R.M., Hernandez-Bermejo, J.E., Royo, S. and Kasimis, N.A. (2009) Analysis of postfire vegetation dynamics of Mediterranean shrub species based on terrestrial and NDVI data. *Environmental Management*, **43**, 876-887.

[42] Van Leeuwen, W.J.D., Casady G.M., Neary, D.G., Bautista, S., Alloza, J.A., Carmel, Y., Wittenberg, L., Malkinson, D. and Orr, B.J. (2010) Monitoring post-wildfire vegetation response with remotely sensed time series data in Spain, USA and Israel. *International Journal of Wildland Fire*, **19**, 75-93.

[43] Ustin, S.L. and Xiao, Q.F. (2001) Mapping successional boreal forests in interior central Alaska. *International Journal of Remote Sensing*, **22**, 1779-1797.

[44] Chen, X., Vierling, L. and Deerling, D. (2005) A simple and effective radiometric correction method to improve landscape change detection across sensors and across time. *Remote Sensing of the Environment*, **98**, 63-79.

[45] Paruelo, J.M. and Lauenroth, W.K. (1995) Regional patterns of normalized difference vegetation index in North American shrublands and grasslands. *Ecology*, **76**, 1888-1898.

[46] Gomez-Mendoza, L., Galicia, L., Cuevas-Fernandez, M.L., Magana, V., Gomez, G. and Palacio-Prieto, J.L. (2008) Assessing onset and length of greening period in six vegetation types in Oaxaca, Mexico, using NDVI-precipitation relationships. *International Journal Biometeorology*, **52**, 511-520.

[47] National Oceanic and Atmospheric Administration (2010)

[48] Zhong, L., Ma, Y., SuhybSalama, M. and Su, Z. (2010) Assessment of vegetation dynamics and their response to variations in precipitation and temperature in the Tibetan Plateau. *Climatic Change*, **103**, 519-535.

[49] Lloret, F., Lobo, A., Estevan, H., Maisongrande, P., Vayreda, J. and Terradas, J. (2007) Woody plant richness and NDVI response to drought events in Catalonian (northeastern Spain) forests. *Ecology*, **88**, 2270-2279.

[50] Kobziar, L.N. and McBride, J.R. (2006) Wildfire burn patterns and riparian vegetation response along two northern Sierra Nevada streams. *Forest Ecology and Management*, **222**, 254-265.

[51] Kavgaci, A., Carni, A., Basaran, S., Basaran, M.A., Kosir, P., Marinsek, A. and Silc, U. (2010) Long-term post-fire succession of *Pinus brutia* forest in the east Mediterranean. *International Journal of Wildland Fire*, **19**, 599-605.

[52] Tarrega, R. and Calabuig, E.L. (1987) Effects of fire on structure dynamics and regeneration of *Quercus pyrenaica* ecosystems. *Ecologia Mediterranea*, **13**, 79-86.

[53] Keeley, J.E., Brennan, T. and Pfaff, A.H. (2008) Fire severity and ecosystem responses following crown fires in California shrublands. *Ecological Applications*, **18**, 1530-1546.

[54] Epting, J., Verbyla, D. and Sorbel, B. (2005) Evaluation of remotely sensed indices for assessing burn severity in interior Alaska using Landsat TM and ETM+. *Remote Sensing of Environment*, **96**, 328-339.

[55] Perry, D.A., Hessburg, P.F., Skinner, C.N., Spies, T.A., Stephens, S.L., Taylor, A.H., Franklin, J.F., McComb, B. and Riegel, G. (2011) The ecology of mixed severity fire regimes in Washington, Oregon, and Northern California. *Forest Ecology and Management*, **262**, 703-717.

12

Note on the vegetation of the mounts of tlemcen (Western Algeria): Floristic and phytoecological aspects

Brahim Babali*, **Abderrahmane Hasnaoui, Nadjat Medjati, Mohamed Bouazza**

Laboratory of Ecology and Management of the Natural Ecosystems, Department of Ecology and Environment, Aboubakr Belkaid University, Tlemcen, Algeria; *Corresponding Author

ABSTRACT

Of the four national hunting reserves in Algeria, the Mounts of Tlemcen Moutas reserve http://reservebio-tlm.com, **characterized by a large area, reliefs and a specific climate, implies significant floristic and faunistic richness. Currently, the coexistence of species, such as *Quercus faginea* subsp. *tlemcenensis* (DC.) M., *Lonicera implexa* L., *Ruscus aculeatus* L., indicates a forest dominant ecological atmosphere, although the region has experienced repeated fires during the 90's. In this research, a phytoecologicaland syntaxonomical analysis is obvious. More than 300 species have been inventoried and indexed in more than 70 families and this shows the importance of phyto-diversity of the studiedregion. In the analysis of the phyto-ecological parameters, we could notice a regression of the vegetal cover in its diversity.**

Keywords: Biodiversity; Floristic inventory; Phytoecological; Anthropozoological action; Climate; Moutas; Tlemcen

1. INTRODUCTION

The currently developed methods of biodiversity extinction have large uncertainties but all converge on acceleration whatever would be the economic models. To assess the loss of biodiversity, we worked on the disappearance and fragmentation of plant life media (inventory of natural habitats).

The reserve is particularly sensitive in terms of plant diversity, it underwent in the past human pressures and significant fires. The ecological landscape includes different habitats moving to a scrub, with considerable variations.

The knowledge of this dynamic and this floristic inventory is an important research path for us. Analyses of biodiversity lead in particular to show that the maximum biodiversity is not in the primitive forest *sensu stricto*, but in the moderately man altered spaces [1].

We will discuss this problem here from floristic inventory formed by tree structures and their stages of degradation as it is at this level that they can be analyzed.

The vegetation of the national parks and natural reserves in the Mediterranean basin have been studied by many authors like Gruber and Sandoz [2]; Véla *et al.* [3]; Hill and Véla [4]; Ibn Tattouand & Fennane [5]... and other works in Tlemcen region like those of Benabadji *et al.* [6]; Mesli *et al.* [7]; Letreuch-Belarouci *et al.* [8]; Medjahdi *et al.* [9] and Bouazza *et al.* [10].

2. METHODOLOGY

Location and structure of Tlemcen hunting reserve:

The study area is located in the western part of Northwest Algeria at about 46 km as the crow flies from the sea and 26 km south-west of the city of Tlemcen. The reserve, part of Hafir forest, occupies the highest and most wooded area of the Mounts of Tlemcen. It is located about 34°41' to 34°49' north and 01°25' to 01°35' west (**Figure 1**).

It occupies an area of 2156 hain a 15 km perimeter; it is characterized by typically mountainous reliefs of the Tamaksalet massif with a remarkable difference in altitude. The altitude is between the extreme points from Ras Torriche 1303 m andthe region of Sidi Messaoud at 1017 m.

It is geographically limited:
• To the east by the town of Aïn Ghoraba;
• In the north-east by the municipality of Sabra;
• To the west and northwest by the municipality of Bouhlou;

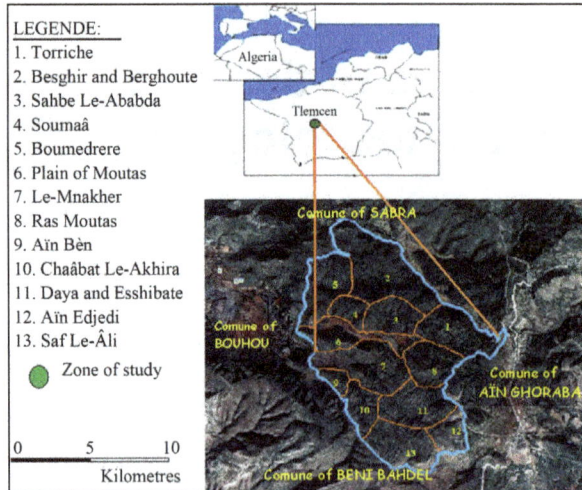

Figure 1. Location map.

• In the south and southwest by the city of Beni Bahdel.

It receives about 500 m mannual rainfall and shows average temperatures between 6.4°C (January) and 26.4°C (July and August). It is classified in the cool winter's sub-humid bioclimatic level with a dry season (5 months) which lasts from June to October.

The soils are varied depending on the topography of the region. On the tops, they are not very deep where of-time the bed rock levels. Sometimes deep towards very deep on all along the principal river basins of the hydrographic network.

In the reserve, the main rocks source is sedimentary: calcareous sandstone, sandy limestones, dolomites and marl [11].

The complete list was made from the following works:

Moutas Herbarium which is made up to now more than 220 species recorded by researchers from the laboratory of plant/botanical ecology from the University of Tlemcen,

Field work during 2010, 2011 and 2012, we have completed 70 surveys using conventional techniques and methods of ecology (inventory, minimum area, transect, flora network...).The method of Braun-Blanquet [12], which is expressed by analytical characteristics: abundance, dominance and sociability on a scale of 1 to 5 to help us to do a thorough analysis of the vegetation.

The basic work used for the identification of taxa collected in the field is from the studies done by Quézel &Santa [13]; Battandier & Trabut [14]; Valdés*et al.* [15]; Blanca *et al.* [16]; Maire [17] and Dobignard [18].

Scientific Name: the word list adopted is that of: *Index synonymique et bibliographique de la flore d' Afrique du Nord* [18] and synonyms from: *Nouvelle flore d' Algérie et des régions désertiques méridionales* [13].

3. RESULTS AND DISCUSSIONS

Ecological zoning of the main forest groupings.

3.1. The Oak Forests

Four major species of the genus *Quercus* were found in the reserve: *Quercus ilex* L., *Quercus suber* L., *Quercus coccifera* L. and *Quercus faginea* subsp. *tlemcenensis* (DC.) Maire and Weiller (=*Quercus faginea* subsp. *Broteroi* (Coutinho) A. Camus).

3.1.1. Evergreen Oak Forest

Dominates almost all of the reserve and is characterized by the evergreen oak presence (*Quercus ilex* subsp. *ballota* (Desf.) Samp) within a 942.4 hectares area.

The major issues, well developed and characterized by a large trunk, are located at the edges of cultivated lands within the reserve where the soil is deep and very rich in organic matter. These characters change as the altitude increases, the soil becomes shallower and bedrock appears at the surface. In addition, North exposures have a significant contribution to water compensation, allowing the taxa development in integrated settlements linked to *Quercetea ilicis* [6,19].

The distribution of Kermes oak (*Quercus coccifera* subsp. *coccifera* L.) is very limited, and in the extreme west and north-west of the reserve close to Tamaksalet (Bouhlou municipamity) and some species are scattered throughout the center of the reserve at Ras Mnakher, Mnakher and Souamaa. These areas are a warmer (xeric).

3.1.2. Zeen Oak Forest

The zeen oak (*Quercus faginea* Lamk) is a deciduous oak of meso-and supra-Mediterranean types [20-22], endemic to the western Mediterranean (Iberian Peninsula, Morocco, Algeria and Tunisia) [23]. It would be represented in the Mounts of Tlemcen by a sub-species: *Quercus faginea* subsp. *tlemcenensis* (DC.) M. (**Figure 2**)

This oak occupies 1/5th of the reserve with an area of 428 ha. It is found mainly in the southern and south-

Figure 2. *Quercus faginea* subsp. *tlemcenensis* formations at Tor- riche (Moutas). Photo. Babali B. September 2011.

western part of the reserve at Torriche, Ras Moutas, Mnakher, Chaâbat La'akhra and Aïn Ben. It is also found in the extreme north at Besghirand Boumedrer.

The Tlemcen zeen oak, ranging in size between 5 and 7 m, prefers deep soils and limestone-rich substratewith-fresh degraded materials and rare silica. This species exists and dominates the valleys and hollows of the reserve.We can consider that this species benefits from the water compensation, despite drought and this can be explained by compensation edaphic-climatic phenomena [24]. It is practically non-existent or so on the summit where the soil depth is less thick, and even if it does, it is most unusual and with a dwarf size that barely exceeds 2 m.

Among the accompanying taxa are: *Cytisus arboreus* subsp. *Baeticus* (Webb) Maire *Cytisus villosus* Pourret, *Hedera algeriensis* Hibberd, *Ruscus aculeatus* L., *Smilax aspera* L., *Viburnum tinus* L., *Lonicera implexa* L., *Pistacia terebinthus* L., *Asplenium ceterach* L., *Umbilicus rupestris* (Salisb.) Dandy, *Phillyrea latifolia* L. *Ampelodesmos mauritanicus* (Poiret) Durand & Schinz...

3.1.3. Cork Oak Forests

Representedby the cork oak relics: *Quercus suber* L. (**Figure 3**), in a very limited area, which does not exceed 20 ha, they are frequently found in Saf-el-Ali, Aïn Djedi and other relics in Torriche, Boumedrerand Ras Moutas south side and finally a few stalks at Mnakher. Their growth is generally less strong after fire. These species are typical of low-intensity fire regimes, but common in the study area [25-27].

The vegetation associated with these cork oak is: *Lavandula stoechas* L., *Anagallis arvensis* L., *Erica arborea* L., *Arbutus unedo* L., *Stauracanthus boivinii* (Webb) Samp *Ampelodesmos mauritanicus* (Poiret) Durand & Schinz, *Asparagus acutifolius* L., *Daphne gnidium* L., *Cytisus villosus* Pourret, *Cistus clusii* Dunal., *Cistus creticus* L., *Cistus salvifolius* L., *Cistus ladanifer* subsp.

mauritianus Pau & Sennen. These plants prefer siliceous substrates.

3.2. Conifers

They are softwood thermophilic with an extremely wide ecological spectrum. We have: Thuja: *Tetraclinis articulata* (Vahl) Masters (**Figure 4**).

Endemic to North Africa [28,29] it colonizes areas with low rainfall (300 - 500 mm) [30]. This species is slightly represented in the Moutas reserve. It occupies, especially the northwest portion of the reserve: southwestern slopes and the southern slopes of Boumedrer, Ras Mnakhert toward Bouhlou, Aïn Ben Soumaâ and Safel-Ali. It is associated with *Pistacia lentiscus* L., *Chamaerops humilis* var. *argentea* Andrew *Globularia alypum* L., *Macrochloa tenacissima* L. (Kunth), *Phyleria angustifolia* L., *Asparagus albus* L.

The Juniper: *Juniperus oxycedrus* Subsp. *Oxycedrus*

It is widespread in the reserve with scattered blankets. This indicates the presence of degradation oak stands.

• Other conifers are represented in the form of plantations in the rest area Torriche and near the forest house Boumedrer as Aleppo pine *Pinus halepensis* Miller, stone pine (*Pinus pinea* L.) cedar (*Cedrus atlantica* (Endl.) Carrière) and cypress (*Cupressus sempervirens* L.)

3.3. Riverine

The reserve is surrounded by natural sources: Aïn-Boumedrer the largest and most common, AïnE-Djedi, Aïn Moutas and Aïn El-Ben. The vegetation, adjacent to these springs and streams, is riparian representing vegetation structure at least partly azonal [31], or indicators of wetlands such as *Rubus ulmifolius* Schott, *Dittrichia viscosa* (L.) Greuter, *Typha latifolia* L., *Carex hispi-*

Figure 3. Relic of cork oak. Sahb El Ababda (Moutas). Photo. Babali B. October 2010.

Figure 4. Degraded forest *Tetraclinis articulata* based at-Tameksalet-Moutas (south side). Photo. Babali B., October 2012.

da Willd., *Populus alba* L., *Salix pedicellata* Desf., *Juncus maritimus* Lam., *Hypericum perforatum* L., *Mentha rotundifolia* L., *Ficus carica* L., *Calamintha nepeta* (L.) Savi, *Potentilla reptans* L., *Zannichellia peltata* Bertol., *Groenlandia densa* (L.) Fourr., *Apium nodiflorum* (L.) Lag, *Arundo donax* L., *Dactylorhiza durandii* (Boiss. & Reuter) M. Lainz, *Ranunculus ficaria* L., *Ranunculus aquatilis* L., *Ranunculus spicatus* Desf. *Sonchus maritimus* L., *Nerium oleonder* L. *Trachelium caeruleum* L., *Jasminum fruticans* L., *Vitis vinifera* subsp. *sylvestris* (D.C. Gmelin) Hegi. *Rorippa nasturtium-aquaticum* (L.) Hayek (= *Nasturtium officinale* R. Br.).

Riparian forests are dominant in this area; unfortunately the foresters do not take this fact into account in their statistics.

3.4. The Scrub

Over a large area of 680 ha, the scrubs are characterized and dominated by more xeric coppices depleted in forest and pre-forest species that occupy the land like thechamaephytes or nanaophanerophytes which prefer gradients and hot spots (southern slopes) belonging to *Pistacio-Rhamnetalia Alaterni* represented by *Chamaerop shumilis*var. *argentea* Andrew *Ampelodesmos mauritanicus* (Poiret) Durand & Schinz, *Pistacia terebinthus* L., *Thymus munbyanus* Boiss. & Reuter, *Fumana thymifolia* (L.) Webb, *Calicotome intermedia* (Salzm.) C. Presl, *Rhamnus lycioïdes* L.

Other taxa grow in an expansionary way after the fire e.g., *Cistus ladanifer* L., *Cistus creticus* L. and *Cistus salvifolius* L. and taxa characterized by their high regeneration such as thujaand evergreen oak that can participate in the formation of scrub landscapes [20].

3.5. Lawns (Figure 5)

"in the short-cycle crops adapted to use a fleeting resource, tolerance and/or need of light (light-demanding species) make them exclusive or preferential plants ofo-

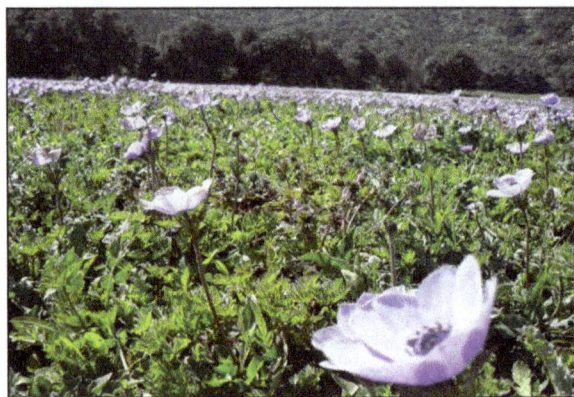

Figure 5. Annual plant lawn with *Anemone coronaria* before Moutas plain crops. Photo. Babali B., March 2011.

ligotrophic dry grasslands or rocks exposed to wind and temperature extremes." [32].

They are spread over about 106 hectares and dominated by annual species (Therophytes) caused by high anthropozoological action and further degradation (fire).

This group belongs to the *Thero-Brachypodietea* in general, is on calcareous substrata; characterized by *Rhaponticum coniferum* (L.) Greuter, *Bombycilaena discolor* (Pers.) Lainz, *Reichardia tingitana* (L.) Roth, *Scorzonera laciniata* L., *Trifolium stellatum* L., *Ajuga iva* var. *pseudoiva* subsp. *Pseudoiva* (DC.) Briq., *Teucrium polium* L., *Ophrys lutea* subsp. *lutea* (Cav.) Gouan, *Carex halleriana* Asso., *Rumex bucephalophorus* L., *Briza maxima* L. *Vulpia ciliate* Dumort…

List of vascular taxa listed by family in Tlemcenhunting reserve.

The list includes 322 species belonging to 72 families: Compositae (Asteraceae), Leguminosae (Fabaceae), Labiatae (Lamiaceae), Gramineae (Poaceae) and Orchidaceae (see palnches 1 - 2 in the annex).

➢ ALLIACEAE
 ***Allium Cupani* Raf.**
 ***Allium roseum* L.**
➢ AMARYLLIDACEAE
 ***Acis autumnalis* (L.) Herb**
 =*Leucojum autumnale* L.
 ***Narcissus cantabricus* DC.**
 ***Narcissus serotinus* L.**
 ***Narcissus tazzeta* L.**
 ***Narcissus tazzeta* subsp. *pachybolbus* (Dur.) Baker**
 ***Pancratium foetidum* var. *oranense* Pomel**
➢ ANACARDIACEAE
 ***Pistacia lentiscus* L.**
 ***Pistacia Terebinthus* L.**
➢ APOCÉNACEAE
 ***Nerium oleander* L.**
➢ ARACEAE
 ***Arisarum simorrhinum* Durieu**
 =*Arisarum vulgare* subsp *simorrhinum* (dur.) M.et W.
 ***Biarum Bovei* subsp. *dispar* (Schott.) Engler**
➢ ARALIACEAE
 ***Hedera helix* L.**
➢ ARECACEAE = PALMACEAE
 ***Chamaerops humilis* subsp. *Argentea* André.**
➢ ARISTOLOCHIACEAE
 ***Aristolochia baetica* L.**
 ***Aristolochia paucinervis* Pomel**
 =*Aristolochia longa* subsp. *paucinervis* (Pomel) Batt.
➢ ASPARAGACEAE
 Anthericum liliago* L. subsp. *algeriense
 ***Aphyllanthes monspeliensis* L.**
 ***Asparagus acutifolius* L.**
 ***Asparagus albus* L.**
➢ ASPHODELACEAE

Asphodelus ramosus **L.**
=*Asphodelus microcarpus* Salzm et Viv

➢ ASPLENIACEAE
Ceterach officinarum **Lamk.**
=*Asplenium ceterach* L.

➢ BORAGINACEAE
Anchusa italica **Retz.**
=*Anchusa azurea* Mill.
Borago officinalis **L.**
Cerinthe gymnandra **Gasparr.**
=*Cerinthe major* subsp. *gymnandra* (Aspar.)M.
Cynoglossum cheirifolium **L.**
Echium vulgare **L.**
Neatostema apulum (**L.**) **I.M. Johnston**
=*Lithospermum apulum* (L.) Vahl.

➢ CAMPANULACEAE
Campanula dichotoma **L.**
Campanula rapunculus **L.**
Trachelium caeruleum **L.**

➢ CAPRIPHOLIACEAE
Lonicera implexa **L.**
Viburnum tinus subsp. *tinus* **L.**

➢ CARYOPHYLACEAE
Dianthus cintranus **Boiss. & Reuter**
=*Dianthus gaditanus* Boiss.
Dianthus serrulatus subsp. *macranthus* **Maire**
Paronychia argentea **Lam.**
Silene vulgaris (**Moench**) **Garcke**
=*Silene inflata* (Salisb.)Sm.
Silene latifolia subsp. *latifolia* **Poiret**
Silene ramosissima **Desf.**
Stellaria media (**L.**) **Vill.**

➢ CISTACEAE
*Cistus clusii*Dunal.
Cistus creticus **L.**
=*Cistus villosus*L.
Cistus ladanifer subsp. *mauritianus* **Pau & Sennen**
=*Cistus ladaniferus*L.
Cistus salvifolius **L.**
Fumana thymifolia (**L.**) **Webb**
Halimium umbellatum (**L.**) **Spach**
Helianthemum cinereum (**Cav.**) **Pers.**
Helianthemum helianthemoides (**Desf.**) **Grosser**
Helianthemum salicifolium (**L.**) **Miller**

➢ COLCHICACEAE
Colchicum lusitanum **Brot.**
=*Colchicum autumnale* L.
Merendera filifolia **Camb.**

➢ COMPOSITAE = ASTERACEAE
Anacyclus pyrethrum (**L.**) **Link**
Atractylis cancellata **L.**
Bellis sylvestris **Cirillo**
Bombycilaena discolor (**Pers.**) **Laínz**
=*Micropus bombycinus* subsp *discolor* Lag.

Calendula arvensis **L.**
Calendula suffruticosa **Vahl**
Carlina gummifera (**L.**) **Less.**
Catananche caerulea **L.**
Centaurea pullata **L.**
Cichorium intybus **L.**
Cirsium echinatum (**Desf.**) **DC.**
Dittrichia viscose (**L.**) **Greuter**
=*Inula viscose* (L.) Ait.
Echinops strigosus **L.**
Filago fuscescens **Pomel**
Glebionis segetum (**L.**) **Fourr.**
=*Chrysanthemum segetum* L.
Helichrysum stoechas (**L.**) **Moench**
=*Elichrysum stoechas* (L.) DC.
Inula montana **L.**
Mauranthemum paludosum (**Poiret**) **Vogt & Ober-prieler**
=*Leucanthemum paludosum* (Poiret) non Bar.
Pallenis maritima (**L.**) **Greuter**
=*Asteriscus maritimus* (L.) Less.
Pallenis spinosa (**L.**) **Cass.**
Phagnalon saxatile (**L.**) **Cass.**
Phagnalon sordidum (**L.**) **Reichenb.**
Reichardia tingitana (**L.**) **Roth**
Rhaponticum acaule (**L.**) **DC.**
Rhaponticum coniferum (**L.**) **Greuter**
=*Leuzea conifera* (L.) DC.
Scolymus grandiflorus **Desf.**
Scolymus hispanicus **L.**
Scorzonera laciniata **L.**
Scorzonera undulata **Vahl.**
Senecio vulgaris **L.**
Sonchus asper (**L.**) **Hill**
Sonchus maritimus **L.**
Staehelina dubia **L.**
Taraxacum obovatum (**Willd.**) **DC.**

➢ CONVULVULACEAE
Convolvulus althaeoides **L.**
Convolvulus arvensis **L.**
Convolvulus cantabrica **L.**
Convolvulus humilis **Jacq.**
Convolvulus tricolor **L.**

➢ CRASSULACEAE
Pistorinia breviflora subsp. *intermedia* (**Boiss. & Reuter**) **Greuter & Burdet**
=*Cotyledon breviflora* (Boiss.) M.
Sedum sediforme (**Jacq.**) **Pau**
Sedum album **L.**
Umbilicus rupestris (**Salisb.**) **Dandy**
=*Cotyledon umbilicus-veneris* subsp. *Pendulina* (DC.) Batt.

➢ CRUCIFERAE = BRASSICACEAE
Alyssum simplex **Rudolphi**

Biscutella didyma **L.**
Erysimum grandiflorum **Desf.**
=*Erysimum bocconei* (all.) Pers.
Lepidium hirtum **(L.) Sm.**
Lobularia maritima **(L.) Desv.**
Nasturtium officinale **R. Br.**
= *Rorippa nasturtium-aquaticum* (L.) Hayek
Raphanus raphanistrum **L.**
Sinapis alba **L.**
Sinapis arvensis **L.**
➤ CUPRISSACEAE
Juniperus oxycedrus **L. subsp.** *oxycedrus*
Tetraclinis articulata **(Vahl) Masters**
Cupressus sempervirens **L.**
CYPERACEAE
Carex halleriana **Asso**
Carex hispida **Willd.**
➤ DIOSCOREACEAE
Diocorea communis **(L.) Caddick & Wilkin**
=*Tamus communis* L.
➤ DIPSACACEAE
Cephalaria leucantha **(L.) Roemer & Schultes**
Lomelosia stellata **(L.) Raf.**
=*Scabiosa stellata* L.
Sixalix atropurpurea **(L.) Greuter & Burdet**
=*Scabiosa atropurpurea* L.
➤ ERICACEAE
Arbutus unedo **L.**
Erica arborea **L.**
Erica multiflora **L.**
➤ EUPHORBIACEAE
Euphorbia helioscopia **L.**
Euphorbia nicræensis **All.**
Euphorbia squamigera **Lois.**
➤ FAGACEAE
Quercus coccifera **L. subsp.** *coccifera*
Quercus faginea **subsp.** *tlemcenensis* **(DC.) Maire et Weiller**
=*Quercus faginea* subsp. *broteroi* (Coutinho) A. Camus
Quercus Ilex **subsp.** *Ballota* **(Desf.) A. DC.**
Quercus suber **L.**
➤ GENTIANACEAE
Centaurium erythraea **Raf.**
=*Centaurium umbellatum* (Gibb.) Beck.
➤ GÉRANIACEAE
Erodium moschatum **(L.) L'Hér.**
Geranium malviflorum **Boiss. & Reute**
➤ GRAMINEAE = GRAMINACEAE = POACEES
Aegilops geniculata **Roth**
Aegilops triuncialis **L.**
Ampelodesmos mauritanicus **(Poiret) Durand & Schinz**
= *Ampelodesma mauritanica* (Poiret) Dur.et Sch.

Anisantha madritensis **(L.) Nevski**
=*Bromus matritensis* L.
Anisantha rubens **(L.) Nevski**
=*Bromus rubens* L.
Arundo donax **L.**
Avena sativa **L.**
Avena sterilis **L.**
Brachypodium sylvaticum **(Huds.) P. B.**
Briza maxima **L.**
Bromus hordeaceus **L.**
Cynosurus echinatus **L.**
Festuca coerulescens **Desf.**
Hordeum murinum **L.**
Lagurus ovatus **L.**
Macrochloa tenacissima **(L.) Kunth**
=*Stipa tenacissima* L.
Vulpia ciliata **Dumort.**
➤ HYACINTHACEAE
Drimia maritima **(L.) Speta**
=*Urginea maritima* var. *pancration* (Stein.) Baker.
Drimia undulata **Jacq.**
=*Urginea undulata* (Desf.) Steinh. subsp *typica M.*
Leopoldia comosa **(L.) Parl.**
=*Muscari comosum* (L.) Mill.
Muscari neglectum **Guss.**
Oncostema peruviana **(L.) Speta**
=*Scilla peruviana* L.
Ornithogalum algeriense **Jord. & Fourr**
=*Ornithogalum umbellatum* L.
Ornithogalum narbonense **L.**
=*Ornithogalum pyramidalis*auct. non L.
Prospero autumnalis **(L.) Speta**
=*Scilla autumnalis* L.
Prospero obtusifolium **(Poiret) Speta**
=*Scilla obtusifolia* Poiret
Uropetalum serotinum **(L.) Ker Gawl.**
=*Dipcadi serotinum* (L.) Medik.
➤ HYPERICACEAE
Hypericum perforatum **L.**
Hypericum tomentosum **subsp** *tomentosum* **L.**
➤ IRIDACEAE
Gladiolus italicus **Mill**
=*Gladiolus segetum* Ker.-Gawl.
Iris planifolia **(Mill.) Dur. et Sch.**
Iris xiphium **L.**
Moraea sisyrinchium **(L.) Ker Gawl.**
=*Gynandriris sisyrinchium* (L.) Parl.
Romulea bulbocodium **(L.) Seb.et Maur.**
➤ JUNCACEAE
Juncus maritimus **Lamk.**
LABIATAE = LAMIACEAE
Ajuga chamaepitys **(L.) Schreber**
=*Ajuga chamaephitis* Schreb.
Ajuga iva **subsp.** *iva* **(L.) Schreber**

Ajuga iva subsp. *pseudoiva* (DC.) Briq. var. *pseudo-Iva*

*Ballota hirsuta*Bentham

Calamintha nepeta (L.) Savi

=*Satureja calamintha* subsp. *Nepeta* correct

Lamium amplexicaule L.

Lavandula stoechas L. subsp. *stoechas*

Marrubium vulgare L.

Mentha rotundifolia L.

Nepeta multibracteata Desf.

Nepeta tuberosa subsp. *reticulata* (Desf.) Maire

Origanum vulgare subsp. *Glandulosum* (Desf.) Iestwaart

=*Origanum glandulosum* Desf.

Phlomis crinita subsp. *Mauritanica* (Munby) Murb.

=*Phlomis crinita* cav.

Phlomis herba-venti L.

Rosmarinus eriocalyx Jord. & Fourr.

=*Rosmarinus tournefortii*de Noé

Salvia verbenaca L.

Stachys ocymastrum (L.) Briq.

Teucrium fruticans L.

Teucrium polium L.

Teucrium pseudochamaepitys L.

Thymus munbyanus subsp. *coloratus* (Boiss. & Reuter) Greuter & Burdet

=*Thymus ciliatus* subsp. *coloratus* (B. & R.) Batt.

➢ LEGUMINOSAE = FABACEAE

Anthyllis polycephala Desf.

Anthyllis vulneraria L.

Argyrolobium zanonii (L.) Link

=*Lotophyllus argenteus* L.

Astragalus caprinus subsp. *caprinus*.

=*Astragalus caprinus* subsp. *Lanigerus* (*Desf.*) M

Astragalus epiglottis L.

Bituminaria bituminosa (L.) Stirton

=*Psoralea bituminosa* L.

Calicotome intermedia (Salzm.) C. Presl

=*Calycotome villosa* subsp. *Intermedia* (Salzm.) M.

Ceratonia siliqua L. (Césalpiniacées)

Coronilla scorpioides (L.) W.D.J. Koch *Cytisus arboreus* subsp. *baeticus* (Webb) Maire

Cytisus villosus Pourret

=*Cytisus triflorus* L'Herit

Erophaca baetica (L.) Boiss.

=*Astragalus lusitanicus* Lamk.

Genista ramosissima (Desf.) Poiret

=*Genista cinerea* subsp. *ramosissima* (Desf.) Maire

Genista tricuspidata subsp. *Duriaei* (Spach.) Batt.

Lathyrus latifolius L.

Lotus hispidus DC.

Medicago italica subsp. *Tornata* (L.) Emb. et Maire

Medicago polymorpha L.

= *Medicago hispida* Gaertn.

Ononis biflora Desf.

Ononis pubescens L.

Ononis spinosa L.

Scorpiurus muricatus L.

Stauracanthus boivinii (Webb) Samp

=*Ulex Boivinii* Webbvar. *webbianus* (Cosson) Maire

Trifolium angustifolium L.

Trifolium stellatum L.

Trifolium tomentosum L.

Vicia onobrychioides L.

Vicia sativa L.

➢ LILIACEAE

Fritillaria lusitanica subsp. *Oranensis* (Pomel) Valdés

=*Fritillaria messanensis* Raf. var. *atlantica* M.

Gagea Durieui Pari.

Gagea granatelli subsp. *chaberti* Terracc.

Tulipa sylvestris subsp. *australis* (Link.) Pamp.

➢ LINACEAE

Linum suffruticosum L.

Linum tenue Desf.

Linum usitatissimum L.

➢ MALVACEAE

Lavatera trimestris L.

Malope malachoides L.

Malva sylvestris L.

➢ MORACEAE

Fucus carica L.

➢ MYRSINACEAE = PRIMULACEES

Anagallis arvensis L.

Anagallis Anagallis monelli L.

➢ OLEACEAE

Jasminum fruticans L.

Olea europea L. subsp. *europaea*

=*Olea europea* var. *oleaster*

Phillyrea angustifolia subsp. *angustifolia* M.

Phillyrea latifolia L.

=*Phillyrea angustifolia* subsp. *latifolia* (L.)M.

➢ ORCHIDACEAE

Aceras pyramidalis (L.) Reichenb

=*Anacamptis pyramidalis* (L.) L.C. Rich.

Anacamptis coriophora subsp. *fragrans* (Poll.) Bateman, Pridgeon & Chase

= *Orchis coriophora* subsp. *Fragrans* (Poll.) G. Camus

Anacamptis morio subsp. *tlemcenensis* (*Batt.*) E.G. Camus

Anacamptis papilionacea (L.) Bateman, Pridgeon & Chase

=*Orchis papilionacea* L.

Dactylorhiza durandii (Boiss.& Reuter) M.Lainz

=*Orchis elata* subsp. *Durandoi* (B.et R.)

Himantoglossum hircinum (L.) Sprengel

=*Himanthoglossum hircinum* (L.) Spreng.

Himantoglossum robertianum (Loisel.) Delforge

= *Himanthoglossum longibracteatum* (Biv.) Sch.
Ophrys atlantica Munby
Ophrys lutea subsp. **Lutea (Cav.) Gouan**
Ophrys speculum L.
Ophrys sphegifera Willd.
= *Ophrys scolopax* Cav. subsp. *Apiformis*
Ophrys subfusca (Reichenb. fil.) Haussknecht
=*Ophrys lutea* subsp. *Subfusca* (Rchb.) Batt.
Ophrys thenthredimifera Willd. subsp. Ficalhoana
Ophrys thenthredimifera Willd. subsp. Thenthredi-mifera
Orchis anthropophora (L.) All.
= *Aceras anthropophorum* (L.) Ait.
Orchis italica Poiret
Orchis olbiensis Reuter.
= *Orchis maculata* subsp. *Obliensis* (Reut.) Asch. et Gr.
➤ OROBANCHACEAE
Bartsia trixago L.
= *Bellardia trixago* (L.) All. (Scrophiliacées)
Odontites purpureus subsp. purpureus (Desf.) G. Don fil.
= *Odontites bolligeri* E.Rico, L. Delgado & Herrero in Rico *et al.* [33]
= *Odontites purpurea*subsp *purpurea* (Scrophiliacées)
Orobanche ramosa L.
Orobanche variegata Wallr
Parentucellia latifolia (L.) Caruel (Scrophiliacées)
➤ PAPAVERACEAE
Fumaria capreolata L. (Fumariacées)
Fumaria officinalis L. (Fumariacées)
Papaver hybridum L.
Papaver rhoeas L.
Roemeria hybrida (L.) DC.
➤ PINACEAE
Cedrus atlantica (Endl.) Carrière
=*Cedrus libanotica* Link
Pinus halepensis Mill.
Pinus pinea L.
➤ PLANTAGINACEAE
Anarrhinum fruticosum subsp. fruticosum Maire (Scrophiliacées)
Antirrhinum majus L. (Scrophiliacées)
Globularia Alypum subsp. alypum L. (Globularia-cées)
Linaria arvensis L. Desf. (Scrophiliacées)
Linaria triphylla (L.) Miller (Scrophiliacées)
Linaria tristis (L.) Miller (Scrophiliacées)
Plantago mauritanica Boiss. et Reut.
Plantago lagopus L.
Plantago serraria L.
➤ POLYGALACEAE
Polygala monspeliaca L.
➤ POLYGONACEAE

Rumex bucephalophorus L.
➤ POTAMOGETONACEAE
Groenlandia densa (L.) Fourr.
= *Potamogeton densus*L.
➤ RAFFLESIACEAE
Cytinus hypocistis subsp. clusii*Nyman
= *Cytinus hypocistis* subsp. *kermesianus* (Guss.) Wettst.
Cytinus hypocistis subsp. hypocistis L.
= *Cytinus hypocistis* subsp. *ochraceus* (Guss.) Wettst.
➤ RENONCULACEAE
Adonis aestivalis*L.
Anemone coronaria L.var. cyanea (Risso) Ardoino
Anemone palmata L.
Clematis cirrhosa L.
Clematis flammula L. var. parviflora Pomel
Delphinium balansae Boiss. et Reut.
Ranunculus arvensis L.
Ranunculus ficaria subsp. ficariiformis Rouy & Fouc.
Ranunculus gramineus L.
Ranunculus macrophyllus Desf.
Ranunculus millefoliatus Vahl
Ranunculus paludosus Poiret
Ranunculus spicatus Desf.
Ranunculus aquatilis L.
➤ RESEDACEAE
Reseda alba subsp alba L.
= *Reseda alba* subsp *eu-alba* L.
Reseda luteola L.
Reseda Phyteuma subsp. collina (Gay) Batt.
➤ RHAMNACEAE
Rhamnus alaternus L. subsp. alaternus
Rhamnus lycioides subsp. oleoides (L.) Jahand. & Maire
➤ ROSACEAE
Crataegus monogyna Jacq.
= *Crataegus Oxyacantha* subsp. *monogyna* (Jacq.) Rouy et Camus
Potentilla reptans L.
Rosa canina L.
Rubus ulmifolius Schott
Sanguisorba minor Scop.
➤ RUBIACEAE
Rubia peregrina subsp. Peregrina L.
➤ RUSCACEAE
Ruscus aculeatus L.
➤ RUTACEAE
Ruta angustifolia Pers.
= *Ruta chalepensis* subsp. *angustifolia* (Pers.) P. Cout
➤ SALICACEAE
Populus alba L.
Salix pedicellata Desf.
➤ SANTALACEAE

Osyris quadripartita **Decne**
➤ SAXIFRAGACEAE
   Saxifraga globulifera **Desf.**
   = *Saxifraga globulüera* Desf.
➤ SCROPHILIACEAE
   *Scrofularia laevigata***Vahl**
   Scrophularia canina **L.**
   Verbascum blattaria **L.**
➤ SINOPTERIDACEAE
   Cheilanthes acrostica **(Balb.) Tod.**
➤ SMILACACEAE
   Smilax aspera **L. var.** *Altissima* **Moris & De Not.**
   Smilax aspera **L. var.** *genuina* **L.**
➤ THYMELAEACEAE
   Daphne gnidium **L.**
➤ THYPHACEAE
   Typha angustifolia L.
➤ UMBELLIFERAE = APIACEAE
   Ammoides pusilla **(Brot.) Breistr.**
   = *A. verticillata (Desf.) Briq.*
   Apium nodiflorum **(L.) Lag**
   = *Helosciadium nodiflorum*Lag.
   Bupleurum rigidum **L.**
   Daucus carota **L.**
   Eryngium tricuspidatum **L.**
   Eryngium triquetrum **Vahl**
   Ferula communis **L.**
   Thapsia garganica **L.**
➤ VALÉRIANACEAE
   Fedia cornucopiae **(L.) Gaertn**
   Valeriana tuberosa **L.**
   Valerianella discoidea **(L.) Loisel.**
   = *Valerianella coronata* subsp. *discoidea* Lois.
➤ VERBENACEAE
   Verbena officinalis L.
   VITACEAE = AMPELIDACEA
   Vitis vinifera **subsp.** *sylvestris* **(C.C. Gmelin) Hegi**
➤ ZANNICHELLIACEAE
   Zannichellia peltata **Bertol.**

4. CONCLUSION AND PERSPECTIVES

One can not speak of plant diversity, of the Moutasreserve, without taking the relative change in the climate of the western part of Algeria into consideration. That's how the phylogenetic potential began a regressive evolution. This latter is accentuated by an increasingly strong anthropo zoological pressure.These ecosystems are fragile and complex and must be tackled in a comprehensive manner. There are many achievements in phytoecology and plant systematics, but little information/data is available regarding in particular the western part of Algeriaendemic species.

Faced with this alternative we insist on developing/expanding plants (aromatic/medicinal and others)

sincethe hunting reserve has a wealth of landscapes and acts as a refuge for sensitive and/or endangered species as *Origanum*, *Ammoides* e.g.

Today, we are moving towards a bank of botanical data to monitor this floristic cover which is close to the ecological break (environmental stress).

For about **30** years we have gone through this area, the changes are significant and we see before our eyes that the regressive evolution has begun. This observation is linked to southern species which are moving to north of Tlemcen Mounts.

REFERENCES

[1]   Blondel, J., (2002) *Problématique de la forêt méditerranéenne*. Hors Scin, **1**.

[2]   Gruber, M. and Sandoz, H. (1995) Inventaire floristique et phytoécologique du parc du Jarret-la Ravelle (Marseille, Bouches-du-Rhône, France). *Bulletin de la Société Linnéenne de Provence*, **46**, 105-118.

[3]   Véla, E., Hill B. and S. Della Casa (1999) Liste des plantes vasculaires du département des Bouches-du-Rhône (France). *Bulletin de la Société Linnéenne de Provence*, **50**, 115-201.

[4]   Hill, B. and véla, E. (2000) Mise à jour de la liste des plantes vasculaires du département des Bouches-du-Rhône. *Bull. Bulletin de la Société Linnéenne de Provence*, **51**, 71-94.

[5]   Ibn Tattou, M. and Fennane, M. (1989) Aperçu historique et état actuel des connaissances sur la flore vasculaire du Maroc. *Bulletin Institut Scientifique de Rabat*, **13**, 85-94.

[6]   Bouazza, M., Mahboubi, A., Loisel, R. and Benabadji, N. (2001) Bilan de la flore de la region de Tlemcen (Oranie-Algérie), forêt méditerranéen, **2**, 130-136.

[7]   Benabadji, N., Benmansour, D. and Bouazza, M. (2007) La flore des monts d'Ain Fezza dans l'Ouest algérien, biodiversité et dynamique. *Sciences & Technologie*, **26**, 47-59.

[8]   Mesli, K., Bouazza, M. and Godron, M. (2008) Ecological characterization of the vegetable groupings of the mounts of Tlemcen and their facies of degradation (west-Algeria). *Environmental Research*, **2**, 271-277.

[9]   Letreuch-Belarouci, A., Medjahdi, B., Letreuch-Belarouci, N. and Benabdeli, Kh. (2009) Diversité floristique des subéraies du parc national de Tlemcen. *Acta Botanica Malacitana*, **34**, 77-89.

[10] Medjahdi, B., Ibn Tattou, M., Barkat, D. and Benabedli, Kh. (2009) La flore vasculaire des monts des Traras. *Acta Botanica Malacitana*, **34**, 57-75.

[11] Anonymous-Bulgarie (1988) Projet d'aménagement cynégétique de la réserve de chasse Moutas-wilaya de Tlemcen. *Lescomplekt-engineering*, **4**, 99.

[12] Braun-Blanquet, J. (1951) Les groupements végétaux de la France méditerranéenne., CRNS, Paris, 297.

[13] Quézel P. and Santa S. (1962-1963) Nouvelle flore d'Algérie et des régions désertiques méridionales. CRNS,

Paris (FR), Tome I: 1-565, Tome II: 566-1170.

[14] Battandier, A. and Trabut, L. (1888-1890) Flore d'Algérie (Dicotylédones). 860.

[15] Valdés, B., Rejdali, M., Kadmiri, A.A.E., Jury, S.L. and Montserrat, J.M. (2002) Catalogue des plantes vasculaires du Nord du Maroc incluant des clés d'ídentification. Consejo Superior d'Investigaciones Científicas, Biblioteca de ciencias, Madrid, Vol. I and II, 1007.

[16] Blanca, G., Cabezudo, B., Cueto, M., Fernández López, C. and Morales Torres, C. (2009) Flora Vascular de Andalucía Oriental, Consejería de Medio Ambiente, Junta de Andalucía, Sevilla.

[17] Maire, R. (1952-1987) Flore de l'Afrique du Nord (Maroc, Algérie, Tunisie, Tripolitaine, Cyrénaïque et Sahara). Le Chevalier, Paris. Vol. 1-16, 5559.

[18] Dobignard, A. (2008) Index, and synonymous bibliographic flora of North Africa, in Press.

[19] Dahmani-Megrerouche, M. (1996) Diversité biologique et phytogéographique des chênaies vertes d'Algérie, Ecologia Mediterranea, **22**, 19-38.

[20] Quezel, P. and Medail, F. (2003) Ecologie et biogéographie des forêts du bassin méditerranéen. Elsevier, Paris, 592.

[21] Laribi, M., Derridj, A. and Acherar, M. (2008) Phytosociologie de la forêt caducifoliée à chêne zéen (Quercus canariensis willd.) Dans le massif d'Ath Ghobri-Akfadou (grande Kabylie, Algérie). *Fitosociologia*, **45**, 1-15.

[22] Messaoudène, M., Tafer, M., Loukkas, A. and Marchal, R. (2008) Propriétés physiques du bois de chêne zéen de la forêt des Aït Ghobri (Algérie). *Bois et Forêts des Tropiques*, **298**, 37-48.

[23] Zine El Abidine, A. (1988) Analyse de la diversité phytoécologique des forêts du chêne zeen (Quercus faginea Lamk.) Au Maroc. *Bulletin Institut Scientifique de Rabat*, **12**, 69-77.

[24] Alcaraz, C. (1989) Contribution à l'étude des groupements à Quercus ilex et Quercus faginea subsp. Tlemcenensis des monts de Tlemcen (Algérie). *Ecologia Medi-*

terranea, **15**, 15-32.

[25] Prodon, R., Fons, R. and Peter, A.M. (1984) L'impact du feu sur la végétation, les oiseaux et les micromammifères dans diverses formations méditerranéennes des Pyrénées Orientales: Premiers résultats. *Revue d'écologie, Terre et Vie*, **39**, 128-158.

[26] Amandier, L. (2004) Le comportement du Chêne-liège après l'incendie conséquences sur la régénération naturelle des subéraies. *Actes du coloque-VIVexpo* 2004: *Le chêne-liège face au feu*, 1-18.

[27] Schaffhauser, A., Curt, T., Véla, E. and Tatoni, T. (2012) Feux récurrents et facteurs environnementaux façonnent la végétation dans les boisements à Quercus suber L. *Et les maquis, C.R. Biologies*, **335**, 424-434.

[28] Hadjadj-Aoul, S. (1995) Les peuplements du Thuya de Berbérie (Tetraclinis articulata (Vahl) Master) en Algérie: phytoécologie, syntaxonomie et potentialités sylvicoles. Thèse d'Etat, Université Aix-Marseille III, Aix-en-Provence and Marseille, 250.

[29] Hadjadj-Aoul, S., Chouieb, M. and Loisel, R. (2009) Effet des facteurs environnementaux sur les premiers stades de la régénération naturelle de Tetraclinis articulata en Oranie, Ecologia Mediterranea, **35**, 19-30.

[30] Quézel, P. (2000) Réflexions sur l'évolution de la flore et de la végétation au Maghreb méditerranéen. Ibis Press, Paris, 117.

[31] Quezel, P. and Medail, F. (2003) Valeur phytoécologique et biologique des ripisylves méditerranéennes. *Forêt Méditerranéenne*, **3**, 231-248.

[32] Véla, E. (2002) Biodiversité des milieux ouverts en région méditerranéenne. Le cas des pelouses sèches du Luberon (Provence calcaire). Phd thesis. University Aix-marseille III, Aix-en-Provence and Marseille, 383.

[33] Rico, E., Delgado, l. and Herrero, A. (2008) Reassessing the Odontites purpureus group (Orobanchaceae) from South-East Spain and North-West Africa. *Botanical Journal of the Linnean Society*, **158**, 701-708.

Appendix

Planche 1. Board color (phot. B. Babali).

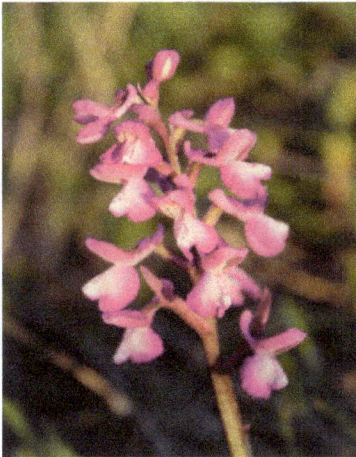

Anacamptis morio subsp. *tlemcenensis*
(Batt.) E.G. Camus

Fritillaria lusitanica subsp. *Oranensis*
(Pomel) Valdés

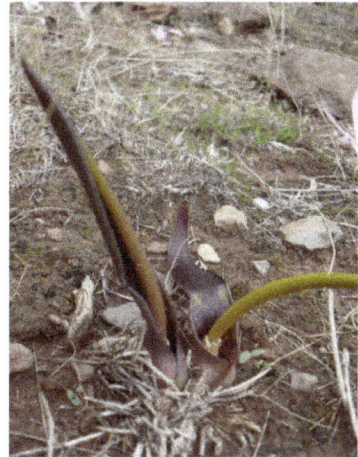

Biarum Bovei subsp. *dispar* (Schott.) Engler

Ophrys atlantica Munby

Iris xiphium L.

Carex hispida Willd.

Briza maxima L.

Pancratium foetidum var. *oranense* Pomel

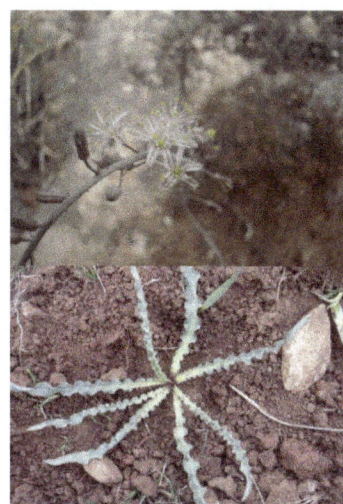

Drimia undulata Jacq.

Planche 2. Board color (phot. B. Babali).

Ajuga iva subsp. *pseudoiva* (DC.) Briq.

Anthyllis polycephala Desf.

Euphorbia nicræensis All.

Cistus ladanifer subsp. *Mauritianus*
Pau &Sennen

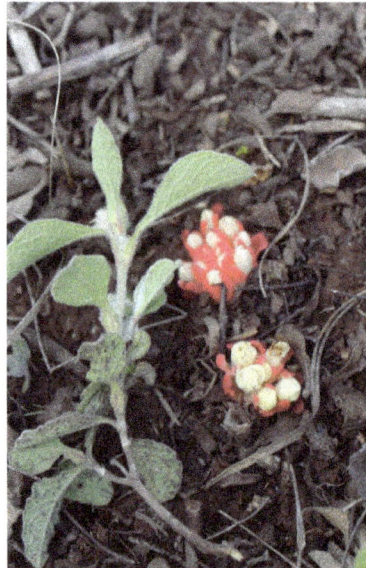

Cytinus hypocistis subsp. *clusii*
Nyman + *Cistus villosus* Pourret

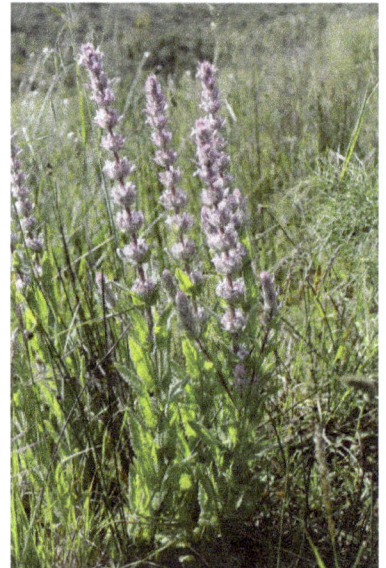

Nepeta tuberosa subsp. *reticulata* (Desf.)
Maire

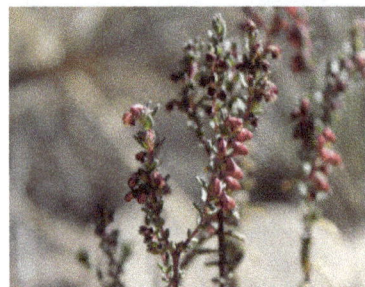

Odontites purpureus subsp.
purpureus (Desf.) G. Don fil.

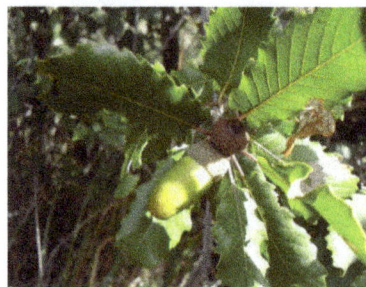

Quercus faginea subsp. *tlemcenensis*
(OC.) Maire et Weiller

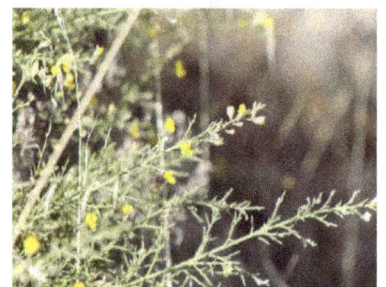

Stauracanthus boivinii
(Webb) Samp

13

Evaluation of habitat quality for selected wildlife species associated with island back channels

James T. Anderson[1], Andrew K. Zadnik[2], Petra Bohall Wood[3], Kerry Bledsoe[4]

[1]Division of Forestry and Natural Resources and Environmental Research Center, West Virginia University, Morgantown, USA;
[2]Wildlife and Fisheries Resources, Division of Forestry and Natural Resources, West Virginia University, Morgantown, USA
[3]US Geological Survey West Virginia Cooperative Fish and Wildlife Research Unit, West Virginia University, Morgantown, USA
[4]West Virginia Division of Natural Resources, Wildlife Resources Section, Fairmont, USA

ABSTRACT

The islands and associated back channels on the Ohio River, USA, are believed to provide critical habitat features for several wildlife species. However, few studies have quantitatively evaluated habitat quality in these areas. Our main objective was to evaluate the habitat quality of back and main channel areas for several species using habitat suitability index (HSI) models. To test the effectiveness of these models, we attempted to relate HSI scores and the variables measured for each model with measures of relative abundance for the model species. The mean belted kingfisher (*Ceryle alcyon*) HSI was greater on the main than back channel. However, the model failed to predict kingfisher abundance. The mean reproduction component of the great blue heron (*Ardea herodias*) HSI, total common muskrat (*Ondatra zibethicus*) HSI, winter cover component of the snapping turtle (*Chelydra serpentina*) HSI, and brood-rearing component of the wood duck (*Aix sponsa*) HSI were all greater on the back than main channel, and were positively related with the relative abundance of each species. We found that island back channels provide characteristics not found elsewhere on the Ohio River and warrant conservation as important riparian wildlife habitat. The effectiveness of using HSI models to predict species abundance on the river was mixed. Modifications to several of the models are needed to improve their use on the Ohio River and, likely, other large rivers.

Keywords: Habitat Suitability Index; Island Back Channel; Model Validation; Ohio River; Riparian Wildlife

1. INTRODUCTION

Over the past 200 years, industrialization and navigational projects have dramatically altered the Ohio River, USA, essentially changing it from a free-flowing river to a series of connected lakes [1]. Many of the islands on the river have been completely eliminated or severely degraded due to these activities [2].

The islands and associated back channels (*i.e.*, areas between an island and mainland not receiving comercial barge traffic) on the Ohio River are generally believed to provide critical habitat features for several wildlife species. However, few studies have quantitatively evaluated the quality of these areas for wildlife [3, 4]. This information is important to assist federal and state resource managers in determining what types of activities are compatible with conserving these areas for wildlife.

A widely accepted method to assess the habitat quality of an area for particular species is the use of Habitat Suitability Index (HSI) models [5,6]. The United States Fish and Wildlife Service (USFWS) originally developed these models as part of the Habitat Evaluation Procedures [7,8]. The models are based on measurements of structural variables necessary for important life requisites of individual species. Each variable is scored from 0 - 1, and then entered into a formula to calculate a final HSI score, also 0 - 1. Higher final HSI scores indicate higher habitat quality for that species [8]. The relation between HSI scores and carrying capacity is assumed to be positively linear [8].

The need to evaluate HSI model performance is commonly recognized [5,9-11]. The preferable means to accomplish this is by testing a model against population measures, such as species density or reproductive success [5,12]. Results of studies attempting such correlations have been inconsistent [9,13-15]. Thus, considering the demand for rapid assessment methods, further validation studies have been encouraged [5,11].

Over 158 HSI models are available from the USFWS in published form. For this study, we chose models for the belted kingfisher (*Ceryle alcyon*; total model) [16], great blue heron (*Ardea herodias*; total model) [17], common muskrat (hereafter muskrat; *Ondatra zibethicus*; freshwater model) [18], snapping turtle (*Chelydra serpentina*; total model) [19], and wood duck (*Aix sponsa*; brood-rearing component and winter model) [20]. Summaries of each of these models can be found in Zadnik [21]. These species are commonly associated with riparian areas and are known to exist along the Ohio River [22]. They also are representative of different taxonomic groups. Furthermore, we believed the variables measured in accordance with these models should provide a thorough representation of the overall habitat quality of the study area.

Our main objective with this study was to evaluate the potential wildlife value, based on HSI scores, of back channel and main channel areas associated with islands on the Ohio River. In addition, to test the effectiveness of these models on the Ohio River, and potentially other large rivers, we related HSI scores and the variables measured for each model with measures of relative abundance for the model species.

2. STUDY AREA AND METHODS

2.1. Study Area

This study was conducted on back channel and main channel areas associated with 10 islands (Captina, Paden, Williamson, Wells, Mill Creek, Middle, Buckley, Muskingum, Neal, and Buffington) on the Ohio River, West Virginia, USA (**Figure 1**). These islands are part of the Ohio River Islands National Wildlife Refuge [22]. They occurred between river kilometer 174 and 349 in 4 separate navigational pools: Hannibal, Willow Island, Belleville, and Racine. Island back channels varied from 0.92 - 39.9 ha in size, 0.63 - 4.13 km in length, and 0.03 - 0.22 km in width. Main channel areas associated with the islands had widths ranging from 0.24 - 0.42 km. A thorough description of the study area can be found in Zadnik [21].

2.2. Methods

2.2.1. HSI Variables

We attempted to evaluate each island back and main channel once during this study using the published HSI

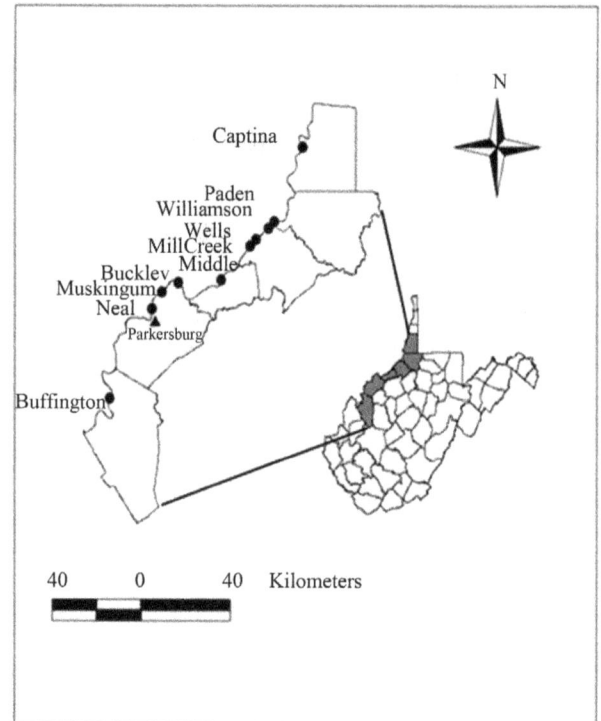

Figure 1. Location of 10 islands (indicated by points) between river kilometer 174 and 349, on the Ohio River, West Virginia, USA.

models for the belted kingfisher, great blue heron, muskrat, snapping turtle, and wood duck. However, Captina Island and Mill Creek Island could not be evaluated using the wood duck brood-rearing component because the back channels associated with those islands did not meet the minimum area required for broods (4 ha) [20]. Similarly, those islands could not be evaluated using the belted kingfisher model because the channels did not meet the minimum shoreline length (1 km) [16]. A total of 27 variables was measured and used in determining HSI scores (**Table 1**). All field measurements were taken from a boat May-September 2001 and 2002. Once measured, each variable was given a suitability index (SI) determined by the specific model. These indices were then used to determine indices for specific life requisites (e.g., cover or food components) and/or final HSI scores (**Table 1**).

We determined the percent canopy cover of aquatic vegetation, herbaceous vegetation, woody downfall, and overhanging vegetation ≤1 m from the water's surface (including coverage ≤15 m from shore) using the line intercept method [23] (**Table 1**). Transects, set at 250-m intervals, were established perpendicular to the flow of water and extending the width of each channel. The number of transects per island ranged from 4 - 19, although an equal number of transects were established per channel across each island. By extending each transect onto shore, the percent herbaceous canopy cover ≤10 m

Table 1. Variables (V) [a] measured and equations used to determine life components (based on variable suitability indices [SI]) [b], and habitat suitability index (HSI) scores for belted kingfisher, great blue heron, muskrat, snapping turtle, and wood duck, on the Ohio River, West Virginia, 2001-2002.

<u>Belted kingfisher [16]</u>

V2 = Water transparency (m) ≤ 15 m from shore	Water component = $(SIV2 \times SIV4 \times SIV5)^{1/3} \times SIV3$
V3 = percent surface obstruction ≤ 15 m from shore	Cover component = SIV6
V4 = percent water area ≤ 60 cm ≤ 15 m from shore	Reproduction component = SIV7
V5 = percent riffles	HSI = lowest of the 3 components
V6 = percent 25-m shoreline subsections with ≥1 potential perch	
V7 = distance from potential perch to potential nesting bank (m)	

<u>Great blue heron [17]</u>

V1 = Distance between potential nesting and foraging areas (km)	Foraging component = $SIV1 \times SIV2 \times SIV3$
V2 = Presence of potential foraging area?	Reproduction component = $(SIV1 \times SIV4 \times SIV5 \times SIV6)^{1/2}$
V3 = Disturbance-free zone ≥100 m at foraging area?	HSI = $(SIV1 \times SIV2 \times SIV3 \times SIV4 \times SIV5 \times SIV6)^{1/2}$
V4 = Presence of potential nesting area ≤250 m from water?	
V5 = Disturbance-free zone (250-m over land or 150-m over water) around potential nest area?	
V6 = Proximity of potential and active nest sites (km)	

<u>Muskrat [18]</u>

V2 = percent of year with surface water present	Cover component = $([SIV2 \times SIV3 \times SIV4]^{1/3} + SIV5)/2$
V3 = percent stream gradient	Food component = $(SIV6 + 2[SIV5])/2$
V4 = percent of channel with surface water	HSI = lowest of the 2 components
V5 = percent of channel with persistent emergent vegetation	
V6 = percent herbaceous canopy cover ≤10 m from water's edge	

<u>Snapping turtle [19]</u>

V1 = water temperature at mid-depth during summer (˚C)	Food component = $(SIV1 \times SIV2 \times SIV3)^{1/3}$
V2 = current velocity (cm/sec)	Winter cover component = $SIV4 \times SIV5$
V3 = percent canopy cover of aquatic vegetation	Reproduction component = SIV6
V4 = maximum water depth > maximum ice depth?	HSI = (food component × winter cover component ×
V5 = percent silt in substrate	reproduction component)$^{1/3} \times SIV7$
V6 = distance to small stream (km)	
V7 = distance to permanent water (km)	

<u>Wood duck [20]</u>

V4 = percent water surface with brood cover	Brood-rearing component = SIV4
V5 = percent water surface with winter cover	Winter habitat component = SIV5

---

[a]Variables followed by a question mark (?) have Boolean answers (yes, no), but all others are continuous variables. [b]Suitability Index (SI) values for a particular variable are scaled from 0 to 1.

from the water's edge was estimated. The percent of the channel with surface water present during typical minimum flow was estimated by measuring the portion of a transect extended across the channel that was not over open water.

We measured the percent silt by collecting 10 random substrate samples using an Ekman dredge and a posthole digger. Each sample was oven dried for >2 days, weigh-

ed, and passed through a 63-micron sieve using a sieve shaker. The material that passed through was weighed and divided by the total sample weight.

We measured average water temperature (˚C) at mid-depth using a temperature sensitive probe. One transect was randomly chosen per channel and temperature readings taken every 5 m. Water velocity was measured as the time it took a neutrally buoyant object (an orange) to travel a measured distance down the center of each channel [23]. Water transparency ≤15 m from shore was measured by randomly taking a Secchi disk reading along each transect [24].

The relation between perches and cover for the belted kingfisher model was based on the number of 25-m channel subsections/km that contained ≥1 suitable perch [16]. Along each channel shoreline, a randomly located 1-km section was divided into 40, 25-m subsections. We then evaluated each subsection to determine if it contained ≥1 potential perch. On-site inspections along with aerial photographs and topographic maps projected with Geographic Information System software (ArcView, hereafter GIS) [25] were used to determine the shortest distance from a subsection containing ≥1 suitable perch to a potential nesting bank [16]. We measured the soil texture of a potential nesting bank using the feel method [23].

We determined the percent of the water area ≤15 m from shore and ≤60 cm deep using a graduated rod [23] along each transect. The percent of riffles in each channel was estimated using a measuring tape [23]. We measured the percent stream gradient in each channel using GIS and topographic maps.

Whether or not a channel contained potential foraging areas for great blue herons was based on the proportion of transects along each channel that we believed met the necessary criteria (presence of a shallow water body with suitable prey population and foraging substrate) [17]. If the majority of transects met the criteria, then the channel was considered a potential foraging area. We assumed the river contained a suitable prey population [4,26]. We used on-site inspections and GIS to determine the following additional variables for the great blue heron model: presence of a disturbance-free zone ≥100 m around potential foraging areas, presence of potential nesting areas ≤250 m from open water, presence of a ≥250-m (land) or ≥150-m (water) disturbance-free zone around potential nest sites, distance between potential nest sites and foraging areas, and proximity between potential nest sites and active nest sites (**Table 1**). Potential sources of disturbance included houses, other buildings, and improved roads [17].

We measured the distance from each channel to a small stream using GIS and on-site inspections. Whether the maximum water depth was greater than the maximum ice depth was determined with winter field observations and local data (**Table 1**). It was not necessary to measure the distance to permanent water or percent of the year with surface water present, as our study area included the permanent water of the river.

2.2.2. Wildlife Population Measures

Belted kingfishers, great blue herons and wood ducks were surveyed with shoreline counts during daylight hours. In 2001, surveys were conducted once in the spring, twice in the summer, and once in the fall. In 2002, surveys were conducted once in the winter, once in the spring, 3 times in the summer, and once in the fall.

Snapping turtles were trapped using commercially available nylon hoop nets (1.5-m long × 0.9-m diameter) with 5-cm mesh (Memphis Net and Twine Company, Inc., Memphis TN, USA) baited with chopped fish [27]. Trapping was conducted once in fall 2001, 3 times in summer 2002, and once in fall 2002. Traps were set for 1 night before being moved. A trap-night was considered as 1 trap found completely intact the day after being set. We completed 376 trap-nights (mean = 75.1 trap-nights per period; SE = 2.5). Captured turtles were shell-notched [28] and released at the capture site.

Muskrat relative abundance was based on direct observation or the observation of appropriate sign (e.g., tracks, cuttings, middens, or burrows). Sign was searched for within a 15-m zone around each HSI transect. Each channel was examined once for appropriate sign. Direct observations within this zone were counted as they occurred. Complete methods for all investigations can be found in Zadnik [21].

2.2.3. Analyses

The relative abundance of belted kingfishers, great blue herons and wood ducks was calculated as the frequency (number/km of channel shoreline) of each observed during the surveys. To avoid double counting from the multiple summer periods, the summer period with the maximum number of individuals of each species observed per year was used in the analyses. Snapping turtle relative abundance was calculated as the number captured/trap-night. The relative abundance of muskrat was calculated as the frequency of transects on which the species was detected.

We used a complete block design analysis of variance (ANOVA) [29], with island as the block, to evaluate final HSI scores and scores for particular life requisites using channel and side (island side and mainland side) as independent variables. A similar ANOVA was used to evaluate species relative abundance using channel, side, year, season, and their interactions as independent variables. Model data and belted kingfisher, great blue heron, snapping turtle, and wood duck relative abundance data were log transformed to meet normal distribution assumptions. Muskrat relative abundance data were power transformed. An alpha level of 0.05 was used for all tests. Presented means and standard errors are untransformed to ease interpretation.

We used simple linear regression to relate final HSI and or life requisite scores with the measures of relative abundance. In this analysis, all data were log transformed. We used multiple regression using the backward variable selection procedure to relate the individual habitat variables for each model with measures of relative abundance. Muskrat model data were power transformed for this analysis. Data for the other models were log transformed. We tested all HSI models, components, and variables against species relative abundance data from the season(s) stated or implied by the model.

3. RESULTS AND DISCUSSION

3.1. Belted Kingfisher

Belted kingfisher density and frequency showed channel x season interactions ($F_{160} \geq 4.96$, $P \leq 0.030$); thus, further analyses were conducted within season. Relative abundance differed only during the summer, with the main channel having a greater mean abundance than the back channel (**Table 2**). The belted kingfisher total HSI model and water component had higher mean scores on the main channel than back channel (**Table 3**). The model and all components failed to show a linear relationship with abundance (**Table 4**). However, results of the multiple regression showed that water transparency and % riffles were positively correlated with belted kingfisher abundance and formed the best-fit model, based on the Akaike Information Criterion (AIC) value (**Table 5**).

These results suggest that the abundance of belted kingfishers on the main channel relative to the back channels is at least partially due to increased foraging opportunities. Kingfishers require clear water to locate potential prey [30]. Though turbidity often occurred throughout the study area, mean transparency was greater on the main channel [21]. Also, kingfishers are known to forage in areas of riffles [30, Brooks and Davis as cited in 16]. Extensive areas of riffles once occurred at the heads of the islands due to gravel beds [22]. However, past activities have degraded these areas, and the few remaining areas of riffles observed within the study area occurred at the mouths of tributaries along the main channel shoreline. The belted kingfisher model could likely be improved for use on the Ohio River by modifying the suitability indices given to water transparency and % riffles.

3.2. Great Blue Heron

Great blue heron abundance did not differ between back and main channels (**Table 2**). The great blue heron reproduction component of the HSI model had a higher mean score on the back channel than the main channel (**Table 3**). A linear association was found between the total HSI model and abundance (**Table 4**). However, no correlation was found between the individual model variables and abundance. The best-fit model included Variable 5 (presence of a disturbance-free zone around potential nest sites) and Variable 6 (proximity of potential and active nest sites; **Tables 1** and **5**).

These results suggest the importance of areas free from human disturbance for great blue herons [17,31]. On the Ohio River, this condition is met most readily on the islands. Herons prefer isolation particularly during nesting [17,32,33]. Indeed, all active nests observed within our study area occurred on islands. The back channels may improve nesting conditions provided on the islands both due to the narrowness of most back

Table 2. Relative abundance of 5 species on back and main channels, averaged across years, associated with 10 islands on the Ohio River, West Virginia, USA, 2001-2002[a].

| | | Channel | | | | | | |
| | | Back | | Main | | | | |
Species	Parameter[b]	$\bar{x}$	SE	$\bar{x}$	SE	F^c	df	P
Belted kingfisher	Spring frequency	0.17	0.05	0.18	0.06	0.00	1, 21	0.993
	Summer frequency	0.35	0.09	0.68	0.10	8.97	1, 21	0.007
	Total frequency	0.26	0.05	0.43	0.06			
Great blue heron	Total frequency	0.52	0.13	0.29	0.05	0.64	1, 27	0.431
Muskrat	Frequency of occurrence	0.68	0.05	0.19	0.04	48.50	1, 27	<0.001
Snapping turtle	no. captured/trap-night	0.17	0.03	0.03	0.01	28.43	1, 27	<0.001
Wood duck	Summer frequency	2.71	0.48	0.78	0.22	14.03	1, 21	0.001
	Fall frequency	1.35	0.44	0.90	0.80	9.81	1, 27	0.004

[a]Belted kingfisher and wood duck summer abundance were only estimated for 8 islands. [b]Frequency = number/km of shoreline. [c]No test indicates test was not performed, due to interaction with season ($P \leq 0.05$).

Table 3. Habitat suitability index (HSI) model scores, and the scores of model components, for 5 species, on back and main channels associated with 10 islands on the Ohio River, West Virginia, USA, 2001-2002[a].

Model	Component	Channel						
		Back		Main				
		$\bar{x}$	SE	$\bar{x}$	SE	F	df	P
Belted kingfisher	Water	0.34	0.01	0.44	0.01	113.06	1, 21	<0.001
	Cover	0.52	0.03	0.47	0.02	1.56	1, 21	0.225
	Reproduction	0.93	0.02	0.95	0.02	0.52	1, 21	0.480
	Total HSI	0.34	0.01	0.42	0.01	54.80	1, 21	<0.001
Great blue heron	Foraging	0.60	0.11	0.55	0.11	0.23	1, 27	0.632
	Reproduction	0.58	0.10	0.39	0.10	5.49	1, 27	0.027
	Total HSI	0.40	0.10	0.33	0.10	0.65	1, 27	0.426
Muskrat	Cover	0.60	0.00	0.60	0.00	0.82	1, 27	0.373
	Food	0.63	0.01	0.55	0.02	21.23	1, 27	<0.001
	Total HSI	0.58	0.01	0.54	0.01	9.89	1, 27	0.004
Snapping turtle	Food	0.01	0.01	0.07	0.02	10.59	1, 27	0.003
	Winter cover	0.34	0.01	0.30	0.02	6.62	1, 27	0.016
	Reproduction	0.98	0.00	0.98	0.00	0.04	1, 27	0.849
	Total HSI	0.04	0.02	0.16	0.04	9.51	1, 27	0.005
Wood duck	Brood-rearing	0.08	0.01	0.02	0.00	106.79	1, 21	<0.001
	Winter habitat	0.10	0.01	0.02	0.01	67.02	1, 27	<0.001

[a]Belted kingfisher model and wood duck brood-rearing component could only be applied to 8 islands due to model parameters.

Table 4. Linear regression models of relative abundance of 5 species modeled with habitat suitability index (HSI) models and model components in areas associated with 10 islands on the Ohio River, West Virginia, 2001-2002[a].

Species[b]	Component	Equation	t	P	df	R^2	AIC
Belted kingfisher	Water	Y = 0.87(water component) − 0.56	0.88	0.384	30	0.02	−71.57
	Cover	Y = −0.31(cover component) − 0.07	−0.54	0.590	30	0.01	−71.06
	Reproduction	Y = 0.47(reproduction component) − 0.66	0.63	0.534	30	0.01	−71.16
	HSI	Y = 0.40(HSI) − 0.37	0.39	0.700	30	0.00	−70.91
Great blue heron	Foraging	Y = 0.11(foraging component) − 1.27	0.33	0.747	34	0.00	2.90
	Reproduction	Y = 0.79(reproduction component) − 1.56	2.23	**0.032**	34	0.13	−1.90
	HSI	Y = 0.89(HSI) − 1.51	2.40	**0.022**	34	0.14	−2.64
Muskrat	Cover	Y = −33.89(cover component) + 19.16	−0.94	0.355	38	0.02	5.78
	Food	Y = 0.74(food component) + 0.79	4.67	**< 0.001**	38	0.36	−203.42
	HSI	Y = 3.73(HSI) − 2.23	3.62	**0.001**	38	0.26	−87.85
Snapping turtle	Food	Y = −1.60(food component) − 1.74	−1.20	0.240	38	0.04	−40.91
	Winter cover	Y = 2.90(winter cover component) − 2.73	2.21	**0.033**	38	0.11	−44.27
	Reproduction	Y = 0.07(reproduction component) − 1.88	0.02	0.988	38	0.00	−39.43
	HSI	Y = −0.71(HSI) − 1.74	−1.25	0.219	38	0.04	−41.04
Wood duck	Brood-rearing	Y = 9.04(brood-rearing component) + 0.11	2.95	**0.006**	30	0.22	22.35
	Winter cover	Y = 8.07(winter cover component) − 3.40	1.00	0.325	34	0.03	72.87

[a]Belted kingfisher model and wood duck brood-rearing component could only be used with 8 islands due to model parameters. [b]Waterbird relative abundance = number observed/km of shoreline; muskrat relative abundance = frequency of sign and direct observations; snapping turtle relative abundance = number captured/trap-night.

Table 5. Backwards selection multiple regression models of relative abundance of 5 species modeled with habitat suitability index (HSI) model variables in areas associated with 10 islands on the Ohio River, West Virginia, 2001-2002[a].

Species[b]	Component	Equation	F	P	R^2	df	AIC
Belted kingfisher	Total HSI	Y = −0.39(water transparency) + 1.20 (% riffles) + 0.13	6.57	**0.004**	0.31	2, 29	−80.71
Great blue heron[c]	Total HSI	If SIV5 = 0, Y = −0.59 − 0.02 (proximity of potential to active nests) − 0.56	4.40	**0.020**	0.22	2, 32	−17.02
		If SIV5 = 1, Y = −0.02 (proximity of potential to active nests) − 0.56					
Muskrat	Total HSI	Y = 0.01(% herbaceous canopy cover) − 0.70	27.05	**<0.001**	0.42	1, 38	−113.35
Snapping turtle	Total HSI	Y = 0.02 (% silt) − 1.66	5.32	**0.027**	0.12	1, 38	−89.46
Wood duck	Brood-rearing	Y = 0.19(brood cover) + 0.10	9.97	**0.004**	0.25	1, 30	−23.39
	Winter cover	Y = 0.34(winter cover) − 6.36	1.17	0.250	0.04	1, 34	116.54

[a]Belted kingfisher model and wood duck brood-rearing component could only be used with 8 islands due to model parameters. [b]Waterbird relative abundance = number observed/km of shoreline; muskrat relative abundance = frequency of sign and direct observations; snapping turtle relative abundance = number captured/trap-night. [c]SIV5 = Disturbance-free zone (250-m over land or 150-m over water) around potential nest area (Yes = 1, No = 0)?

channels compared to the main channel, as well as greater woody canopy closure along the back channel island shorelines relative to the main channel island shorelines [21]. These conditions may provide not only disturbance-free zones but also potentially new nesting sites.

3.3. Muskrat

Muskrat relative abundance was greater on the back channel than main channel (**Table 2**). The muskrat total model and food component had higher mean scores on the back channel than main channel (**Table 3**). There were linear associations between the total HSI score, the food component and relative abundance (**Table 4**). Results of the multiple regression showed a positive correlation between % cover and relative abundance (**Table 5**).

According to the HSI model, muskrat cover can be provided by persistent emergent vegetation within the river channel itself or herbaceous canopy cover along the shore [18]. In our study area, a complete lack of persistent emergent vegetation meant that cover was primarily provided by shoreline herbaceous vegetation, which was most abundant on the back channel island shoreline. Additional cover was likely provided by woody debris and undercut banks [34], conditions provided in more abundance on the back channels (measured as % brood cover for the wood duck model) [21] but not included as variables in the muskrat model.

3.4. Snapping Turtle

Snapping turtles had a greater mean relative abundance on the back channel than the main channel (**Table 2**). However, the snapping turtle food component and total HSI model had higher scores on the main channel

than back channel (**Table 3**). Conversely, scores for the winter cover component were higher on the back channel than the main channel. There was a linear association between the winter cover component and relative abundance (**Table 4**). Similarly, % silt in the substrate, one of the variables included in the winter cover component, was positively correlated with relative abundance (**Table 5**)

The obvious discrepancy between snapping turtle total HSI and food component scores compared with relative abundance is primarily due to the weight given aquatic vegetation by the model. The model assumes that a complete absence of aquatic vegetation removes all food value for the area being evaluated [19]. Hence, since the aquatic vegetation observed within our study area occurred primarily on the main channel [21], that area was considered higher quality habitat as compared with back channels. The model states that snapping turtles are primarily carnivorous in early spring, and then switch to a more herbivorous diet later in the spring and summer [19]. If that was the case in our study area, we should have found a higher abundance of turtles on the main channel than the back channel, particularly in the summer when the few areas of aquatic vegetation were visible (A. Zadnik personal observation). While trapping was not conducted in early spring, back channels had a higher abundance of turtles throughout the 2 seasons in which trapping was conducted, summer and fall. Furthermore, main channel turtle abundance stayed consistently low during those seasons. Due to the overall scarcity of aquatic vegetation on the Ohio River, it is probable that snapping turtles remain primarily carnivorous throughout the year, with back channels likely providing more foraging opportunities than the main channel. The HSI model would be more effective for use on the Ohio River if it put less weight on abundance of aquatic vegetation. However, additional research is needed to evaluate food preferences and identify any seasonal shifts in area use.

The one variable that was positively correlated with relative abundance, % silt in the substrate, reflects the value of back channels as overwintering sites for snapping turtles. Snapping turtles are known to escape from harsh winter conditions by burying themselves deep in the substrate [35]. The model assumes that substrate with a greater % of silt is of higher quality for the species [19]. Mean % silt was greater on the back channel than main channel [21]. Of course, while silt in the substrate appears to positively affect snapping turtles, suspended silt can have negative impacts on the overall habitat quality for other species [36].

Though significant, the correlation between % silt and relative abundance is rather weak ($r^2 = 0.12$; **Table 5**), indicating additional variability not accounted for by the model. Current velocity was another variable included in the model (**Table 1**). Snapping turtles are typically associated with slower waters [35,37]. While it appeared that mean velocity was less on the back channel than main channel [21], our analyses did not find a correlation between velocity and relative abundance. Even with all model variables included, the correlation with relative abundance remained weak ($r^2 = 0.18$; [21]). That still leaves 82% of the variation in turtle relative abundance unaccounted for by the model. Additional research is needed to identify other factors contributing to turtle abundance.

3.5. Wood Duck

Wood duck summer relative abundance (corresponding to the brood-rearing component) and fall abundance (corresponding to the winter model) was greater on the back channel than the main channel (**Table 2**). There were no wood ducks observed during the winter survey.

Both wood duck models had higher scores on the back channel than the main channel (**Table 3**). In addition, the wood duck brood-rearing component showed a linear correlation with relative abundance (**Table 4**). Not surprisingly, the only variable in the model, % brood cover was found to be positively correlated with relative abundance (**Table 5**).

Brood cover is defined as woody downfall, emergent vegetation, and overhanging branches ≤1 m from the water's surface [20]. With the exception of emergent vegetation, which failed to occur during the sampling, these characteristics were >3 times more prevalent on the back channel than main channel [21]. The importance of woody cover for wood ducks in riverine systems is commonly recognized. Cottrell *et al.* [38], in a Tennessee study, found that wooded shorelines and fallen trees were 2 features that determined use of areas by ducklings.

Similarly, Minser [39] found that wood duck brood density (number/km) was positively correlated with woody debris in the water and large overhanging trees. In general, woody overhead cover is considered an essential component of good brood habitat [40].

Nonbreeding habitat for wood ducks is considered similar to high quality brood habitat [41-43], a common feature being the presence of woody vegetation [44]. Hence, due to the lack of emergent vegetation in our study area, winter cover was considered identical to brood cover. The lack of correlation between the winter model and relative abundance is likely due to the absence of wood ducks during the winter survey. This is not surprising, however, as the West Virginia portion of the Ohio River is at the far northern edge of the species wintering range [45]. Furthermore, although the back channels in particular were used for brood-rearing [21], the overall lower quality wood duck habitat on the river compared to other wetland types [46], may have limited use of the study area by this species.

4. MANAGEMENT IMPLICATIONS

With the continuing pressure to develop Ohio River island back channels and back channel mainland shorelines, the need to better understand the value of these areas for wildlife is clear [27]. We found that certain wildlife species appeared to be more abundant in back channel than main channel areas due to particular habitat characteristics not as prevalent on the main channel, including herbaceous vegetation, woody downfall and overhanging limbs, and a substrate with high silt content. Other back channel characteristics potentially benefiting wildlife include protection from human disturbance and slower water current. Conversely, species on the main channel appeared to benefit from overall less turbid water and presence of riffles. Due to the uniqueness and rarity of back channels on the Ohio River, their conservation and restoration are warranted [21,27].

We found that the total HSI model for the muskrat and the brood-rearing model for the wood duck can both be used to successfully predict habitat quality for those species on the Ohio River and possibly similar large rivers. In addition, the great blue heron total model can be used with limited success. Our results indicate the snapping turtle total HSI model needs to be modified to better account for the likely year-round carnivorous habits of this species. As published, the winter cover component of the model can be used to some extent to predict use of the river by snapping turtles. However, further research is needed to identify other variables affecting snapping turtle relative abundance. Finally, our results indicate the belted kingfisher model needs to be modified to better predict use of the river, possibly by changing the suitability indices to include measures of water transparency and riffle area.

5. ACKNOWLEDGEMENTS

We thank the West Virginia Division of Natural Resources, United States Environmental Protection Agency, United States Geological Survey West Virginia Cooperative Fish and Wildlife Research Unit, and the West Virginia University Davis College of Agriculture, Natural Resources, and Design for funding, resources, and support. We are grateful to the late G. Seidel for his assistance and expertise with statistical analysis and W. S. Kordek for project support. We thank M. Christ, W. Kordek, J. Pitchford, and S. Welsh for comments on this manuscript. We also thank the many individuals who assisted in the field and laboratory. Turtle trapping and handling protocols were approved by the West Virginia University Animal Care and Use Committee (protocol #001209). This is scientific article number 3160 of the West Virginia University Agricultural and Forestry Experiment Station. Use of trade names or products does not constitute endorsement by the U.S. Government.

REFERENCES

[1]   Frost, S.L. and Mitsch, W.J. (1989) Resource development and conservation history along the Ohio River. *Ohio Journal of Science*, **89**, 143-152.

[2]   United States Army Corps of Engineers. (2000) Ohio River ecosystem restoration program, integrated decision document and environmental assessment. United States Army Corps of Engineers, Great Lakes and Ohio River Division, Cincinnati.

[3]   Thorp, J.H. (1992) Linkage between islands and benthos in the Ohio River, with implications for riverine management. *Canadian Journal of Fisheries and Aquatic Sciences*, **49**, 1872-1882.

[4]   Millard, M.J. (1993) Nearshore habitat use by larval fishes near two islands in the upper Ohio River. Dissertation, West Virginia University, Morgantown.

[5]   Brooks, R.P. (1997) Improving habitat suitability index models. *Wildlife Society Bulletin*, **25**, 163-167.

[6]   Morrison, M.L., Marcot, B.G. and Mannan, R.W. (1998) Wildlife habitat relationships: concepts and applications. University of Wisconsin Press, Madison.

[7]   United States Fish and Wildlife Service. (1980) Habitat evaluation procedures (HEP). United States Fish and Wildlife Service, ESM 102.

[8]   United States Fish and Wildlife Service. (1981) Standards for the development of habitat suitability index models. United States Fish and Wildlife Service, Release No. 1-81, 103 ESM 103.

[9]   Thomasma, L.E., Drummer, T.D. and Peterson, R.O. (1991) Testing the habitat suitability index model for the fisher. *Wildlife Society Bulletin*, **19**, 291-297.

[10] Bender, L.C., Roloff, G.J. and Haufler, J.B. (1996) Evaluating confidence intervals for habitat suitability models. *Wildlife Society Bulletin*, **24**, 347-352.

[11] Roloff, G.J. and Kernohan, B.J. (1999) Evaluating reliability of habitat suitability index models. *Wildlife Society Bulletin*, **27**, 973-985.

[12] Prosser, D.J. and Brooks, R.B. (1998) A verified habitat suitability index for the Louisiana waterthrush. *Journal of Field Ornithology*, **69**, 288-298.

[13] Clark, J.D. and Lewis, J.C. (1983) A validity test of a habitat suitability index model for clapper rail. *Proceedings of the Annual Conference of the Southeastern Association of Fish and Wildlife Agencies*, **37**, 95-102.

[14] Cook, J.G. and Irwin, L.L. (1985) Validation and modification of a habitat suitability model for pronghorns. *Wildlife Society Bulletin*, **13**, 440-448.

[15] Robel, R.J., Fox, L.B. and Kemp, K.E. (1993) Relationship between habitat suitability index values and ground counts of beaver colonies in Kansas. *Wildlife Society Bulletin*, **21**, 415-421.

[16] Prose, B.L. (1985) Habitat suitability index models: Belted kingfisher. United States Fish and Wildlife Service Biological Report 82 (10.87).

[17] Short, H.L. and Cooper, R.J. (1985) Habitat suitability index models: Great blue heron. United States Fish and Wildlife Service Biological Report 82 (10.99).

[18] Allen, A.W. and Hoffman, R.D. (1984) Habitat suitability index models: Muskrat. United States Fish and Wildlife Service, FWS/OBS-82/10.46.

[19] Graves, B.M. and Anderson, S.H. (1987) Habitat suitability index models: Snapping turtle. United States Fish and Wildlife Service Biological Report 82 (10.141).

[20] Sousa, P.J. and Farmer, A.H. (1983) Habitat suitability index models: Wood duck. United States Fish and Wildlife Service, FWS/OBS-82/10.43.

[21] Zadnik, A.K. (2003) Wildlife use and habitat quality of back channel areas associated with islands on the Ohio River, West Virginia. Thesis, West Virginia University, Morgantown.

[22] United States Fish and Wildlife Service. (2002) Ohio River Islands National Wildlife Refuge, comprehensive conservation plan. United States Fish and Wildlife Service, Parkersburg.

[23] Hays, R.L., Summers, C. and Seitz, W. (1981) Estimating wildlife habitat variables. United States Fish and Wildlife Service, FWS/OBS-81/47.

[24] Orth, D.J. (1983) Aquatic habitat measurements. In: Nielsen, L.A. and Johnson, D.L., Eds., *Fisheries techniques*. American Fisheries Society, Bethesda, 61-84.

[25] ESRI, ARC/INFO. (1994) Version 7.01. Environmental Systems Research Institute, Redlands.

[26] Van Hassel, J.H., Reash, R.J., Brown, H.W., Thomas, J.L. and Matthews Jr., R.C. (1988) Distribution of upper and middle Ohio River USA fishes, 1973-1985 I., associations with water quality and ecological variables. *Journal of Freshwater Ecology*, **4**, 441-458.

[27] Zadnik, A.K., Anderson, J.T., Wood, P.B. and Bledsoe, K. (2009) Wildlife use of back channels associated with islands on the Ohio River. *Wetlands*, **29**, 543-551.

[28] Cagle, F.R. (1939) A system of marking turtles for future identification. *Copeia*, **1939**, 170-173.

[29] User's Guide. Version   SAS Institute. (1990) SAS/STAT 6, 4th Edition, Vol. 2, SAS Institute, Cary.

[30] Davis, W.J. (1982) Territory size in *Megaceryle alcyon* along a stream habitat. *Auk*, **99**, 353-362.

[31] Vance, J.A, Angus, N.B. and Anderson, J.T. (2012) Riparian and riverine wildlife response to a newly created bridge crossing. *Natural Resources*, **3**, 213-228.

[32] Gibbs, J.P., Woodward, S., Hunter, M.L. and Hutchinson, A.E. (1987) Determinants of great blue heron colony distribution in coastal Maine. *Auk*, **104**, 38-47.

[33] Watts, B.D. and Bradshaw, D.S. (1994) The influence of human disturbance on the location of great blue heron colonies in the lower Chesapeake Bay. *Colonial Waterbirds*, **17**, 184-186.

[34] Errington, P.L. (1937) Habitat requirements of stream-dwelling muskrats. *Transactions of the North American Wildlife Conference*, **2**, 411-416.

[35] Ernst, C.H., Barbour, R.W. and Lovich, J.E. (1994) Turtles of the United States and Canada. Smithsonian Institution Press, Washington DC.

[36] Bridges, C.M. and Semlitsch, R.D. (2002) Linking xenootics to amphibian declines. In: Lannoo, M., Ed., *Status and Conservation of United States Amphibians, Conservtion Essays*, Vol. 1, University of California Press, Berkeley, in press.

[37] DonnerWright, D.M., Bozek, M.A., Probst, J.R. and Anderson, E.M. (1999) Response of turtle assemblage to environmental gradients in the St., Croix River in Minnesota and Wisconsin, U.S.A. *Canadian Journal of Zoology*, **77**, 989-1000.

[38] Cottrell, S.D., Prince, H.H. and Padding, P.I. (1990) Nest success, duckling survival, and brood habitat selection of wood ducks in a Tennessee riverine system. In: Fredrickson, L.H., Burger, G.V., Havera, S.P., Graber, D.A., Kirby, R.E. and Taylor, T.S., Eds., *Proceedings of the North American Wood Duck Symposium*, St. Louis, 20-22 February 1988, 191-197.

[39] Minser, W.G. (1993) The relationship of wood duck brood density to river habitat factors. *Proceedings of the Annual Conference of the Southeastern Fish and Wildlife Agencies*, **47**, 112-122.

[40] Webster, C.G. and McGilvrey, F.B. (1966) Providing brood habitat for wood ducks. In: Trefethen, J.B., Ed., *Wood Duck Management and Research: A Symposium*, Wildlife Management Institute, Washington DC, 70-75.

[41] McGilvrey, F.B. (1968) A guide to wood duck production habitat requirements. Bureau of Sport Fish and Wildlife Resources Publication, Washington DC, 60.

[42] Gilmer, D.S., Kirby, R.E., Ball, I.J. and Riechmann, J.H. (1977) Post-breeding activities of mallards and wood ducks in north-central Minnesota. *Journal of Wildlife Management*, **41**, 345-359.

[43] Haramis, G.M. (1990) The breeding ecology of the wood duck: A review. In: Fredrickson, L.H., Burger, G.V., Havera, S.P., Graber, D.A., Kirby, R.E. and Taylor, T.S., Eds., *Proceedings of the North American Wood Duck Symposium*, 20-22 February 1988, St. Louis, 45-60.

[44] Hein, D.A. and Haugen, A.O. (1966) Autumn roosting slight counts as an index to wood duck abundance. *Journal of Wildlife Management*, **30**, 657-668.

[45] Bellrose, F.C. (1980) Ducks, geese and swans of North America. Stackpole Books, Harrisburg.

[46] Balcombe, C.K. (2003) An evaluation of vegetative communities and wildlife habitat use in mitigated and natural wetlands of West Virginia. West Virginia University, Morgantown.

14

Establishing landuse/cover change patterns over the last two decades and associated factors for change in semi arid and sub humid zones of Tanzania

Amos Enock Majule

Institute of Resource Assessment (IRA), University of Dar es Salaam, Dar es Salaam, Tanzania

ABSTRACT

This study investigated landuse cover change patterns and established potential environmental and social factors that have contributed to changes in two zones namely sub humid and semi-arid found in southern highland and central parts of Tanzania respectively. The overall objective was to understand change patterns; the process evolves and clearly isolates various factors that have contributed to the changes over the last 20 years. A total of four villages, two in each zone were involved whereby historical land use cover changes were analysed using remote sensing techniques. To do so satellite imageries for 1991 and 2011 and those of 1986 and 2009 for sub humid and semi-arid zones respectively. Factors for changes were established through focus group discussions (FDGs) with a total of 80 participants (20 per village) and household (HH) interviews subjected to 10% of the total number of HH per village. Both woodlands and bush lands decreased in the expense of mixed farming in both sub humid and semi-arid zones to a maximum of 121% and 146.8% respectively. Wetland farming also increased particular in sub humid zone. In general, both environmental and social factors were found to have contributed to LUCC in various magnitudes in both zones. Such observed change on landuse will continue and it is recommended that there is a need to have in place and implement proper landuse plan also have capacity building programs on climate and land management issues for both livelihood and ecosystem sustainability need to in place.

Keywords: Agriculture; Climate Change; Landuse; Mixed Farming; Woodlands; Vinyungu

1. INTRODUCTION

The majority of communities living in rural areas of Sub-Sahara Africa (SSA) depend their livelihood on the utilization of natural resources particularly from forest, woodlands, wetlands and agricultural land for crops and livestock production [1]. Such resources have been dwindling in many places but in some places where sustainable management are in place and not transformed to other emerging new landuse (LU) both non wood and wood products play a significant role in terms of providing both products and other services [2,3]. A number of studies in the region have reported for example how different livelihood system interacted with natural resources in the past where natural resources were not under pressure and what is recently happening [4,5]. For example over the last 50 years ago in East African ecological gradients of for example Mounts Kenya and Kilimanjaro natural resources were used in a sustainable way because population was low in such areas and economical pressure was low [6,7]. Also during that time much of the land was suitable to support various crops and livestock production and hence there was no much need to acquire new land. Apart from forest resources other landuse types such as woodlands, wetlands, crop lands, and bush lands supported different ecosystems which then to a large extent maintained both social and natural systems [3].

Over the last 30 years a situation has changed significantly and this is likely to continue changing if some measures are not taken. Due to changing landuse in different agro-ecological systems in Tanzania, different new landuse cover (LUC) types have emerged with various implications to both community livelihoods and upon environment [8,9]. Various reasons have been proposed for observed changes but yet more studies are still needed since factors for change are sometime spatially determined. For example there are limited reports which show that long term climate change may result to landuse cover change due to changes in rainfall patterns and rise in temperature favouring other land uses to emerge with consequences on community livelihoods [10]. On the other hand, [11] made an attempt to link the impact of various environmental factors including climate change and landuse and findings showed that the link is somehow indirect which suggests further investigation.

On the other hand, in Tanzania a study by [12] has revealed that there has been an increase in cultivated land due to population increase and investment in agriculture sector. In addition, a study by [13] found the root causes of the land use and cover changes that have taken place on the southern slope of Mount Kilimanjaro are many and multifaced. They include demographic factors, colonial and post-independence government policies, institutional factors, legislation, as well as socio-cultural, economic and environmental factors. [14] added two more factors on the causes of agricultural land expansion in East Africa namely; access to markets and changes in land tenure arrangement. Further more studies by William [15,16] revealed that liberalized economic relations, such as free trade agreements and economic globalization change people's relations to the physical environment hence induce changes in LU. The intensity and scale of land use change has increased drastically in recent decades, partly due to expansion of farmlands, settlements and climate stresses that force agro-pastoral communities to search for virgin pastures and croplands.

Basing on these trends, land use cover change and degradation of the natural resources such as land, water, forests, woodlands, grasslands and wildlife in Tanzania have been common in various ecosystems in particularly slopes of Kilimanjaro [13]; Usangu basin and slopes of Mount Rungwe in the volcanic region [8,9,17] and thus exposing rural communities to vulnerable ecosystems [18,19]. Although many studies have been able to establish changing landuse cover types at a wide scale in various agro ecological systems. Very few of them have been able to isolate various biophysical and social factors that have contributed to such changes and where the changes observed have positive impacts. The overall aim of this study was to assess and analyse different landuse cover types over the last 2 decades in two agro ecological zones of Tanzania characterised by semi humid and semi arid conditions. The study also made an attempt to isolate both biophysical and social factors that have played a significant role in causing landuse cover change.

2. METHODOLOGY

2.1. Description of the Study Area

The study was conducted in two climatic zones namely sub humid and semi arid zones. In sub humid zone the study was conducted in Mufindi district in Iringa region which forms the part of the southern highland zone of Tanzania (**Figure 1**). Mufindi is one of four districts in Iringa Region located in Southern highlands of Tanzania lying between 8°00' and 9°15' South and longitude 34°35' to 35°55' East. The district is situated about 80 km from Iringa Municipality and is bordered by Iringa district to the North, Morogoro region to the East, Njombe region to the South and Mbeya region to the West. The area received a minimum and maximum rainfall of 1000 and 2000 mm/yr respectively and both average annual minimum and annual temperatures remains to be 16°C and 23°C on average respectively [20].

In the semi arid zone this study was carried out in Nzega district in Tabora region, Tanzania in Undomo and Mbutu villages (**Figure 2**). Nzega is one of the six districts of Tabora region, which covers the area of 9226 km^2, and lies between longitudes 36°30' and 33°30'E and latitudes 3°45' and 5°00'S. The District is characterized by unimodal type of rainfall, which falls between November and April. The distributions are unreliable with annual average usually less than 700 mm. The temperature ranges from 28°C to 30°C. The highest temperature is experienced in October just before onset of rainfall. A dry spell between normally occurs between mid January and February [19].

2.2. Historical Analysis of Land Use Cover Changes

Processing and analysis of remote sensing and GIS data on establishing change detection (land cover or land use changes in the sample study wards involved five key procedures: image interpretation, image classification, image digitization, overlying analysis and generalization of error matrix. In this case overlaying analysis was employed so as to derive new information from two land cover layer covering the same area. The process was done by using satellite images available of year 1986 and 2009 for semi arid zone and that of 1991 and 2011 for sub humid zone to discover land cover changes based on the computed area. Change detection for each landuse category was then performed using standard procedure used by [8].

2.3. Establishing Factors for Landuse Cover Change

In order to establish factors of change and implications of new emerging land upon environment and community

Establishing landuse/cover change patterns over the last two decades and associated factors for change in semi arid and sub humid zones of Tanzania

111

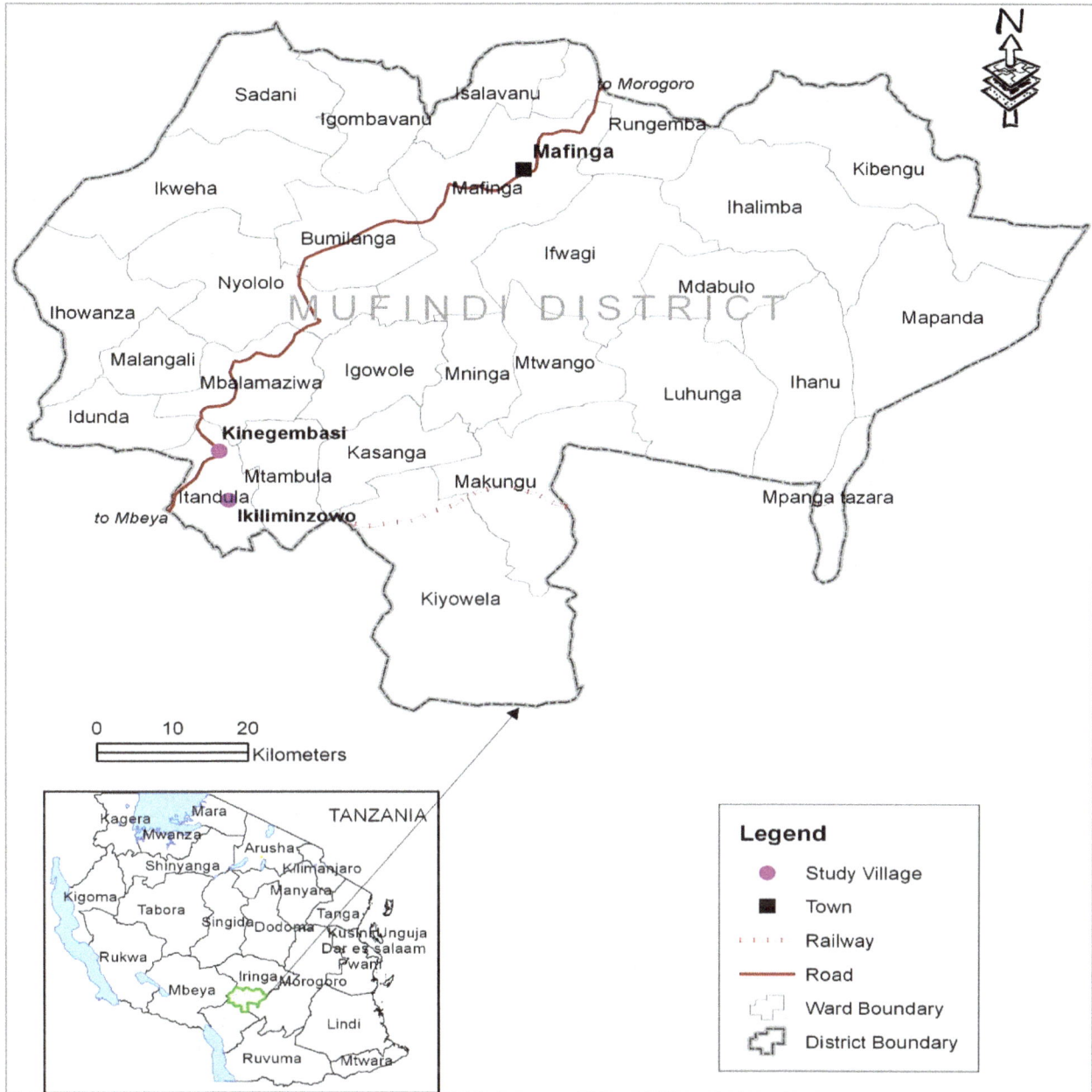

Figure 1. Location of study area and villages in sub humid zone, Tanzania.

livelihoods data were collected through literature review and through focus group discussions in respective study villages using a checklist administered to a total of 20 people per village strategically selected. This was supplemented by information collected through household interviews in respective zones per villages presented in **Tables 1** and **2**. In this case a checklist with both open and closed questions was adopted.

A similar approach was also used by [18] when studying land degradation in semi arid area of central Tanzania. Where the information was not satistisfactory on certain issues discussion with key informants was also employed. Both qualitative and quantitative data analysed using the

Statistical Package for Social Science (SPSS) software version 16.0 and spread sheet and results were presented in different forms including Tables and Bar charts.

3. RESULTS AND DISCUSSION

3.1. Historical Analysis of Landuse Cover Changes in Two Zones

3.1.1. Sub-Humid Case Study Zone

Major landuse/cover types found in the study area is presented in **Tables 3** and **4** for Kimilinzowo and Kinegembasi villages respectively. In Kimilinzowo (**Table 3**) results indicate that there has been a substantial

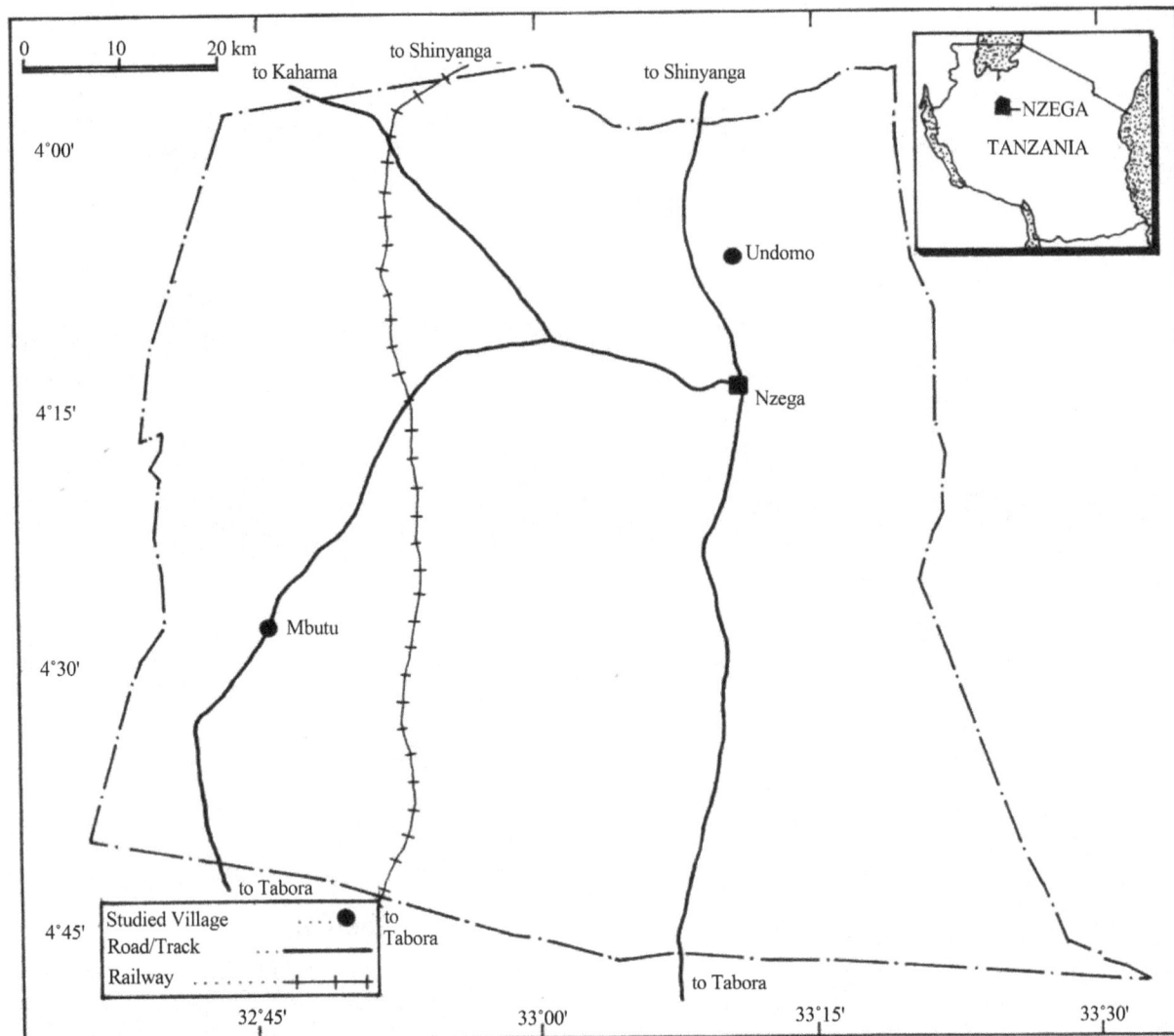

Figure 2. Location of study area and villages in semi arid zone.

Table 1. Distribution of household sample size in semi arid zone.

Village name	Number of households	Number of respondents	% of sample
Mbutu	512	52	10
Undomo	570	57	10
Total	**1082**	**109**	**10**

Table 2. Distribution of household sample size in sub-humid zone.

Village name	Number of households	Number of respondents	% of sample
Kinegembasi	560	50	9.2
Kimilinzowo	762	70	9.2
Total	1322	120	18.4

Source: Ward office (2012).

change of land-use/cover in the following manner; bush land with scattered crop land has decreased by 16%; bare soil has declined by 3%. On the other hand closed woodland has decreased by 11% while cultivated land has increased by 121%. Other landuse cover types in particular grassland with scattered crop land have increased by 5%, open grassland has decreased by 6% and open woodland has increased by 10% while woodland with scattered cropland has also decreased by 10%. In general there is no natural forest detected suggesting that deforestation occurred in the area beyond 1991 and what is happening is dynamic change of other landuse cover types on the expense of expanding crop land in the area.

It is evident that open grasslands which mostly were wetlands has decreased due to expansion in valley bottoms cultivation traditionally known as vinyungu as these are considered as adaptation strategy to drought due to increasing rainfall unreliability [20,21]. Analysis

Establishing landuse/cover change patterns over the last two decades and associated factors for change in semi arid and sub humid zones of Tanzania

113

Table 3. Landuse cover change analysis in Kimilinzowo.

Land use/cover types	Years		Changes	
	1991	2011	1991-2011	1991-2011
	Area (ha)	Area (ha)	Change Index (ha)	Change (%)
Bush land with scattered cropland	2378	1435	−943	−39.6
Bare soil	194	000	−194	−100
Closed woodland	437	404	−33	−7.6
Cultivated land	1053	2322	+1269	120.5
Grassland with scattered cropland	802	1125	+323	40.3
Open grassland	535	146	−389	−72.7
Open woodland	000	582	+582	+100
Woodland with scattered cropland	615	000	−615	−100
Total	**6014**	**6014**		

Table 4. Landuse cover change analysis in Kinembasi village.

Land use/cover types	Years		Changes	
	1991	2011	1991-2011	1991-2011
	Area (ha)	Area (ha)	Change Index (ha)	Change (%)
Bush land with scattered cropland	1133	964	−169	−14.9
Closed woodland	10	16	6	60.0
Cultivated land	612	1293	681	111.3
Grassland with scattered cropland	1638	319	−1319	−80.5
Open grassland	0	669	669	100.0
Open grassland seasonally inundated	99	145	46	46.5
Open woodland	571	656	85	14.9
Total	**4063**	**4062**		

of satellite image for Kinegembasi village on landuse/cover changes for the same period did not differed substantially if you compare with that of Kiliminzowo village (**Table 4**). For example, bush land with scattered crop land also decreased by 14.9% while bare soil, closed woodland have decreased in the order similar to that of Kiliminzovo. It was also interesting to note that cultivated land has increased by 111.3% in the order similar to that of Kiliminzowo village. Other land cover types including grassland with scattered crop land decreased by 80.5% suggesting a conversion to crop land particular wetlands. Finding revealed that open woodland was not significant in such area (**Table 4**) although it increased by 14.9% suggesting positive impact of conservation initiatives.

3.1.2. Semi Arid Case Study Zone

Tables 5 and **6** present an analysis of historical lan-

duse cover change for selected case study villages in semi arid zone. In general there are more landuse cover types as compared to case study villages in sub humid zone (10 vs 7) and this suggests that there a diversification in landuse practices in semi arid zones as compared to sub humid. This is true because in semi arid zones people are sparsely populated allowing them to diversify their economic activities due large land available as compared to sub humid zones which are densely populated and have limited land. **Table 5** shows that dense and open bush lands, bushed grass land have decreased by 100%, 96.5% and 94.4% respectively on the expense of land under mixed cropping which increased by 64.9%. In general this indicates a significant expansion of agriculture activities in the area due to high demand of food crops by people. The area is very potential for food production in the region due to high fertility nature of vertisol which dominate the area.

Table 5. Landuse cover change analysis for Mbutu case study area.

Land use/cover types	Years		Changes	
	1986	2009	1986-2009	1986-2009
	Area (ha)	Area (ha)	Change Index (ha)	Change (%)
Dense Bushland	4913	0	−4913	−100
Open Bushland	27,227	954	−26,273	−96.5
Open Bushland (Lowland)	43	366	323	751.2
Mixed Crop Land	68,332	112,664	44,332	64.9
Bushed Grassland	23,007	1287	−21720	−94.4
Bushed Grassland Seasonally Inundated	283	0	−283	−100
Open Grassland	9886	4117	−5769	−58.4
Open Grassland Seasonally Inundated	269	416	147	54.6
Wooded Grassland	2012	4146	2134	106.1
Water	34	243	209	614.7
Mixed Crop land-Lowland	849	7786	6937	817.1
Settlement	260	216	−44	−16.9
Closed Woodland	121	459	338	279.3
Open Woodland	2902	7484	4582	157.9
Totals	**140,138**	**140,138**		

Table 6. Landuse cover change for Undomo case study.

Land cover types	Years		Changes	
	1986	2009	1986-2009	1986-2009
	Area (ha)	Area (ha)	change Index (ha)	Change (%)
Dense Bushland	0	178	178	100
Open Bushland	943	861	−82	−8.7
Open Bushland-Lowland	635	0	−635	−100
Mixed Cropland	2619	6464	3845	146.8
Bushed Grassland	2611	245	−2366	−90.6
Bushed Grassland Seasonally Inundated	262	278	16	6.1
Open Grassland	1227	0	−1227	−100
Open Grassland Seasonally Inundated	2168	0	−2168	−100
Wooded Grassland	486	1134	648	133.3
Water	275	34	−241	−87.6
Settlement	67	58	−9	−13.4
Closed Woodland	1332	724	−608	−45.6
Open Woodland	1877	4526	2649	141.1
Totals	**14,502**	**14,502**		

Establishing landuse/cover change patterns over the last two decades and associated factors for change in semi arid and sub humid zones of Tanzania

115

Other landuse cover change of interest is an increase of both closed and open woodlands suggesting that there is a positive response towards conservation of woodlands due to various campaigns including REDD+ interventions and presence of a forest reserve. FGD revealed that there has been an increase of cultivation in wetlands and this is shown by a reduction of both grassland including seasonally inundated areas (**Table 5**). At Undomo Village (**Table 6**) patterns and trend on landuse cover change is quite similar to that reported for Mbutu village but with the exception of decreasing area under closed wood land. In general the land area is also small but this does not necessarily affect the pattern on LUCC observed.

The area under mixed cropping (**Table 6**) has increased on the expense of other landuse types due to various factors. It also apparent that water bodies has decreased and this was explained to be due to drying and lack of dams as compared to Undomo where there is a number of artificial dams.

3.2. Factors for Land Use Cover Change

3.2.1. Environmental Related Factors

To discuss environmental factors that have contributed to landuse cover change in case study zones one should consider both driving forces and the process it self. Various scholars have suggested that poverty, changes in policies and strategies are among the key driving forces to landuse cover change (LUCC). Findings showed that in all four case study villages three major wealth groups namely the rich, middle and poor groups are common and their distribution at village level ranged on average in the order of 10%, 30% and 60% for the rich, the middle and the poor respectively. It was clearly noted that the majority of people are poor and most of them live under poverty and this has contributed to degradation or natural resources contributing to LUCC. It was also clearly revealed that changes and introduction of new polices in the country contributed to LUCC changes for example the introduction of villagilization policy in Tanzania in mid 1970's whereby people were clustered in villages called "Ujamaa" or "live close together" contributed to deforestation in new areas where they moved into for settlement and opening of new farms. A similar explanation for LUCC in semi arid areas of Tanzania has been documented by [8,18] for sub humid zone.

Environmental change in particular changing rainfall pattern and other rain related factors are shown in **Figure 3** for semi arid zone were found to have contributed directly or indirectly to LUCC. Similar observations were also reported for sub humid zone where by 81% of the total respondents indicated so. However 19% of responded were not able to clearly the pattern and trend of rainfall some perceiving that it has increased. This observation is not surprising and it could be explained in

terms of existing spatial variability of rain which is common in such areas.

Increasing temperature was also among factors mentioned during FGD that has contributed to LUCC in both sub humid and semi arid zones. In this case increasing temperature over decades was perceived by communities in all case study villages and this has been validated by empirical data on long term temperature trends by both Gwambene [17] and Kaijage [22] sub humid and semi arid zones respectively. A snap short on communities' perception with regard to temperature pattern at local level is shown in **Figure 4**.

In general the majority of communities reported that temperature has increased over that last 20 years and this trend is increasing. A paper by [19] provides a detailed analysis on the implication of both increasing temperature and changing rainfall pattern on both crops and livestock production. It is evident from the paper that crop pests and diseases have increased leading declining crop by nearly 40% over the last 20 years. Livestock production has also decreased and farmers by clearing woodlands and other LUC with understating that new land is fertile and can compensate loss incurred. In reality declining crop and livestock productivity could be associated with more than one factors and this calls for capacity building to farmers through research and training.

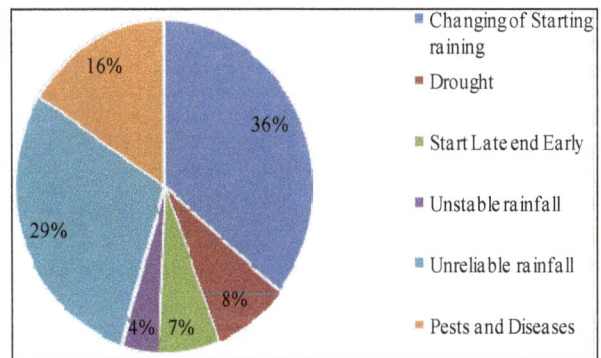

Figure 3. Rainfall related factor for LUCC in semi arid zone.

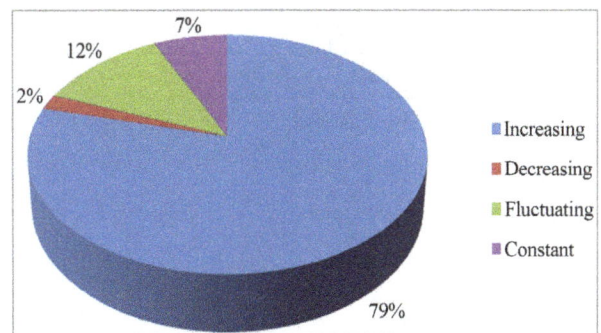

Figure 4. A snap shot of communities' perception on temperature trend in semi arid zone.

Reported by 80% of the respondents was declining soil fertility in most dry land areas and this has forced farmers to open new areas which are more fertile particularly in semi arid areas. In sub humid zones due land scarcity a tendency has been to expand wetland to take advantage of moisture and accumulated soil fertility and this allow multiple crops and vegetable to be grown and it has been considered to be an adaptation option. Similar cases were also reported by individuals who participated in FGDs in both zones. Kiunsi and Meadow [2] also observed a similar case when studying land management issues in Monduli, Tanzania.

3.2.2. Socio-Economic Related Factors

Findings from household interview indicated that most of the people migrated in the study villages due to various reasons mainly being due to agricultural activities and following up their relatives as presented in **Table 7**. The role of agricultural expansion on landuse cover change is not surprising since it has also been reported by [23,24].

Migration of people in case study villages to follow their relatives and marriages is among major factors contributing to increased human population and hence pressure on natural resources (**Table 7**). There are also those who migrated for other social economic related activities including looking suitable land with pasture for livestock keeping, seeking employment and other business opportunities. This also contributed to population increase in case study areas and hence LUCC.

3.2.3. Demographic Factors

Of importance in this study was an increase on human population at the rate of 4.5% and 3.2% respectively in sub humid and semi arid zone. This has increased pressure on natural resources in both zones resulting changes on landuses. In semi arid zone it was found that about 53% of the HH were Sukuma whereby 40% were Nyamwezi in tribes. The two tribes are native and commonly found in most parts of the region. Other ethnic

Table 7. Social economical reasons for in migration in study areas in % per zone.

Major reasons	Sub humid	Semi arid	Mean
Search for agricultural crop land	38.5	60.0	49.0
Looking for new business opportunities	7.5	9.0	8.25
New employment opportunities	NA	9.0	9.0
Look for pasture land	NA	2.0	2.0
Following relatives and marriages	44.0	20.0	32.0
Total	100	100	100

NA: Not Applicable.

groups constituting 7% were found to be Gogo, Rangi, Nyakyusa and Nyiramba which originated from neighbouring regions.

4. CONCLUSION

Land use cover types differs in numbers and sizes in both sub humid and semi arid zones being larger in size and more in semi arid zone. However both cover and use types have changed over the last 20 years into different directions to absorb shock associated with both environmental changes and social economic, demographic and other related demands which are also growing. It is evident that climate change which is also triggering other factors such as land degradation has played a significant change and this is coupled by migration of people into new potential areas suitable for agriculture which was found to be expanding and being intensified. Two recommendations can be made. One is to have proper landuse plan such areas and implement them and second is to design and implement proper capacity building programs on climate and land management issues for both livelihood and ecosystem sustainability.

5. ACKNOWLEDGEMENTS

The author wishes to thank CORUS through the Rungwe Environmental Scientific Observatory Network and Rockefeller Foundation for providing financial assistance to undertake this important study. The University of Dar es Salaam management is thanked for granting permission to researchers to implement this study. Tabora and Iringa regions authority is thanked for allowing this study to be conducted in their regions. All technicians at GIS laboratory working at the Institute of Resource Assessment are thanked their analytical work. Finally I would like to thank research assistants namely Nassib Muzo and Lucy Kassian for their assistance on data collection in the field.

REFERENCES

[1]   Adger, W.N., *et al.* (2007) Successful adaptation to climate change across scales. *Global Environmental Change*, **15**, 77-86.

[2]   Kiunsi, R.B. and Meadow, M.E. (2006) Assessing land degradation in the Monduli District, northern Tanzania. *Land Degradation Development Journal*, **17**, 509-525.

[3]   Majule, A.E., *et al.* (2009) Natural resource contribution to community livelihoods. Experience from selected case studies in Tanzania. Dar es Salaam University Press LTD, Dar es Salaam.

[4]   Benard, F.E., *et al.* (1989) Carrying capacity of the eastern ecological gradient of Kanya. *National Geographic Research*, **5**, 399-421.

[5]   Campbel, D., *et al.* (2003) Root causes of land degradation in the Loitoktok area, Kajiado. LUCID Working Paper No. 19, International Livestock Research Institute, Nai-

Establishing landuse/cover change patterns over the last two decades and associated factors for change in semi arid and sub humid zones of Tanzania

117

robi.

[6] Maitima, M.J., *et al.* (2009) The linkages between land use change, land degradation and biodiversity across East Africa. *African Journal of Environmental Science and Technology*, **3**, 310-325.

[7] Majule, A.E., *et al.* (2009) Ecological gradients as a framework for analysis of land use change in East Africa. *African Journal of Ecology*, **47**, 55-61.

[8] Muganyizi, J.M. (2009) Land use changes within agricultural systems and their implications on food security in Rungwe district, Tanzania. M.Sc Dissertation, University of Dar es Salaam, Dar es Salaam.

[9] Majule, A.E., *et al.* (2010) Underlying threats on forest reserves in Tabora Region, Western Tanzania: The case of Igombe River and Simbo forestry reserves. *Journal of Studia Universtatis Babes-Bolyai*, pp. 137-150.

[10] IPCC (2007) Working group II fourth assessment report. Climate change: Climate change impacts, adaptation and vulnerability.

[11] Muzo, N. (2012) Climate change and its impacts on agricultural land use practices in selected villages of Nzega District, Tanzania. MSc. Dissertation, Dar es Salaam University Press, Dar es Salaam.

[12] FAO (2008) The state of food and agriculture, biofuels prospects, risks and opportunities in Tanzania. FAO, Rome.

[13] Mbonile, M.J., *et al.* (2003) Land use change pattern and root causes on the slope of mount Kilimanjaro, Tanzania. LUCID Working Paper 25, International Livestock Research Institute, Nairobi.

[14] Olson, J.M., *et al.* (2004) The spatial pattern and root causes of land use changes in East Africa. LUCID Project Working Paper 47, International Livestock Research Institute, Nairobi.

[15] William, C.M.P. (2003) The implications of land use changes on forests and biodiversity; A case of the 'half mile strip' on mount Kilimanjaro, Tanzania. LUCID Working Paper Series Number 30, International Livestock Research Institute, Nairobi.

[16] Rowcroft, P. (2005) Gaining ground: The socio-economic driving forces behind decisions regarding land use and land use change. An overview. Working Paper 16, Vientiane.

[17] Gwambene, B. (2007) Climate change and variability and adaptation strategies and its implications for land resources in Rungwe District, Tanzania. A Master of Science Dissertation, University of Dar es Salaam, Dar es Salaam.

[18] Kangalawe, R.Y.M., *et al.* (2005) An analysis of land use dynamics and land degradation process in the great rift valley, Central Tanzania: A case of Iramba District. OSSREA Publications, Addis Ababa.

[19] Majule, A.E., *et al.* (2013) Impact of climate change on natural resources and community livelihood in Tanzania: Experiences from semi arid areas of Tanzania. Transworld Research Network, Kerala.

[20] Majule A.E. and Mwalyosi, R.B.B. (2005) Enhancing Agricultural productivity through sustainable irrigation. A case of Vinyungu farming system in selected zones of Southern Highlan, Tanzania. In: Sosovele, H., Boesen, J. and Maganga, F., Eds., *Social and Environmental Impacts of Irrigation Farming in Tanzania: Selected Cases*, Dar es Salaam University Press, Dar es Salaam.

[21] Kassian, L. (2012) Socio-economic and environmental implication of valley bottoms farming as adaptation strategy to climate change: A case of Mufindi District in Iringa Region. M.Sc Dissertation, University of Dar es Salaam, Dar es Salaam.

[22] Kaijage, H.R. (2012) Impact of climate change on groundwater dynamics in Nzega District, Tanzania: Triangulation of indigenous knowledge and empirical method scenario analytical approach. M.Sc Dissertation, University of Dar es Salaam, Dar es Salaam.

[23] Rowell, D.L. (1994) Soil sciences: Methods and applications. Longman, London.

[24] Nelson, V. and Stathers, T. (2009) Resilience, power, culture and climate and new research directions: A case study from Semi-Arid Tanzania. *Gender and Development Journal*, **17**, 81-94.

15

Urban surface water system in coastal areas: A comparative study between Almere and Tianjin Eco-city

Tao Zou[1*], Zhengnan Zhou[2]

[1]Beijing Tsinghua Tongheng Urban Planning and Design Institute, Beijing, China; *Corresponding Author
[2]School of Architecture, Tsinghua University, Beijing, China

ABSTRACT

In the purpose of defining typical urban water management challenges in coastal lowlands in the context of global climate change, a comparative study was conducted between two coastal new towns respectively located in the Netherlands and Northern China. Comparative method is applied to define main functioning patterns of urban water systems in the two cases, then computer simulations were used to further compare drainage capacity in order to reveal the trends of urban water management. Major result has shown that Almere in the Netherlands generally more advanced in urban water management as multiple functioning patterns are available. Strong dykes maintain competence for land subsidence and sea level rise. Open water system decreases local runoff and increases water retention level. Systematic control of sluices and locks which serve for shipping and waterfront landscaping are simultaneously isolating contaminants from outer water body. Tianjin Eco-city in China has shown both strengths and weaknesses. It takes large amount of reclaimed water as main landscaping water source, which adapts to local water pollution and shortage while requires highly centralized facilities. Large water body is reserved and huge scale underground drainage system built, but it is still vulnerable to heavy storms due to the lack of efficient surface water drainage system. Coastal line control does not adequately prevent from increasing storm surge risks in the future. SWMM simulations have supported the viewpoint of distributed surface water with a higher efficiency for storm drainage. Meanwhile, surface water system returns more added values to urban development. The study is corresponding well with the theory of water sensitive city. As a conclusion, urban water system should always incorporate methods to achieve higher system resilience based on multiple functioning patterns.

Keywords: Urban Surface Water System; Urban Water Management; Coastal Areas; SWMM

1. INTRODUCTION

1.1. Vulnerability of Coastal New Towns

Coastal zones are highly vulnerable to sea level rise in the context of global climate change. Stronger and higher frequency of storm surges poses an increasing threat to coastal urban area, and especially to the infrastructure systems. The main issues include flooding, water resources, land loss and water quality as well. Sea level rise may cause immense economic loss due to the direct or indirect damage to the urban areas, sewers, ports and other infrastructure. In China, many coastal lowlands have limited or no human-built protection against impacts from sea level rise or storms. Simultaneously the quality of water is likely to decrease, with the fluctuation of runoff that either increased sediments and pollutants or decreased flushing that leads to higher salinity levels. In north China coastal cities, increased salinity is vastly threatening water quality for residential users, and affecting crops as well.

Cities located in coastal regions are very vulnerable to extreme weather events. While in recent years, large number of new commercial and residential development in China accelerates the concentrations in coastal areas. In megacities like Shanghai and Tianjin, or city clusters located around them, dozens of new-towns with a planned population from 100,000 to 500,000 emerges along coastal low-lying belts, among which Sino-Singapore Tianjin Eco-city is one of the most known projects in China.

Without adequate consideration on the rising risks, the vulnerable costal lowland area has been developed into new settlements. The phrase "Wrong location" is commonly used when arguing on coastal new towns' planning. But decision maker's faithful submission to the rule of urban concentration has led to the adaption of these "wrong places" unavoidable.

Looking back into the near history, European cities had also experienced such fast growing period decades before. Almere is one of them. As an extension of Amsterdam city, Almere is still considered to be located wrongly due to geological reason [1]. But fortunately enough, Dutch seems far more experienced in coping with coastal risks.

1.2. Introduction on the Study

A comparative study was conducted between two coastal new towns respectively located in China and the Netherlands (**Figure 1**). The purpose is to define typical challenges and the differences between these urban developments in coastal lowlands. Local strategies for planning the urban surface water system and their functioning patterns differ greatly according to local climate and other social-economic situations. In order to further discuss on possible future adaption to extreme weather events, a brief SWMM simulation was applied to each case, so as to compare the two methodologies coping with the same water issues. The comparison intentionally implicates the better orientation of urban water management in different phases of urban planning.

2. PRELIMINARY STUDIES AND METHODS APPLIED

2.1. Cases General Information

2.1.1. Almere of the Netherlands

Initially built as a satellite town of the capital Amster-

Figure 1. Same scale satellite images of Tianjin Eco-city and Almere.

dam, Almere is located in Flevoland of the Netherlands, 20 km away from Amsterdam. The city covers 248.77 km^2 in total, including water area of 118.29 km^2. As the largest city of Flevoland, Almere has become the seventh largest city in the Netherlands with the population over 180,000 by July 2008. Owing to the expansion plan of Almere, its population is expected to reach 350,000 by 2030.

Almere is a typical polder city. The sea reclamation project in Flevoland is initially planned for agricultural purpose. After the end of World War II, two new polder cities were planned because of the rapid population growth and the increasing house demand. One is Lelystad, and the other one is Almere. As one of the latest built cities in the Netherlands, with the first building emerged in 1976, Almere was formally established as a municipality until 1984.

2.1.2. Sino-Singapore Tianjin Eco-City

Sino-Singapore Tianjin Eco-city (SSTEC, or Tianjin Eco-city) is an international cooperated new town development started from 2008. The 34.2 km^2 of coastal salinized lowland was chosen in consideration of preventing from taking up farmlands in the region. The first phase covers 7.8 km^2 of development capable of accommodating 105,000 residents in 2015, then 350,000 residents or 110,000 families by 2030 when fully built.

Aiming at a demonstration project of sustainable urban development in China, Tianjin Eco-city is seeking for environmental friendly and low-impact development. Work-life balance and compact-mixed land-use structure, as well as sustainable mass transit system has been largely considered. Water recycling due to regional water shortage is also taken as one of the most important strategies. Large scale reclaimed water plant has been built, and storm water drained through huge scale underground piping system.

In the past years, a few families moved into the new eco-city, while national level economic policy has lowering down the pace.

2.2. Climate and Topographical Backgrounds

Both located in the temperate zone in northern sphere, Almere has higher latitude of 52.37N, while Tianjin Eco-city locates 39.15N. Although much to the south, Tianjin has an annual average temperature not much higher than Almere. The major difference is a colder winter and a much hotter summer in Tianjin.

Annual precipitation in Almere is a bit higher than Tianjin. With a lower annual actual evaporation, Almere shows a wet climate in general. In Tianjin, more than 70% precipitation is received during the hot summer period, and throughout remaining months of the year it appears very dry. Tianjin is now facing severe water shortage due to high population density and general dry climate as in other northern cities in China (**Table 1**).

Topographically, Almere is a polder city lying several meters lower than sea-level. The polder is prevented from flooding by strong dykes along the periphery of inside lowlands. Tianjin Eco-city was historically consisting of salt fields and a large contaminated reservoir. Dykes for defensing storm surges lie near its border. The whole area is quite flat and mostly around 1 m above sea-level, which will be raised to around 3 m according to its master planning.

2.3. Future Challenges for the Two Cases

Future dilemma in sustainable development for both coastal new-towns is obvious. Among many of the main problems encountered, the following ones are the most harmful and urgent.

2.3.1. Storm Tide

Storm tide is one of the most challenges in the context of global climate change. As about two thirds of land in the Netherlands is highly vulnerable to flooding, it is very important to develop on the reclaimed land following the huge scale damming. Luckily enough, the risk of storm tides for Almere has been tremendously overtaken by the two dams of Afsluitdijk and Houtribdijk. In contrast, Tianjin Eco-city is situated much adjacent to Bohai Bay with one and only dam about 500 m away from its southeastern border. Strong tide had historically hit part of the site, and future extreme weather event is threatening this area still.

2.3.2. Land Subsidence

Land subsidence is another important issue. Three quarters of the Netherlands is subsiding, the young and flexible soil of Almere is subsiding even more rapidly.

This process is even intensified by drainage and construction, and can only be stopped by the construction of large water storage facilities. In Tianjin, as part of the most densely populated area in Northern China, deep ground water over-exploitation has led to severe land subsidence. This problem overlaps with sea level rise, and has enormously downturn capability of dams originally built for 100 years return period to a much lower level.

2.3.3. Water Shortage

Freshwater resource might not be an issue for Almere due to abundant and evenly distributed precipitation, while Tianjin is facing severe shortage of freshwater during most months in a year and suffers from heavy storms in summer period. Water pollution and saline soil for Tianjin Eco-city are also higher risks for keeping a good urban living quality.

2.3.4. Extreme Weather Event

Almere is naturally a part of a polder that totally relies on the largest pumps in the world, while Tianjin Eco-city is more densely developed with much heavier rains during the summer period. Extreme weather event, an unprecedented rainfall, could be a rigorous challenge to both cities.

2.4. Comparative Methods Applied

2.4.1. General Research Framework

When studying on urban surface water system, it will naturally relate to other urban water management issues. Ground water and urban infrastructures immensely affects surface water system, while only the latter one shows more obvious clues to all who concerns urban form. This might also be the reason why urban planners, in a conventional manner, takes care more on surface water system, and feel less responsible on other unseen

Table 1. Climate and topographical information for both cases.

	Tianjin Eco-city (Tanggu)[a]	Almere (Amsterdam)[b]
Longitude	117.75E	5.22E
Latitude	39.15N	52.37N
Annual average temperature	13.0°C	9.6°C
Coldest month average temperature (Jan.)	−3.0°C	2°C
Hottest month average temperature (Jul.)	26.8°C	18°C
Annual precipitation	545.4 mm	755 mm[c]
Annual actual evaporation	-	563 mm[c]
Annual evaporation from water surface	2025.4 mm	-

[a]Climate data source: Tanggu station from China Meteorological Data Sharing Service System; [b]Climate data source: http://www.amsterdam.climatemps. com/; [c]Reference [2].

issues. The comprehensive framework of the general program excludes the possible bias. Surface water is directly supported by ground water and infrastructure system (**Figure 2**).

2.4.2. Information-Based Mapping and Structuring

In order to understand the urban water systems accurately and concisely, structures of these urban water systems have been firstly divided into single elements. These elements were reproduced into ArcGIS and SWMM, thus to form the whole information-based picture of the urban water systems. Data had been collected mainly via site visits and satellite image interpretation, while contribution from local planning authorities and institutes greatly helps to build up a more credible and precise mapping.

2.4.3. Theory of Water Sensitive City

Based on the general urban planning philosophy, urban water management issues are always complex and case sensitive. Thus initial works on the comparison of the operational modes and functions for the two urban water structures are largely related with local situation, in spite of considering the incomparable preconditions that might confuse the results. Introduction of the theory of water sensitive city has apparently clarified the comparative framework. A dimension of time has been added, and the two urban water management systems have been seen evolving.

2.4.4. SWMM Modeling

SWMM modeling focusing on storm water drainage helps further to compare the different structures. As many parameters had been hypothetically setup due to the lack of precise topographical data, the touch-and-go type of modeling is far from accurate. Thanks to the high similarity of for both cases though, the model fulfills the basic requirements of an urban study at this specific level.

The basic concept is to study on how spatial structure of the systems would affect the storm water drainage efficiency, rather than how pipelines or waterways should be designed. Modeling for both cases is limited to a "proper scale". In the case of Almere, only main waterways are modeled, not any capillary surface water networks or underground storm water pipes have been taken into account. Same in Tianjin Eco-city, drainage pipelines within main sub-catchment area are not included.

Almere Stat, the center part of Almere, has been simulated instead of the whole large region. Rainfalls have been assumed to be the same event cluster, which is 5, 10 and 50 years return period in Tianjin.

3. STRUCTURE AND FUNCTIONS OF THE URBAN SURFACE WATER SYSTEMS

3.1. The Case of Almere

Almere is planned with several semi-independent core areas with blocks, public facilities and location markers. These core areas are connected to the urban center through the public infrastructures. Ditches and various kinds of water bodies intersect in the six urban areas of Almere. The highly efficient surface water system of the polder city is protecting the land from damage, and maintaining social and economic development within this region.

3.1.1. Structure of Surface Water System

1) Lakes

Almere is surrounded by water in three sides. The lake Markermeer in the north is the twin lake of IJsselmeer built in the early Zuiderzee Works. The Markermeer covering 700 km^2 is 3 to 5 m in depth. Due to economic and ecological concerns, the area that initially planned to build polder turns to form a lake after building a dam. Nowadays, the Markermeer functions as fresh water reservoir, buffering for drought and flood control, and ecological conservation as well. IJmeer in the west converges with the Markermeer; lake in the south is Gooimeer. The three external lakes' water level is very close to sea level.

The entire Almere is below the sea level, and the road elevation is −6.2 to −3.15 m N.A.P. The overall terrain of the planning area is relatively flat; most high-lying areas are located in the southwest; low-lying areas are located in the north. Lake Weerwater in the central part covers

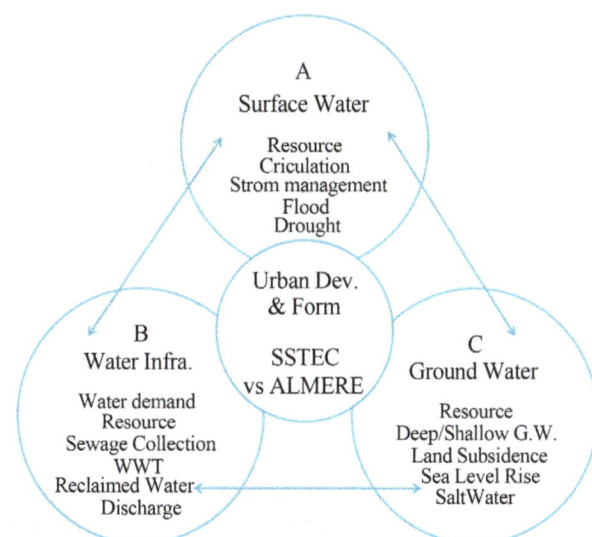

Figure 2. General research framework.

1.5 km^2; Lake North (Noorderplassen) covering 2 km^2 is a wetland protection park.

2) Drain and Ditches

The polder in which Almere locates carries dense river network that are mainly functioning as drain and ditches, including high canal, low canal, normal river ways and capillary water net.

High canal, of 50 m in width, collect seepage water and rainwater from higher ground in the south and direct to Ketelmeer in the east. Owing to the large amount of seepage water here, regular flow of high canals is significant to security, ecology and recreation. Low canals, of 45 m in width, collect excess surface water from lower flow rate area in the middle of polder and use it to adjust the water level. The water quantity of canal is under control, and the water quality is ensured by continuous refreshing water. The large part of Almere is located at west of the two canals. The staggered network of water in the region plays a significant role to maintain the quantitative and qualitative balance of local water system.

3) Pumping Stations

Pumping station de Blocq van Kuffeler is one of the largest pumping stations in the world. It is located on the dam where high and low canals meet in the northeast of Almere. The huge structure is put into use in 1967, which is diesel-powered. It maintains the polder's water elevation with other three pumping stations in Flevolands.

The pumping station comprises four pump units. It works at 85 rpm under normal conditions, with processing capacity from 700 to 850 m^3/min. The pump unit can work for 600 to 900 hours during the dry season, and 1200 hours during the rainy season.

4) Weirs and Sluices

Weirs are mainly used to adjust the water level. Movable weirs are adopted to better control the height of the regulated water. This kind of weir, with a movable flap valve, allows excess rainwater passing through the weir in the condition of heavy rainfall, and then the water surface elevation gets back to the present level. Most of the weirs are automatic or semi-automatic. Water is directly discharged into the canal through the weirs.

3.1.2. Operational Modes and Function Analysis

As a typical polder city, the basic goal for Almere is to maintain a long-term stability of surface water level in the polder. This is essential not only to ensure safety of the city with an expected population of 350,000 in the future, but to exert the functions of urban water system in ecology, recreation and transportation. Water quality must be kept in good conditions to fully achieve these functions.

A key climate feature of the region is the relatively even distributed rainfall all the year round, with a sig-

nificant difference of evaporation between winter and summer. Therefore, "flood in winter and drought in summer" emerges regularly, and the urban water system can be roughly defined as two "operational modes": "winter mode" and "summer mode". In colder seasons, excess surface water is discharged from the polder to the outside water body, so as to prevent flooding from occurring in reclaimed land for urban construction and agricultural use; in warmer seasons of the year, external water is introduced to the polder to ensure stability of regional hydrological system, and to meet agricultural water demand and protect dams from damage as well. In different management regions, the internal water level control in winter and summer are mostly the same, with some slight difference in certain area. Throughout the year, "winter mode" is generally longer because the average annual rainfall (755 mm) is exceeding the total evaporation (563 mm) up to 192 mm [2] (**Figure 3**). In addition, when rainstorm comes, the water system in the polder will turn to "flood relief mode" which pose greater challenge to the system's draining efficiency.

Almere and the polder's water system implements three important functions in these operating modes, which show great hints for future urban development of coastal areas.

1) Damming System: Defense Seawater Intrusion

Almere is under the protection of three dams, respectively built in the projects of the Ijsselmeer formed in the early Zuiderzee Works, the Markermeer formed in the Markermeer project, and the south Flevoland polder project. These dams have completely prevented seawater intrusion.

The standards on hydraulic engineering design and construction in the Netherlands are extremely high. The

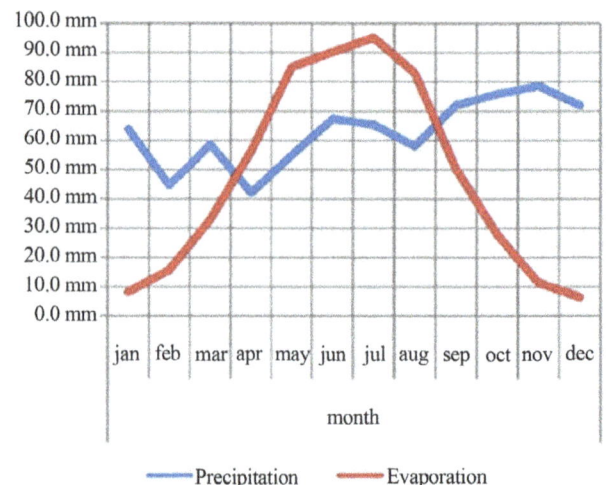

Figure 3. Averaged monthly rainfall and precipitation in millimeters (1971-2000) over the period of one year in the Netherlands. Well visible is that the evaporation is exceeding the precipitation during the summer months [2].

32 km-long Afsluitdijk was completed in 1932 under the design standard of flood with recurrence interval of ten thousand years. It is 3.5 m higher than the given highest tide, and 90 m wide at the sea level, equipped with strict impermeable measures. The dam is still intact even it has experienced maritime storm surges for many times [3].

The Markermeer project has not been fully completed, but the 28 km-long Houtribdijk dam becomes the second line of defense. Markermeer is also a key buffer and protection element for this region. In 2003, the Netherlands suffered a severe drought and many dams were threatened. Water was transferred from Markermeer to surrounding polders imminently, which effectively kept the soil moist and ensured the dam safe.

The multi-layer dam system under scientific and strict management is quite sufficient to handle the storm surges that may increase in the future, and to deal with coastal problems caused by land subsidence and sea level rise.

2) Highly Expanded Open Water System: Higher Resilience Dealing with Floods

The standard drainage capacity of the Netherlands can be converted as 14 mm rainfall drainage per day, and the designed drainage capacity of Flevoland polder is 11 to 18 mm per day [4]. This drainage standard is very high, but still difficult to resist occasional heavy rain. For example, 24 hours rainfall in parts of the Netherlands had reached 130 mm in September 1998, which exceeded the drainage capacity and resulted in flooding in extensive agricultural and urban areas. Over the past few years, such situations made the Netherlands lose billions of euros [5].

These make people realize that a climate proof rainwater management method has to ensure a sustainable and flexible rainwater management system in the polder. The system may achieve selective storage in clean rainwater season, and realize rain flood regulation and storage under extreme precipitation. Although maintaining larger open water area may limit urban development and building use, it is almost impossible to result in rainwater infiltration and underground storage because of the higher water table, small drying height and non-permeable soil structure in the polder. Therefore, expanding use of open water to capture and store rainwater is widely recognized. Based on the Netherlands' guidelines for water resources management in the future, local water affairs management institutes keep the following idea: in the new urban development, open water shall account for 10% of the total land area [6]. In the Flevoland polder, the early open water accounts for only 1%, this is currently expanded to 4%. In the region of Almere with concentrative construction, nearly 4 km² lakes within the polder in the northern, central and southern parts and the ditches all over the city become important buffer water to deal

with storm disaster. Large water storage facilities and extensive open water system greatly reduce surface runoff, as well as construction and operational costs of urban infrastructure.

3) Dynamic Closed Water System: Self-Cleaning Internally

Water management system for Netherlands polder is based on considering each polder as an independent water system, which is mainly to control the stability of the groundwater and surface water elevation. Because of climate reason, the stability of the water level within the polder cannot be self-sustaining. In winter with less evaporation, the excess rainwater will be pumped into the canal at higher terrain, and ultimately discharged into the external lake. The process is reverse in summer, water must be introduced from external lakes, so as to add water and prevent the water level lowered. The control process is effective to ensure stable water quantity, maintain boating conditions, coastline landscaping and ecological services. Simultaneously, it is able to separate water within the polder from pollutants out in the external water bodies.

Nevertheless, researchers are still working on the sustainability of water management system. The system is still lacking of sufficient inherent resilience, a large amount of relatively clean rainwater is discharged, but much polluted external lake water is also introduced inward in the opposite case. Thus, there is certain risk of water quality decrease. So there comes a challenge to avoid cleaner internal water being discharged, and to prevent external pollutants from canals and rivers. In order to achieve this goal, expanding the open water is an important way to collect clean rainwater, as much as possible. However, for the purpose of maintaining good inter-seasonal water quality in a closed system, sewage plant should be fully used and its drainage standards should be improved. Besides, hygrophyte around the lake shall be used to purify the water, such as reed in ponds and ditches.

At the same time, the so-called "circulation model" [7] for sub-areas is vastly established in Almere in accordance with the water level control. This method is not only conducive to water quality maintenance at low cost, but also improve the quality of urban landscape.

3.1.3. SWMM Simulation Overview: The Case of Almere

Total sub-catchment area of Almere Stat is 26.17 km². It has been divided into 58 sub-catchment areas. 2 retention water bodies have been settled. Main waterways width from 35 to 40 m, branch waterways width from 15 to 27 m, distributes with a total length of 29.5 km (**Figures 4** and **5**).

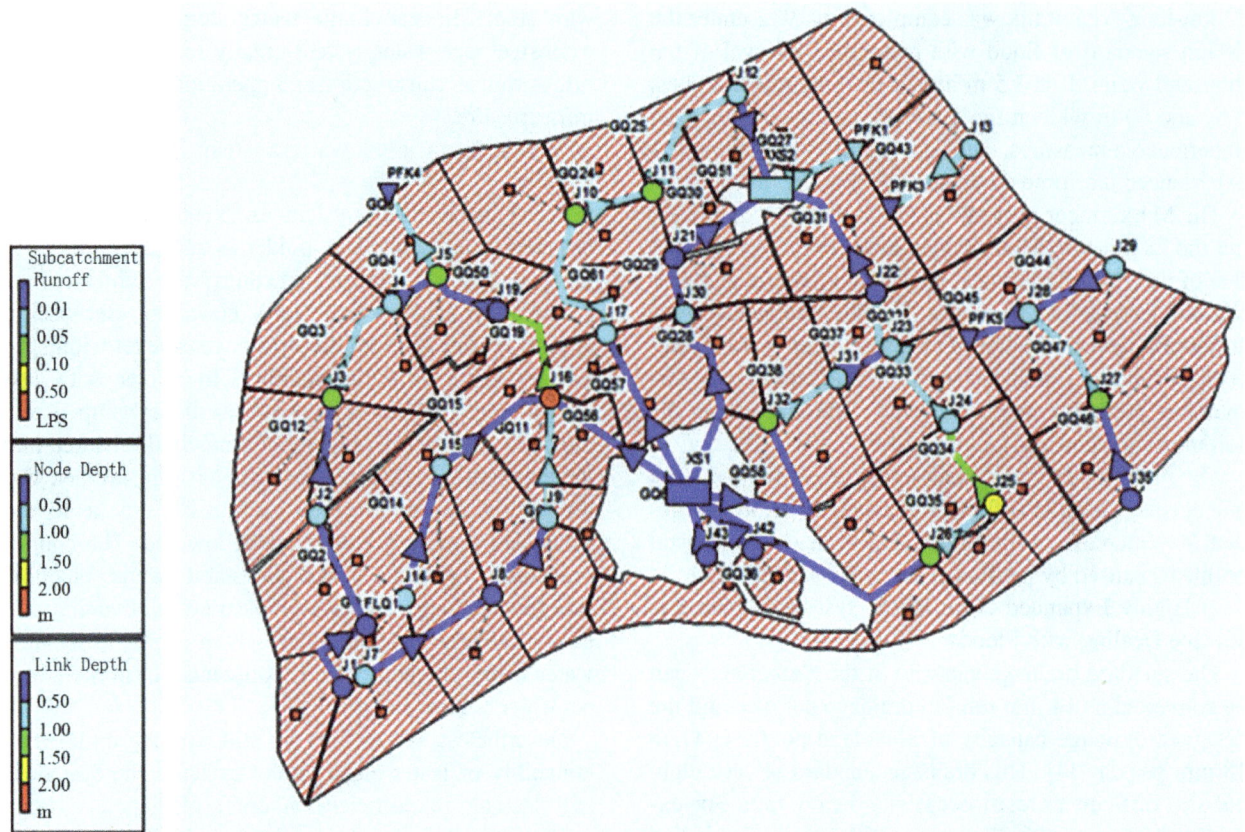

Figure 4. Overview of a SWMM simulation for Almere stat.

Figure 5. Simulation based on different storm events in Almere Stat. Red dots and lines indicates overflow where flooding will occur. In Almere, nowhere turns red under 10-year return period rainfall event. Small proportion of waterways will overflow under 50-year return period rainfall event.

3.2. The Case of Tianjin Eco-City

3.2.1. Structure of Surface Water System

1) Lake and Old River Course

Qingjing Lake is located in the center of Tianjin Eco-city. The original reservoir area covered 7.7 km² in total. The original water body named Yingcheng Reservoir was artificially built on low-lying shallow water with an average depth of 3 m. It had been used as an industrial waste water basin in the past decades. The water covering area varies greatly between rainy season and dry season. When reconstructed and transformed to Qingjing Lake according to the master plan, the total area is reduced to only 1.3 km².

Jiyunhe old river course flows through Tianjin Eco-city. It was part of Jiyunhe River till a curve cut-off project had been implemented for the purpose of flood control in the near history. As a result, Jiyunhe old river course winds around the central water body for 11 km in the planning area, and the new watercourse defines the site border. Its water quality is among the worst before being reconstructed as planned. The width of Jiyunhe old river course varies between 150 to 350 m.

The water body consisting of Qingjing Lake and Jiyunhe old rover course is around 4 km², nearly 12.8% of total site area, which is one of a feature that most other similar type of urban development not likely to exceed.

2) WWTP

With the key target of water conservation, Tianjin Eco-city attaches great importance to the establishment of water recycling system. Through the construction of systems for sewage treatment, water reclamation, rainwater collection and seawater desalination, the non-traditional water use ratio is planned to reach 50% by 2020. The initial capacity of the water reclamation plant is 21,000 tons per day, and the designed capacity is 42,000 tons per day.

3) Pumping Station

Pumping stations were planned to build at both end of the Jiyunhe old river course, in the purpose of balancing interior water amount and quality control as well. The whole interior water body is thoroughly connected without weirs or sluices.

Considering reasonable circulation, another solution is also under discussion: a two-way pumping station built in upper end of the Jiyunhe old river course, and a sluice placed at the lower end.

4) Isolated Ponds

The whole urban surface water system consists predominantly with large water body, and there's only two planned isolated ponds loosely connected with Jiyunhe old river course. These waterway-like ponds were planned to be ecological corridors connecting inside water body to the peripherals. But since they are locating higher, serving like a catchment border, rain water will not be collected through natural processes. Their actual function is more visually driven rather than a green ecological corridor.

5) Rain Water Piping System

Rain water is generally collected via piping systems divided in 4 sub-regions. Within each sub-region, very large section pipes and tunnels are densely distributed and deep buried into ground. The largest section can be up to nearly 16 m^2, and bury more than 3 m deep at the outlet of each sub-region. Storm water is then pumped out into the lake and old river course. The whole system is built like a waterway system underground.

3.2.2. Operational Modes and Function Analysis

The case of Tianjin Eco-city has chosen specific measurements in coping with the urban water functioning patterns. The basic logic can be seen as highly concentrated system with large scale planning and design.

1) Water Reclamation as Main Water Resource

The first and the most featured functioning pattern is taking large amount of reclaimed water as main water resource, thus simultaneously transforming a waste water basin into a clean water body for urban landscaping and recreational purpose. This substantially adapts to water pollution and shortage situation that might as well exists in other regions. But it requires highly centralized waste water treatment facilities and strictly monitored outlet water quality in accordance with the scale and sustainable target.

In late summer, water quality in Jiyunhe River is the best of a year owing to abundant precipitation in the region. It is the best period of time to pump water from external water body to Jiyunhe old river course so as to replenish water for the interior water bodies. In the rest of a year, reclaimed water from the water reclamation plant will be used for replenishment. Water from the Qingjing Lake will also be directed to the water reclamation plant for processing when its quality degrades.

Presently, water output from the WWTP meets First Class B, which will be improved to First Class A in the future. According to First Class A, the N, P concentration limits will be able to achieve that of the National Reclaimed Water Quality Standard, but the salinity could still be very high. Therefore, WWTP outlet water should then send for membrane treatment processes in the water reclamation plant to meet Reclaimed Water Quality Standard.

The high salinity of water output from sewage treatment plant is another challenge. It is primarily caused by the following reasons: high salinity of underground water; serious leakage problems of old drainage pipe systems in Hangu and surrounding areas; pipelines passing through many abandoned salt fields, which make the high-salinity underground water leak into drainpipes; high-salinity effluent from surrounding chemical plants also connecting to drainage pipe system, and eventually discharged to the sewage treatment plant.

2) Circulation model as a Way against Eutrophication

At present, the water body of Qingjing Lake circulates counter clock, with the water reclamation plant in-taking water in the north and discharging in the south. The water body of Jiyunhe old river course circulates clockwise. Qingjing Lake and Jiyunhe old river course are connected by culverts and pumping stations. However, the bottom of Qingjing Lake is high in north and low in south, which is not conductive to reverse circulation by itself. On the other hand, for the southwest section connecting Jiyunhe River to Jiyunhe old river course, water quality tends to degrade due to poor liquidity.

In master planning, two connecting ends of Jiyunhe old river course and Jiyunhe River will be built with pumping stations for water exchange. However, the current strategic tendency is that the north end being equipped with a two-way pump which is able to drain and replenish, and the south end built with a sluice.

3) Large Area of Water Body Serve for Visual Landscaping

Large volume of water body is planned, but the designing purpose is originally intended to be more visually rather than ecological functions. As the main water body is relatively concentrated, the site is still vulnerable to heavy storms due to the lack of efficient surface water drainage system.

3.2.3. SWMM Simulation Overview: The Case of Tianjin Eco-City

Total sub-catchment area of Tianjin Eco-city is 19.04 km^2, which has been divided into 4 main sub-regions according to the planning. Main pipelines diameter ranges from 1.2 to 2.2 m, piping tunnels section area from 5 to 16 m^2 with a total length of 63.2 km (**Figures 6 and 7**).

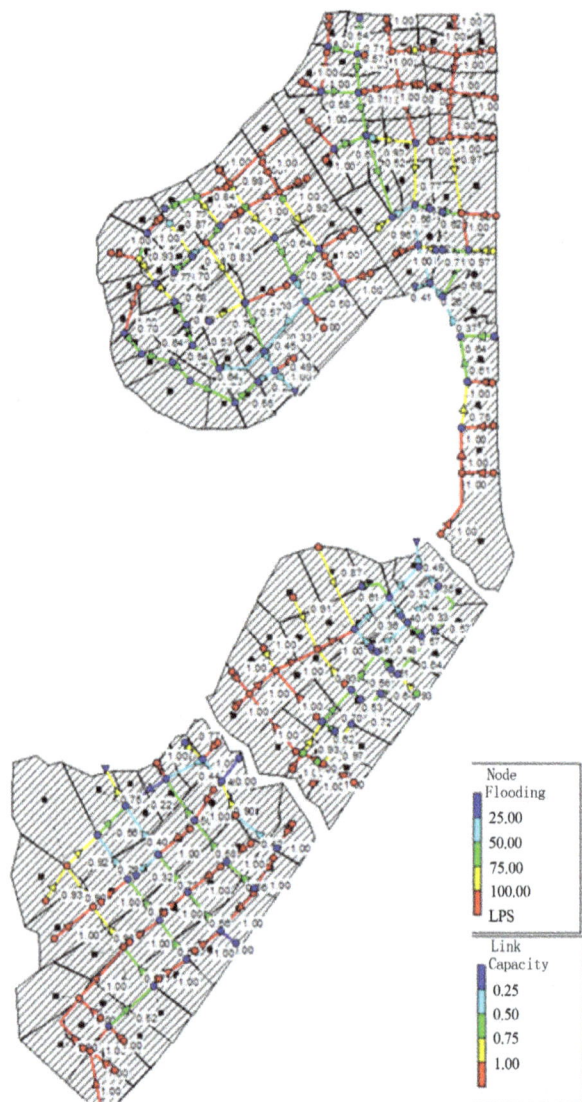

Figure 6. Overview of a SWMM simulation for Tianjin Eco-city.

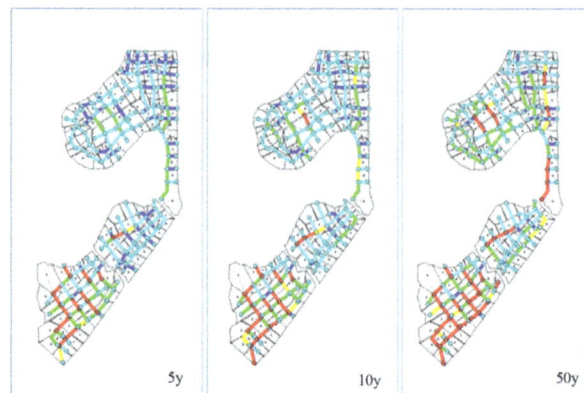

Figure 7. Simulation based on different storm events in Tianjin Eco-city. Red dots and lines indicates overflow where flooding will occur. A large proportion of drainage pipes and tunnels turn red under 10-year return period rainfall event. Massive overflow occurs under 50-year event.

4. DISCUSSION

4.1. Largely Distributed Surface Water System Gains Higher Drainage Efficiency

Storm water drainage system in Tianjin Eco-city is one key feature in its master planning. The scale and design size of the piping networks are surprisingly huge. However, deeply buried pipelines and tunnels have apparent bottleneck effect in terms of the drainage efficiency. Namely, they couldn't drain properly once a pipeline is fully loaded, blocked or poorly drained.

This will not occur to open watercourses which are more flexible. Most waterways in Almere are built with rectangular cross-section and trapezoid revetment. Therefore, their watercourses have certain degree of flexi-

bility capacity in case of heavy rains. Especially, revetments formed on the basis of natural slope protection have inhibiting effects on water surge problems.

It can be seen from the simulation for the above two cases, the widely distributed surface water system of Almere is more efficient with less overloading problem in the drainage path under the same rainstorm reoccurrence period. Furthermore, the two cases are extremely different in utilization effects, although both of them have large open water body. In Tianjin eco-city, converge of water in sub-catchments ultimately flows to the deeply buried pipes, and eventually pumped to the lake. This poses main challenge to the drainage efficiency.

The widely distributed open water network shows obvious advantages to improve drainage efficiency, increase flexibility and ensure stability.

4.2. Surface Water System Takes up Some Land but Returns Added Value

Open surface water system will take up some of the urban space, which may be an overly extravagant choice for high-density urban area. However, parks and greenbelts are essential for planning for most proportions of the urban development areas. These open spaces can be combined with water or low-lying land, and will form attractive urban space. In this case, the urban space occupied by the open water system will return added value in another good way. It can be seen from Almere, the open water system combined with the lake forms a waterfront space for the city, which is more comfortable and interesting than a vast open water area.

4.3. Both Cases Are Specifically Developing towards Water Sensitive City

The theory of water-sensitive city provides us a way to

understand the development course of urban water system management. Tianjin Eco-city, as a case of emerging economy, can be considered as the development stage just entering the drained city from a sewer city. Almere, also as a typical polder city, has developed into a waterway city in advance. From this point of view, Tianjin Eco-city shall take Almere as an example to learn from.

Tianjin Eco-city is suffering serious water scarcity which restricts its development. It has put forward the planning program of taking full use of reclaimed water, which to some extent is transforming to a water cycle city. In this aspect, Tianjin Eco-city seems more advanced. It may also be a reference to the future city development.

5. CONCLUSIONS

With higher frequency of extreme climate events, global climate change is challenging the urbanization process and population concentration in coastal zones. Although there are many differences in premises and backgrounds comparing with Asian coastal cities, Dutch polder cities have demonstrated the potential dealing logic and operation methods under extreme cases.

Almere with its typical layout of polder city, dispersed settlements, dense river network, large area of open water body, precise control of water level, efficient drainage system, sound damming system, provides us with many experiences in coping with the potential coastal risks in low-lying areas. For the many coastal new town construction projects in emerging economies, such as Tianjin Eco-city, more skills must be acquired to prevent seawater intrusion, to maintain internal water balance, and to achieve good water quality, so as to lay a more solid foundation for urban sustainable development locally and globally.

Climate challenges will be a long-term issue for coastal cities that are seeking for larger and denser developments. In the coming decades, the necessities for learning how to adapt to larger range of uncertainty are crucial requirements for urban planners. As a conclusion,

the urban water system planning should always incorporate methods to achieve higher system resilience based on multiple functioning patterns.

6. ACKNOWLEDGEMENTS

Thanks to H. M. de Brauw for discussions on related models. Also thanks to Mingyi Li, Jiexin Cheng for their help on calculation and simulation studies, thanks to Jianjia Wang for her translation work.

The project is supported by International Science & Technology Cooperation Program of China (2010DFA74490).

REFERENCES

[1] Hooimeijer, F.L., Meyer, H. and Nienhuis, A. (2005) Atlas of Dutch water cities. Sun Architecture, Amsterdam.

[2] Schuetze, T. (2011) Climate adaptive urban design with water in Dutch polders. *Water Science and Technology*, **64**, 722-730.

[3] Zhou, C. (1987) Helan shuili guanli gaishu. *Guangdong Hydropower Technology*, **1**.

[4] Schuetze, T. and Chelleri, L. (2013) Integrating decentralized rainwater management in urban planning and design: Flood resilient and sustainable water management using the example of coastal cities in the Netherlands and Taiwan. *Water*, **5**, 593-616.

[5] Dam, P.J. (2001) Sinking peat bogs: Environmental change in Holland, 1350-1550. *Environmental History*, **6**, 32-45.

[6] Tjallingii, S. (2008) The water issues in the existing city. In: Hooimeijer, F. and Toorn Vrijthoof, W.V.D., Eds., *More Urban Water*: *Design and Management of Dutch Water Cities*, Urban Water Series, Taylor & Francis/Balkema, Leiden.

[7] Tjallingii, S. (1996) Ecological conditions. Strategies and structures in environmental planning. Vol. 2, IBN Scientific Contributions, DLO Institute to Forestry and Nature Research (IBNDLO), Wageningen.

16

Selective use patterns of woody plant species by local communities in Mumbwa Game Management Area: A prerequisite for effective management of woodland resources and benefit sharing

Chansa Chomba[1*], Vincent Nyirenda[2], Mitulo Silengo[1]

[1]School of Agriculture and Natural Resources, Disaster Management Training Centre, Mulungushi University, Kabwe, Zambia;
*Corresponding Author
[2]Zambia Wildlife Authority, Directorate of Research P/B 1 Chilanga, Zambia

ABSTRACT

Selective patterns of human uses of woody plants in Mumbwa Game Management Area were investigated using quantitative survey methods. Major causes of human encroachment into the wildlife zone were assessed so that appropriate management actions could be taken to ensure continued supply of goods and services to the local community. Woody plant species were found to be diverse with 93 species recorded in the study area. Of these, the community utilized 92 (99%) in different ways. Trees were cut for various reasons, major ones being; building poles, fire wood, fibre, fruit collection, medicine, bee honey collection, house hold tools and utensils and clearing for agriculture. Clearing for agriculture was the most damaging, because it involved removal of below and above ground woody biomass of all sizes and suppression of their regeneration during cultivation in subsequent years. Of the recorded human uses, 2366 kg of woody plant material was consumed per head/$^{yr-1}$ as fire wood. The day to day consumption of firewood varied with season. In the cold season (May-August), a 26 kg (mean weight) bundle of firewood was consumed in three days while in the warm season it lasts five days. Certain species were particularly selected; *Julbernardia paniculata, Pericopsis angolensis, Brachystegia speciformis, Brachystegia boehmii, Julbernardia globiflora, Brachystegia longifolia, and Pteleopsis anisoptera*. In building and construction, differences were observed in the species and size of poles was used. The mean sizes of roofing poles were 3.5 metres long and 0.18 metres mid-length girth. Wall poles were 2.4 metres long and 0.40 metres mid-length girth. For the main house of about two rooms each, there were an average number of 48 poles in the roof (45,859.2 cm^3) and 28 (284,653.6 cm^3) in the wall. Clearing for agriculture was the main cause of damage to woody plants in the Game Management Area. The extension of human settlements into the wildlife zone and towards the Itezhi-tezhi road is likely to increase loss of woody vegetation, and will have a negative impact on the habitat for wildlife.

Keywords: Utilization; Woody Plants; Local Community; Agriculture

1. INTRODUCTION

Woody plants are central to the livelihood systems of thousands of rural and urban dwellers in Africa [1-8]. A range of products from woodlands support rural living by way of medicines, food, building timber and fuel (**Figure 1**). Ecologically, trees play an important role in controlling soil erosion by intercepting rain drops, providing a habitat to many organisms and also creating a micro climate which both man and beast utilize (**Figure 2**). They are also central to the spiritual needs of people, with specific trees even blocks of woodland being conserved by local communities for cultural reasons, for example, burial groves. Sacred groves associated with spirits of the dead or with territorial deities are found not only in Zambia but also in other countries throughout the miombo

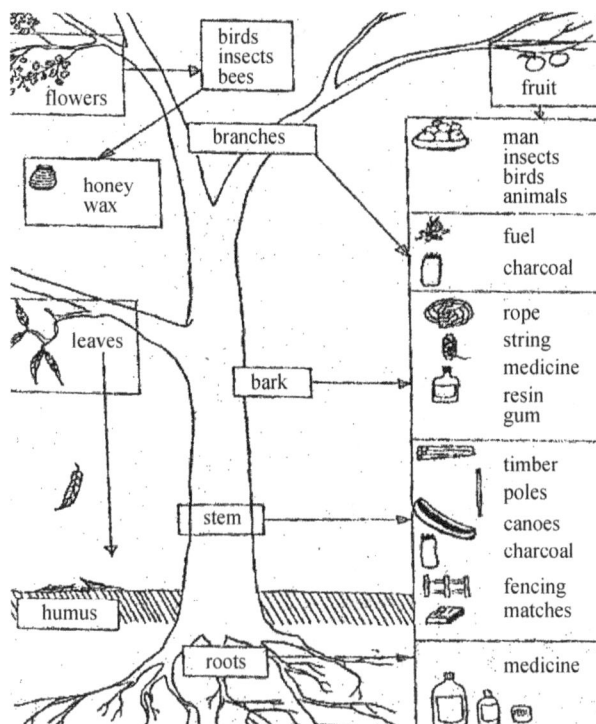

Figure 1. Different human uses of woody plants (source: Storrs, 1968).

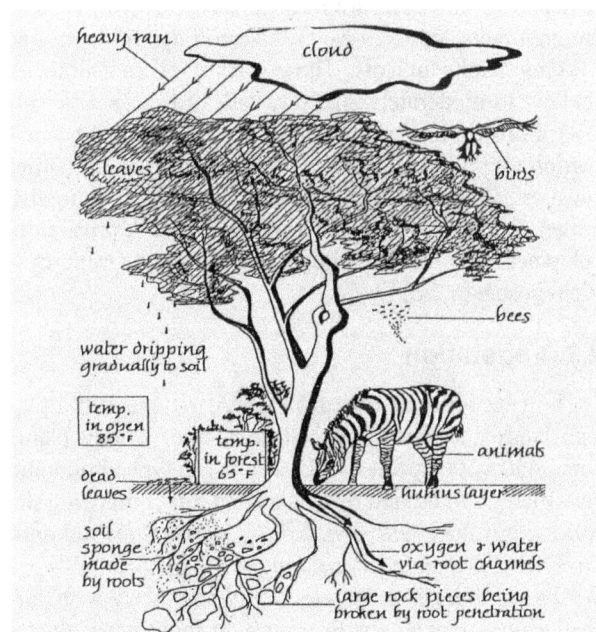

Figure 2. Different ecological uses of woody plants (source: Storrs, 1968).

eco-region of east and southern Africa [7].

Woodlands also provide products for towns and cities in Zambia. The most important of these is fuel, but also a source of income are non-fuel timber products including honey, bees wax, edible fruits, edible insects, vegetables, game meat, mushrooms, traditional medicine, and wood

for handicraft [8-10]. These goods and services create a demand for woody plants and even when the woodlands are located in an inhabitable escarpments and mountainous landscapes of rugged and rocky nature, or are protected by law such as forest reserves, zoned Game Management Areas or National Parks. They are still legally or illegally accessed for collection of various products [11].

In Zambia, there is a growing concern about deforestation. An estimated 0.5% of the country's woodlands were lost each year [12]. The figure could even be higher at the moment. Around large human settlements, this loss may be considerably higher. Although fires, especially those late in the season, are one of the principal causes of deforestation, and it is accentuated by other anthropogenic factors. The cutting of trees for firewood and charcoal, combined with clearing for agriculture for instance, are additional causes of deforestation around large human settlements, towns and major fishing areas [12]. Traditionally, firewood was collected from dead wood on the ground, often as old trees that have died of natural causes or as a product of land clearance for agriculture, but with the increase in human population, this is becoming less practicable and live trees are now being cut for fuel wood. Firewood cutting however, does not amount to complete clearance of woody biomass, as some of the species are preferred to others, so the true extent of the problem is not immediately obvious, and this has led to lack of awareness among the public. Other problems include declining fallow periods and the clearance of land for both small and large scale agriculture [12-16].

In protected areas such as Mumbwa Game Management Area, where conditions of multiple property rights exist with the state through ZAWA legally co-opting traditional/customary tenure to create a statutory system overlaid on a traditional one, the exploitation of resources is often antagonistic. This is not unique to Zambia, as in many parts of Africa, governments lay claim to greater amounts of land and other natural resources than they can effectively control. The result is that authority and management continue to be compromised and open access tendencies thrive.

The local community in Mumbwa GMA, however, benefit from the natural resources, through the provision of social facilities such as schools, clinics as well as employment facilitated by revenue accrued from trophy hunting. Such revenues are shared on 50:50 basis between ZAWA and Community Resource Boards (CRBs). With regards to forest products in Protected Areas however, management is generally centralized under ZAWA and the Department of Forestry (FD). As a result, local people tend to share the proprietorship of the forests with the state in conflicting circumstances. In many instances, local communities permit timber cutting in GMAS without seeking ZAWA's authorization which is contrary to the Zambia Wildlife Act No. 12 of 1998. The other indication of resentment is the increasing levels of illegal settlements in the wildlife zone. The increased numbers of illegal settlers and the accompanying clearing for agriculture and other extractive uses of vegetation exacerbate habitat transformation/loss, weakening the status of Mumbwa GMA as a buffer to Kafue National Park. An effective GMA as a buffer zone, is intended among other things to; 1) extend those habitats contained within the National Park, 2) provide for the needs of threatened and endangered species of animals, 3) act as additional habitat outside the National Park boundary thereby enhancing the capacity to contain wildlife species likely to move out of the National Park [17,18], and 4) sustain trophy hunting and limited photographic safaris from which local communities benefit. The GMA is also supposed to serve other environmental protective functions such as soil and water conservation [18,19]. If a buffer zone is encroached by human settlements, its ecological functions wane. This is because human settlements in most instances have predictable needs for firewood and building timber, such that even when settlements are on the periphery they would eventually expand deeper into the wildlife zone and overtime, the GMA would fail to meet their requirements and settlers would seek such resources from the National Park itself [8,19,20].

The aim of this study therefore, was to assess human use of the woody plants in Mumbwa GMA, focusing on clearing for agriculture, wood fuel and building poles and to a lesser extent other uses such as honey collection, medicines and house hold tools. This study targeted woody plants because they take longer time to grow than herbaceous plants. Popular woody species that are habitat specific slow growing and slow reproducing and which are debarked or dug out for dyes, medicine, food or horticultural purposes are particularly vulnerable to over harvesting. Studies carried out in Tanzania showed that effects of harvesting of trees for building materials or fuel wood would be more apparent if there was; complete removal of woody biomass, low levels of recruitment and suppression of coppices [11,21,22]. Harvesting of poles and laths for instance utilize much smaller stem size classes than the timber industry, but nevertheless can impact on tree recruitment and woodland structure [21, 23].

In order to develop appropriate management strategies that would ensure continued supply of goods and services on a sustainable basis, the human use of woody plants as well as their ecological importance in maintaining soils and biota should be understood. **Figures 1** and **2** demonstrate the different ecological and human uses of woody plants which should form the basis for the need to protect them, which is why this study is important.

2. METHODS AND MATERIALS

2.1. Study Area

Mumbwa GMA is situated on the eastern boundary of Kafue National Park. It lies between 14°58' and 15°18' south and 25°58' and 26°58' east and covers an area approximately 3370 km^2 in extent (**Figure 3**).

Relief and drainage comprise of an undulating landscape with a general elevation of approximately 1000 m above sea level broken only by a ridge of over 1200 m a.s.l. in a SW-NE direction. The Nansenga and Nagoma Rivers dominate the drainage system.

Mean annual rainfall ranges from 1000 mm with mean annual rainy days of 80 except for the southern tip which has 70 days. The mean date for the onset of rains is 10th November and the mean date for retreat is 20th March. The evapouration rates reach the peak of 295 mm in October, falling to 115 mm in February. These figures may however, change in view of the threats of global warming and changing weather patterns.

The geology is underlain predominantly by the Karroo-Granite complex flanked by a series of Katangan meta—sediments, shales, sandstones and siltstones to the northwest and southwest. An inlier of Katanga rocks is found towards the northwest corner of the GMA [8,24]. The GMA soils, however, are of different types with 88% of the area being covered by Orthi-eutric dystric leptosols, Chromic-haplic uxisols. These consist of well drained, shallow to moderately shallow, dark brown to dark yellowish brown, rocky, fine loamy to clayey soils. A small portion (approximately 12%) in the north western corner consists of moderately shallow, brown to yellowish brown, friable, clayey soils with a very weak profile development and in places imperfectly drained Orthic-dystric regosols [8,24].

2.2. Vegetation

The vegetation of Mumbwa GMA is typical of a plateau landscape, with the total number of woody plants being 450 - 500 species [8,25-27]. It is a typical miombo classified as woodland of plateau and hill comprising the following; Pure *Julbernardia paniculata* woodland, rarely with *Brachystegia longifolia* of *Brachystegia floribunda*; pure *Brachystegia longifolia* rarely with *Julbernardia paniculata* or *Brachystegia spiciformis*; almost pure *Brachystegia utilis* woodland occasionally with a little *Julbernardia paniculata*, *Brachystegia longifolia* or *Brachystegia spiciformis*; almost pure *Brachystegia boehmii* woodland occasionally with a fair amount of *Julbernardia paniculata*; pure *Brachystegia* spp woodland with little *Julbernardia paniculata* or *Brachystegia longifolia*; pure *Brachystegia bussei* woodland with *Sterculia* spp; almost low pure *Brachystegia spiciformis* woodland on hill tops; *Julbernardia paniculata*—

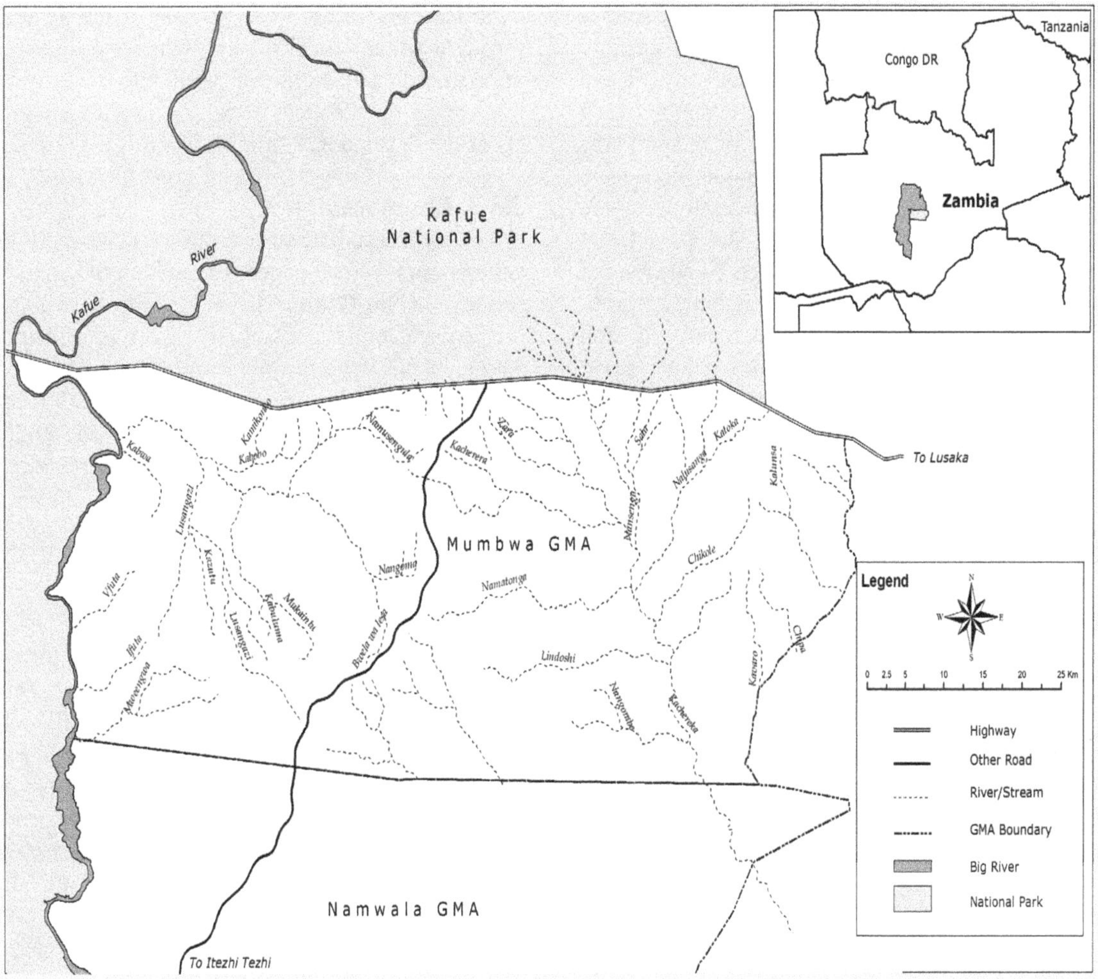

Figure 3. Location of study area, Zambia.

Brachystegia longifolia woodland with varying percentages of the two co-dominants; mixed miombo woodland and various mixtures of almost all plateau species of *Brachystegia* with *Julbernardia paniculata*.

The chief characteristic of the miombo woodland in the study area was its openness below canopy. Small trees and shrubs were scattered so that visibility within the woodland was high. The chief associates of *Julbernardia* and *Brachystegia* species were *Acacia macrothyrsa*, *Albizia antunesiana*, *Azanza garckeana*, *Bauhinia petersiana*, *Clerodendrum glabrum*, *Cordia* spp, *Grewia flavescens*, *Margaritaria* spp, and *Ximenia caffra*. *Acacia scheinfurthii*, *Capparis tomentosa*, *Clematis brachiata*, *Hippocratea indica*, *Jasminum stenolobum*, *Turbina shirensis* and *Vernonia aurantiaca* were the most prominent climbers.

The *Acacia-Albizia* woodland was mainly found on the alluvial plains where it formed very open woodland with a few shrubs but abundant tall grass. The dominant species were *Acacia* and *Albizia* spp, which occurred in pure or almost pure stands accompanied by *Ficus capen-*sis, *Combretum imberbe*, *Lonchocarpus capassa*, *Combretum ghasalense*, *Piliostigma thonningii*, and *Zizyphus abyssinica*.

2.3. Human Settlements

Settlers were either randomly distributed as determined by water availability or linearly distributed along main streams, Lusaka-Mongu highway and access roads. The population was densest between the Lutale River and Nansenga stream. Since the 1990s however, settlements have been spreading towards Chungu stream into the wildlife zone. The total human population in the GMA was estimated at 6000 in 2000 but had increased to more than 10,000 by 2011.

2.4. Field Methodology

2.4.1. Vegetation Use Assessment

Transects 1 km long each were laid out at randomly located points along the Nansenga River which forms the boundary between settlement and wildlife zones. At each of the random points along the boundary, transect lines were laid out perpendicular to each other for 1 km into the settlement and wildlife zones respectively. There were ten transects in each zone and ten plots along each transect line. The modified Whittaker plot method was used [28]. Along each transect line, sample plots were located at random [29,30]. The plots were used to sample woody plants greater than 10 cm dbh. The diameter at breast heigh was taken to be 1.3 metres above ground [29,30] The variable measurements included; girth, height and crown diameter [29,30]. In measuring dbh circumference was measured and converted later using the formula; $C = D\pi$. Thus $D = C/\pi$. The basal area (BA) being calculated as πr^2 while volume was calculated as $\pi r^2 h \times 1/3$; where C = circumference, D = diameter, H = height, r = radius and pie ($\pi = 3.14$) is a constant.

Where tree stems branched below dbh, the individual stems were counted and their measurements combined as also applied in other similar studies [8,29]. If the branching was at ground level, each stem was treated as an individual tree. This was because measurements used to calculate basal area were taken at 1.3 metres above ground level. Tree height was measured using a tree height measuring rod. Canopy cover was measured using a 100 metre measuring tape laid on the ground from one edge of the crown to the other D_1, the second measurement being taken perpendicular to the first one as D_2.

All trees of circumference greater than 10 cm dbh in each 25 m × 25 m plot were examined for any human induced damage. Human damage categories were developed as described below to give an indication of the frequency with which a species was the most diverse in its uses and whether it may be in danger of over exploitation. Damage categories were classified as; coppice when the stem was cut below 50 cm above ground level; pollard when cut at above 50 cm; lop when branch was cut; strip when bark had been removed as in fibre collection and scarred when there are identifiable axe or other human made tool marks on the stem or branch. Difficulties were encountered where vernacular names referred to large genera and many species with one common name for different species. This was the case with *Combretum* and *Terminalia* species. Specimens of plant species not identified in the field were taken to the herbarium for verification. Data collected was analysed using a Minitab 14.0 computer software programme. This survey was carried out in 1998 and 2011/12 seasons.

2.4.2. Questionnaire Survey

The survey covered villages located in the settlement zone of the GMA. The villages sampled were within 1 km of the wildlife zone boundary. This closeness to the wildlife zone boundary allowed the assessment to determine the influences that villagers might have on the

wildlife zone. Households resident at the same locality for no less than 12 calendar months were interviewed. A pre coded questionnaire was used with the last two questions being open ended, designed to reveal the many facets of the respondents' attitude towards the GMA.

2.4.3. Enumerators

To ensure gender balance, three female and three male enumerators were engaged from within the communities' resident in the GMA. The six enumerators took measurements of buildings and fire wood as well as participating in the Participatory Rural Appraisal. The advantage was that they were familiar with local species and were knowledgeable of and known to the community. Prior to starting work, the enumerators were trained for a period of two weeks during which trial questionnaires were conducted to determine their suitability for the local community.

2.4.4. Participatory Rural Appraisal

Groups of people resident in the GMA of different age groups and gender depending on the topic being discussed were gathered together and information collected from them by asking questions regarding natural resource utilization. Interviewees were assured that any answer given would not lead to any criminal proceeding. The PRA took advantage of Sunday masses, wedding parties and other public gatherings as organizing meetings solely for PRA were non responsive. The number of questions was limited depending on the occasion as it was found that asking too many questions led to inaccurate responses.

2.4.5. Building Materials and Firewood Survey

For a given human use, some plants were preferred by local people to others and are therefore, sought out and used. This provided the researchers with an indication of the relative importance of each species to the local community. Species preferences were also determined from the house hold consumption surveys, which gathered data on the frequency with which species were used for certain purposes, the preferred species for each human use and reasons for their selection.

2.4.6. Measuring Wooden Structure

Building materials were divided into two major categories: roofing poles and wall poles for houses and floor pole, horizontal and upright poles for grain racks (**Figure 4**). Measurements of 32 houses randomly located in different villages were made. In each building; house, grain rack or granary, species used were identified, mid length girth taken in cm, and length of pole measured in cm. For walling poles, the examiner excavated (with the permission of the owner) the pole to expose the part buried in

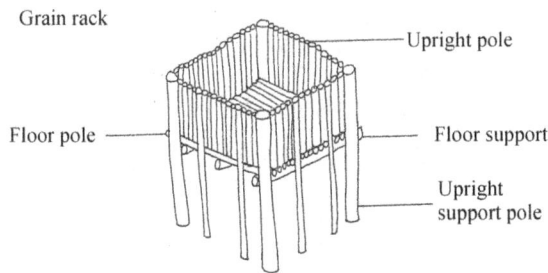

Figure 4. Conceptual use of woody plants in grain rack construction, Zambia 2012 (modified after: Chansa, 2000).

the ground to measure the length of the pole and how deep the poles were buried in the ground. Permission was sought from each house hold to enter their house and take measurements of roofing and wall poles.

2.4.7. Measuring Firewood

Units of head loads, wheelbarrow, bicycle and scotch cart towed by oxen or donkeys were measured. Bundles of firewood were weighed and the number of poles counted. In order to investigate the methods of fire wood collection used, the three female enumerators accompanied women on firewood collection trips. This is because women were the main firewood collectors as recorded during PRA and questionnaire survey. The enumerator allocated all the fire wood collected into categories as, felled trees, taken from woodpiles made during clearing of fields for agriculture, from standing dead trees, or gathered from the ground. Group meetings were also held to consolidate the commonest firewood collection methods. In each village, species preferences and uses of firewood were recorded from the bundles of firewood examined. All bundles of firewood examined were weighted to determine weight in kg.

2.5. Limitations

Determining species preference and classifying damage categories involved in some instances, relatively arbitrary approaches and subjective assignation of numerical values to different criteria. Although this approach has been criticized by some researchers, the aim in this study was to obtain relative and not absolute values of species' preferences and damage and was therefore effective in achieving this goal. The boundary between the wildlife and settlement zones was also often arbitrary except where it was demarcated by a prominent physical feature such as a river.

3. RESULTS

3.1. Use of Woody Plants by Local Communities

In this study 93 plant species were recorded and of

these 92 (99%) were used by the local community. In fact, each one of them had more than one use (**Table 1**). Fourteen were recorded for medicinal and magical uses, 21 for tools and utensils, 25 for wall poles, 33 for roofing poles, seven for firewood, 13 for fruit trees and five were reserved for hanging bee hives.

3.2. Differences in Woody Plant Density between Settlement and Wildlife Zones

There were more stems and trees per hectare in wildlife zone than settlement zone. Three species were recorded to have equal relative frequency of 5% in both settlement and wildlife zones. These were; *Julbernardia panniculata*, *Brachystegia boehmii* and *Erythrophleum africanum*. *Brachystegia boehmii* was the most abundant in both settlement and wildlife zones, with relative sighting frequency of 14% in settlement zone and 20% in wildlife zone. *Diplorynchus condylocarpon* was more abundant in the settlement zone than wildlife zone 9% and 5% respectively. The mean and total basal area for all species was higher in the wildlife zone than in the settlement zone, 8420.20 cm^2; 117,549 cm^2 for wildlife zone and 4346.82 cm^2; 30,764 cm^2 for settlement zones respectively. Many species had been eliminated from the settlement zone as their sighting frequency was low. Species which had been eliminated in the settlement zone showed low relative sighting frequency of less than 5% in the wildlife zone. Local communities extended their use of woody plants into the wildlife zone seeking species of preference which had been eliminated in the settlement zone.

3.3. Building Materials

Thirty two buildings were surveyed, in which 36 species were recorded as being used in roofing and 25 in walling. Mann-Whitney U test showed a significant difference in the number of species used for roofing and walling (P < 0.001). Woody plant species were selected depending on the human use type and special plant characteristics befitting that human use. Depending on the type and part of building some species were used as roofing poles and others as wall poles while others were used in both walling and roofing. *Diplorynchus condylocarpon*, *Erythrophleum africanum*, *Parinari curatellifolia*, *Pericopsis angolensis* and *Burkea africana* were the five most commonly used species for roofing timber. In wall construction, *Pericopsis angolensis*, *Erythrophleum africanum* and *Burkea africana* were the three most commonly used species (**Table 1**).

3.4. Pole Size Class Selection in Building Materials

Among the 32 houses surveyed, girth size and length

Table 1. Human uses of woody plants in Mumbwa Game Management Area, Zambia, 1998-2011.

Species name	Human uses					
	Firewood	Roof pole	Wall pole	Tools	Fruits/Medicinal	Bee honey/Other
Acacia polyacantha	na	*	na	na	na	na
Acacia tortilis	na	*	na	na	na	na
Afzelia quanzensis	*	na	na	*	na	*
Albizia adianthifolia	na	na	na	na	na	*
Albizia antunesiana	na	na	na	na	na	*
Albizia harveyi	na	na	na	na	na	*
Albizia versicolor	na	*	*	na	na	*
Anisophyllea boehmii	na	na	na	na	*	na
Azanza garckeana	na	na	na	na	*	na
Balanites aegyptiaca	na	na	na	na	*	na
Bauhinia petersiana	na	*	na	na	na	*
Bobgunnia madagascariensis	na	*	*	*	na	*
Brachystegia boehmii	*	*	*	na	na	*
Brchystegia longifolia	*	*	na	na	na	na
Brachystegia manga	*	na	na	*	na	*
Brachystegia spiciformis	*	na	na	*	na	*
Bridelia micrantha	na	*	*	na	na	na
Burkea africana	na	*	*	*	na	*
Capassa violacea	na	na	na	na	na	*
Combretum collinum	na	na	*	na	na	na
Combretum molle	na	na	*	*	na	na
Combretum fragrans	na	na	*	na	na	*
Combretum imberbe	na	*	*	na	na	na
Combretum spp	na	na	*	na	na	na
Combretum zeyheri	na	*	*	na	na	na
Commiphora mollis	na	*	*	na	na	na
Dalbergia melanoxylon	na	*	*	*	na	na
Dalbergia nitidula	na	*	*	*	na	na
Dalbergia nyasae	na	*	*	na	na	na
Dichrostachys cinerea	na	na	na	na	*	na
Diospyros kirkii	na	na	na	na	*	*
Diospyros mespiliformis	na	na	na	na	*	*
Diospyros spp	na	na	na	na	na	*
Diplorynchus condylocarpon	na	*	*	na	na	*
Ekbergia benguelensis	na	na	na	na	na	*
Entada abyssinica	na	na	*	na	na	na
Etythrophleum africanus	na	*	*	*	na	*
Faurea intermedia	na	na	na	na	na	na
Faurea speciosa	*	*	*	na	na	na
Flueggea virosa	na	na	na	na	na	*
Ficus sur	na	na	na	na	na	*
Ficus spp	na	na	na	na	na	*
Gardenia spp	na	na	na	na	na	*
Hexalobus monopetalus	na	na	na	na	na	*
Hymenocardia acida	na	na	na	na	na	*
Julbernardia paniculata	*	*	*	na	na	*
Julbernardia globiflora	*	na	na	na	na	*

Continued

Kigelia pinnata	na	na	na	na	*	*
Lannea discolor	na	na	na	na	na	*
Markhamia obtusifolia	na	*	na	na	na	na
Monotes africanus	na	*	*	na	na	na
Monotes glaber	na	na	na	na	na	*
Monotes spp	na	na	na	*	na	na
Ochna pulchra	na	*	na	na	na	na
Olax obtusifolia	na	na	na	na	na	*
Olfiedia dactylophylla	na	na	na	na	na	*
Ozoroa insignis	na	na	na	*	na	na
Parinari curatellifolia	na	*	*	*	*	na
Peltophorum africanum	na	*	na	na	na	na
Pericopsis angolensis	*	*	*	*	na	*
Pilostigma thonningii	na	na	*	na	*	*
Pteleopsis anisoptera	*	*	*	na	na	na
Pterocarpus angolensis	na	*	*	*	na	na
Pseudolachnostylis maprouneifolia	na	*	*	*	na	na
Rauvolfia caffra	na	na	na	na	*	*
Rhus longipes	na	na	na	na	*	*
Senna abbreviata	na	na	na	na	*	*
Sclerocarrya caffra	na	na	na	na	*	na
Schrebera trichoclada	na	*	na	na	na	na
Staganotaenia araliacea	na	na	na	na	na	*
Sterculia quinqueloba	na	na	na	*	na	na
Strychnos cocculoides	na	na	na	*	*	na
Strychnos innocua	na	na	na	na	*	*
Strychnos pungens	na	*	na	*	na	*
Strychnos spinosa	na	na	na	*	na	*
Syzygium cordatum	na	na	na	na	*	*
Syzyguim guineense	na	*	*	na	*	na
Tamarindus indica	*	na	na	na	*	*
Terminalia sericea	na	*	*	na	na	na
Terminalia spp	na	na	na	na	na	*
Trichilia emetica	na	*	*	*	*	na
Uapaca kirkiana	na	*	*	*	*	na
Uapaca nitida	na	*	*	na	*	na
Uapaca sansibarica	na	na	na	na	*	na
Uvariastrum hexaloboides	na	na	na	na	na	*
Vitex doniana	na	na	na	*	*	na
Ximenia americana	na	na	na	*	*	na
Ziziphus abyssinica	na	na	na	na	na	*
Zanha africana	na	na	na	na	na	*

Notes: Tree species whose uses were not properly defined by the local community were grouped together under other uses. Species which had more than one use or received greater damage were probably most useful to the community.

of poles varied between roofing and walling poles. Roofing poles were longer than walling poles. The average length for roofing poles was 3.5 metres (n = 1856 poles) and mid length girth was 0.18 metres. The walls consumed shorter but thicker poles than roofs. The average wall pole length was 2.4 metres (n = 896 poles) and the mid length girth was 0.40 metres. The average number of poles in the roof was 58 and 28 for the wall. The average number of species per roof was 4 and 2 for the wall. *Pericopsis angolensis*, *Burkea africanum* and *Erythro-*

pleum africanum were commonly used both in roofing and walling units. For these species, relative sighting frequency was below 5% implying that their numbers were declining.

3.5. Methods of Collecting Building Poles

Different methods were used to collect building poles and these were; cutting down the whole tree of desirable size (60%), climbing a tree and lopping or pollarding branches of desirable girth and straightness (30%) and re-using old poles (10%).

3.6. Firewood

3.6.1. Pole Size, Species Selected and Transport Used

As for firewood of the total 60 bundles sampled, the average weight per bundle was 26 kg and 91 bundles or 2.3 tonnes (2366 kg) were consumed per head per year. Seven species were particularly selected for firewood. Chi-square test showed a significant difference in the selection of species for firewood (P < 0.001). The reasons given for the selection of species were those yielding good fire with embers lasting a long period of time while producing little smoke. *Pericopsis angolensis* in particular was chosen because it gave a hot fire and *Julbernardia paniculata* because it gave hot embers and did not spark (**Figure 5**). As regards transport, head loads method of fire wood transportation, was the commonest 60%), bicycle 27% and wheel barrow was the least 13%.

Division of labour was based on age group and gender. Women collected most of the firewood (68%, n = 600) and of these, female children comprised 31% (n = 600).

This was probably the reason why men seemed generally unconcerned about fuel needs.

3.6.2. Methods of Collection

Four common methods of firewood collection were recorded based on 360 respondents:

1) Dead wood collected from the ground. Women and children said that this was an easy method as it did not require cutting down a tree. This method though preferred by women and children was not common (15%) because it was difficult to find dead wood in the settlement zone. Late fires usually consumed above ground dry biomass.

2) Standing dead trees are identified and cut down. This was one of the commonest methods (30%). For hard trees such as *Pericopsis angolensis*, cutting was done by men, usually the husband or older male children. The mother and children would then return as demand arises to collect branches from the fallen tree. Usually only the branches were removed and large stems were abandoned unless the tree was near the homestead, in which case it would slowly be cut up until finished. Villagers confessed (83%, n = 180) entering the restricted wildlife zone or the adjacent National Park in search of dead trees to cut down. One of the respondents whose name could not be revealed for fear of being arrested explained that when they wanted to poach firewood from the National Park, they normally did not cut it using an axe unless it was very far (>5 km) from a Wildlife Scout's camp as doing so nearby would alert them. Instead they used fire which was left overnight burning, returning the following day to increase the volume of fire if the tree was still standing or to break off branches if the tree had fallen.

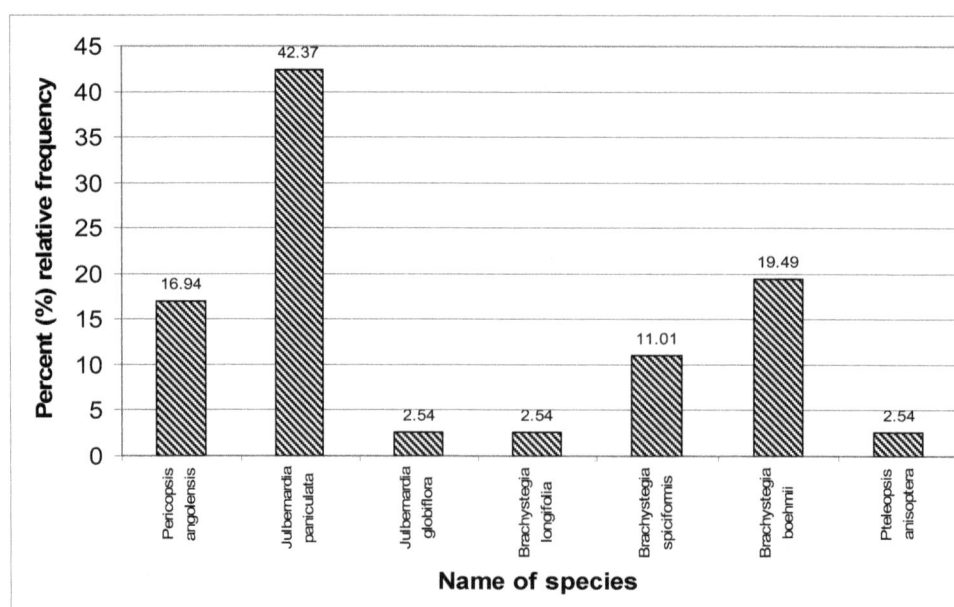

Figure 5. Firewood preferences in the study area, Zambia, 1998-2011/12.

3) The other method (30%) was the collection from wood piles. These woodpiles are made when fields are being cleared in readiness for cultivation. In this method there was less species selection as all the woody plants were cleared prior to cultivation. The sizes of poles ranged from saplings with circumference less than 10 cm (usually used in rekindling a fire) to big stems. If the field was near home, the big logs were slowly reduced to nothing by slicing off with an axe starting with the bark until the whole log has been taken. This method was most popular during the cultivation period (late October-December) when households were busy with farm work and little time was available for other activities. It saved a lot of time said one of the female respondents.

4) In this method, a preferred species was felled down and left to dry for many days. Though destructive, it was the second commonest (25%) method. Firewood collectors, mainly women and children returned to collect as need arose. When a tree has been felled for this purpose then it belonged to that particular household and no other person would knowingly collect the firewood without permission from the rightful owner. This method was highly selective of the most desirable species such as *Julbernardia paniculata* which was chosen because it can burn and give good embers even before it becomes completely dry.

3.6.3. Seasonality of Firewood Consumption

All the respondents in the survey (100%, n = 600) used firewood for heating, cooking and lighting. All the respondents in the survey (100%, n = 600) reported using more firewood in the cold season May-August during which time an average 26 kg bundle of firewood is consumed in three days time while the same quantity lasted 5 days in the warm season. In the cold season, firewood in addition to cooking was used for heating. This was also the season when maize (*Zea mays*), sorghum (*Sorghum vulgare*) and finger millet (*Eleusicine coracana*) are harvested. Eighty five of the respondents indicated that most wedding parties and rituals that require beer brewing were postponed to this period of the year when food was plentiful. Beer brewing, wedding parties for instance, consume more firewood per unit time than the ordinary day to day usage and bigger logs of circumference usually greater than 50 cm are used. Single women or widows with school going children reported brewing beer more often as away of raising money for school fees. Some reported to brew up to 200 litres per week. In the rainy season, firewood is piled under shelter of the roof of a kitchen or main house to keep it dry, but one respondent said this way of keeping firewood provides shelter for snakes and would rather have it lean against a big tree so that rain water drains off quickly.

3.7. Other Uses of Woody Plants

Honey hunting was a common practice in the local community. Many male respondents (45%, n = 280) confirmed having collected honey from bees (*Apis mellifera*) at least once every year from either wildlife zone or the National Park. The honey collections seasons were; March-May and August-November.

Three common ways of locating bees were recorded. Keeping a look out on trees and the ground for bees; blowing a homemade instrument (usually made from a dry fruit of *Strychnos cocculoides*). The instrument produced a whistle-like sound which then attracted then greater honey guide (*Indicator indicator*), and chance encounters of the honey guide as they walk in the bush. It was also found that bees selected host trees and were most popular in five of the recorded species (**Table 1**).

3.7.1. Fibre Collection

Fibre was mainly collected for domestic use. Commercial production of mats and baskets was not recorded. Households engaged in mat making preferred palm leaves.

Fibre was used when collecting firewood to tie it up in bundles. Fibre was not required for big logs as these would be carried by pickup truck, wheel barrow or Scotch-cart. For head loads, strong fibre was needed so that it could not break during transportation. *Brachystegia boehmii* was the main source of fibre, but when this species was not available other members of the genera *Brachystegia* or *Julbernardia* were used. Sometimes women recycled old fibre by soaking it in water to make it soft. Saplings of 10 cm in girth and above were used by either stripping the bark or by felling and debarking. The debarked branches were later picked up as firewood. During fibre collection, strips of 4 - 5 cm wide and about 1 m long were made and thus about 50% of the circumference of young trees was debarked but sometimes the whole stem was stripped. Fibre for building was collected by men and usually targeted larger trees.

3.7.2. Medicinal Plants, Wild Fruits and House Hold Items

1) Medicinal and Magical Value

Most respondents (75%) were reluctant to reveal medicinal and magical uses of trees. Where names of trees were given, only the disease or condition treated was indicated with no details of how the concoction was prepared and how the treatment was done. Trees recorded for medicinal value are shown in **Table 2**.

2) Wild Fruits

Many respondents (67%, n = 120) indicated that wild fruits were important food items. Thirteen wild edible species were recorded (**Table 3**). All fruits were collected free of charge from the bush. *Anisophyllea boehmii* and

Table 2. Trees of medicinal and magical values recorded in the study area, 1998-2012, Zambia.

Species	Disease treated	Magical value
Kigelia africana		The fruit is used to enlarge male private parts
Olax obtusifolia		Used as a sexual stimulant
Piliostigma thonningii		Bark is used as a lucky charm
Combretum zeyheri	Roots are used for treating bilharzia	
Rhus longipes		Charm for attracting women
Senna abbreviata	Root concoction mixed with parts from other trees is used to treat venereal diseases	
Diplorhynchus condylocarpon	Roots and leaves are used to treat snake bites	
Balanites aegyptiaca	Kills bilharzia snails	Root concoctions mixed with animal parts especially warthog tail, is used as a sexual stimulant and also cures male impotence
Rauvolfia caffra		Used to repel visiting witches
Bauhinia petersiana	Root concoctions used to treat stomach upsets	
Oldfieldia dactylophylla	Root concoctions used to treat venereal diseases	
Dalbergia nitidula	Used to treat venereal diseases and pneumonia	
Ekbergia banguelensis	Root concoctions are used to promote female fertility and also cures gonorrhea	
Sclerocarya caffra		Used to regulate sex of the child to be born as required by the couple

Table 3. Indigenous fruit trees and nutrient composition of some selected species in the study area, 1998-2011, Zambia.

Species	Popularity	Sold	Nutritional value					
			Energy (KJ)	Energy (Cal.)	Water (g)	Prot. (g)	Fats (g)	Ca (g)
Anisophyllea boehmii	**	+	1345	322	74	0.60	1.0	45.0
Azanza garckeana	*	+						
Diospyros mespiliformis								
Parinari curatellifolia	*	+	1474	353	70.8	2.5	0.3	
Strychnos cocculoides	*	+	1382	331	76	0.3	0.3	45.0
Strychnos innocua								
Syzygium guineense			1407	337	80	5.9	5.5	600.0
Tamarindus indica								
Uapaca kirkiana	**	+	1420	340	58	0.2	1.0	40.0
Uapaca nitida								
Uapaca sansibarica								
Ximenia americana								

Key: En. KJ = Energy in kilojoules, En. Cal = Energy in kilocalories, Prot. = Protein; *Popular, **very popular, + sold.

Uapaca kirkiana were the most popular. *Parinari curatellifolia, Anisophyllea boehmii* and *Uapaca kirkiana* were occasionally sold at the road side. However, half of the respondents (50%, n = 120) indicated that the ever-growing preference for exotic fruits like orange (*Citrus* spp), mango (*Mangifera indica*) and others relegates wild fruits to poor families or for using during periods of drought. Fruit trees however, were preserved in all farm plots, around homes for shade and food.

3) Household Items, Species Used and Replacement Period

Eight house hold items were recorded to have made from woody plants. There were a strong species selection for most tools and utensils. These were; mortar, pestle, cooking stick, axe handle, hoe handle, canoe making, stool and drums of various size (**Table 4**). The durability of the tree species used and/or the treatment made during the preparation and curative process of the item determined the replacement period. Tools that were made from strong species and that were treated by for instance passing through a fire or soaking in water for a given length of time lasted longer than those that did not re-

Table 4. Household items, woody plant species used and replacement period in years, 1998-2012, Zambia.

Name of household tool or utensil	Tree species used	Replacement period in years
Motar	*Albizia antunesiana*	20
	Pterocarpus angolensis	20
	Terminalia spp	20
	Parinari curatelifolia	20
Pestle	*Monotes* spp	18
	Burkea africana	20
	Bobgunnia madagascariensis	20
	Terminalia spp	18
	Albizia antunesiana	15
	Pericopsis angolensis	20
Cooking stick	*Strychnos cocculoides*	5
	Uapaca kirkiana	8
	Ximenia Americana	5
	Ozoroa reticulata	8
	Pterocarpus angolensis	8
	Faurea speciosa	7
Axe handle	*Brachystegia spiciformis*	2
	Dalbergia nitidula	5
	Julbernardia globiflora	2
	Bobgunnia madagascariensis	5
	Pterocarpus angolensis	5
Hoe handle	*Julbernardia globiflora*	2
	Brachystegia spiciformis	4
	Pterocarpus angolensis	4
	Bobgunnia madagascariensis	4
	Albizia antunesiana	5
Canoe	*Afzelia quanzensis*	3
Stools	*Albizia antunesiana*	>20
	Afzelia quanzensis	>20
	Brachystegia spp	>20
Drums	*Albizia antunesiana*	>20
	Sterculia quinqueloba	>20

ceive any treatment. The species of the tree used and the treatment used were found to be the most important factors determining the replacement period. This did not however, discount the extent and frequency of use which would also reduce the replacement period.

3.8. Incidences of Cut Categories

Cut categories 1 - 3 which involved cutting of the stem or branch were combined to make one category and cut categories 4 - 5 which do not involve cutting of stem or

branch formed another category. A significant difference was found between the two classes, with category 4 and 5 being the commonest. Comparison of the five categories showed that stripping was the commonest form of damage Mann-Whitney U test $0.02 < P < 0.05$ (**Figure 6**).

4. DISCUSSION

4.1. Use of Woody Plants

Local communities in Mumbwa GMA utilize woody plant species in many ways to provide for their subsistence. The study proved that despite lower species diversity in miombo woodlands compared with tropical rain forests, the miombo woodland species probably provide more useful types per species than the tropical rainforest species [31]. In their study of the ethno botany in Amazonian Terra-Firme rain forest, found between 34 and 76 tree species which were useful in at least one way or another per hectare plot which was between 49% and 79% of all species present. In this study conducted in a miombo woodland, 92 (99%) of the 93 species recorded were used by local communities and many of the species had more than one human use per tree species.

Other studies on miombo woodlands in central Zambia indicate that most of the research works on vegetation and use has often over emphasized the great diversity of tropical rain forests [38]. Yet the miombo woodlands provide a great diversity of uses even if the number of species per unit area may be less than the tropical rain forest.

4.2. Differences in Plant Density between Settlement and Wildlife Zone

Of the 93 species recorded during the study, 64 were recorded in the wildlife zone and 39 in the settlement zone. The differences in species diversity are attributed to clearing for agriculture in the settlement zone. Other uses such as collection of building poles and firewood may also selectively eliminate some species which cannot tolerate the impact. Clearing for agriculture for instance, involves uprooting trees and subsequent suppression of regeneration during cultivation. Some species however, such as *Bracystegia boehmii* and *Diplorynchus condylocarpon* are still common in the settlement zone probably because of the high regeneration potential and fire tolerance for the latter.

It is also important to note that species that provide more than one use where for instance, the roots, bark, stem and leaves are useful are likely to disappear earlier than those with fewer uses. Form the results of this study, it would appear that when preferred species are eliminated in the settlement zone, rather than turning to alternatives or planting fast growing exotic species, villagers

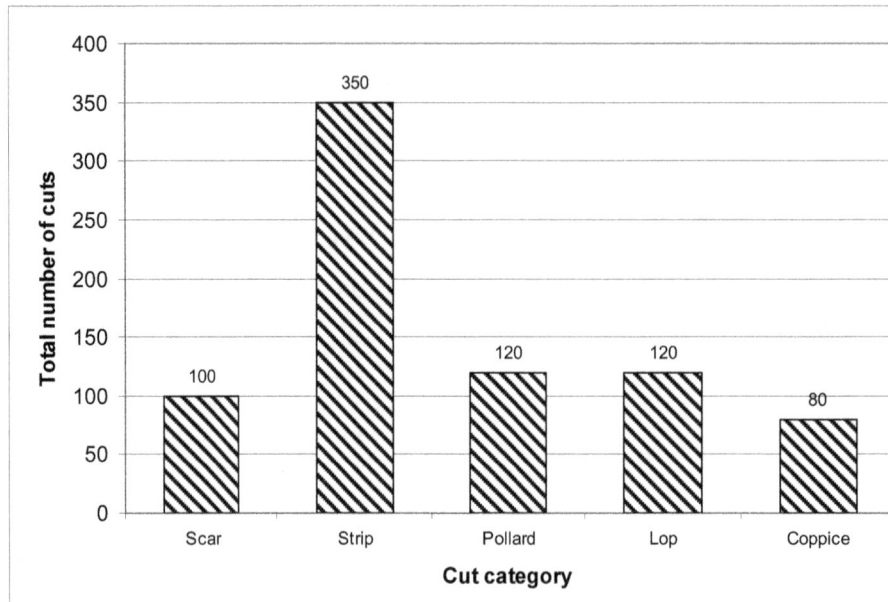

Figure 6. Comparisons of cut categories; coppicing, pollarding, lopping, stripping and scarring in the study area, 1998-2012, Zambia.

chose to collect from the wildlife zone and National Park. This trend is expected to increase as settlements expand. The declining sighting frequency of some species in the settlement zone is an indication that they are being sought after for their particular characteristics. After exploring different ways in which woodland resources are utilized, [8,32,33] showed that it was apparent that the increasing scarcity of miombo woodland resources in settled areas is largely a function of clearance of woody species for cultivation. This study also proves this fact.

In Mumbwa GMA, considerable tenural conflicts arise because rural household believe they have the right to use the land and the resources by virtue of proximity or historical connection to the land often leading to people extending their settlements into the wildlife zone where they access virgin land for agriculture and game meat for protein. Respondents 35 years and older especially women showed great attachment to the resources of the area. Such conflicts are also experienced in other countries in the southern African sub region. In Malawi, for example, [8,32] reported encroachment by smallholders on state lands for poles, firewood, thatch, grazing and even cultivation. The frequency of such cultivation suggests that the creation of estates has restricted small holders' access to forest and other resources as well as access to arable land needed to meet their food requirements. In fact National Forestry Policies are directed towards limiting commercial exploitation of woodlands. The focus is on regulating woodland use and protection against wanton cutting of trees and encroachment rather than on sustainable utilization and management.

These limitations are extended to woodlands nontradi-

tional lands through control regimes, licensing, traffic check points, roadblocks and increased royalty collection from users of woodlands. This means that rural households have no right to exploit trees growing on their land except to cut them when clearing for agriculture [8,34, 35]. There is also a tendency by foresters to consider local communities' use of woodland for non commercial purposes a major cause of deforestation when in actual fact many human uses unless commercialized, do not have deleterious effects on regeneration of miombo woodlands. It was further documented that the dominant trend in regenerating miombo woodland in the absence of frequent hot fires or other intense disturbances is towards woodland [14,36]. Unless plants have been thoroughly uprooted during the initial disturbance, most of the subsequent development of woodland derives from re-growth of coppice from surviving stems and rootstocks. However, he also noted that marked changes in composition and very slow if any, recovery to the original state is likely in areas where *Brachystegia*, *Julbernardia* and other Caesalpiniodeae have been eradicated because these trees have extremely low ability for seed dispersal and their seeds are short lived. It is not easy, however, to eradicate these species anyway. In this survey, contrary to the general perception that all human uses of woody plants cause deforestation, the most destructive was clearing for agriculture. Other human uses that were found to be destructive was the uprooting of *Rhynchosia insignis* (*Munkoyo* roots), which are used in the preparation of a local beer called "Munkoyo" which has led to local extermination of this species in the GMA.

The other destructive human use of woody plants which was revealed during the survey was the debarking of trees for making beehives. One of the popular species for bark bee hives is *Julbernardia paniculata*. Debarking kills trees but the frequency of debarking is not known. Whatever the frequency when trees die after debarking, small gaps are created in the forest, such gaps are beneficial to the regeneration of valuable timber species such as *Pterocarpus angolensis*, so the extent to which this practice of killing trees by debarking may be destructive or beneficial requires to be investigated [37]. What this study gathered was that honey hunting activities are accompanied by debarking of trees for containers to carry honey, but this does not often involve debarking, although debarked patches are later burnt by fires making the tree hollow and reducing their value for timber [8].

Commercial charcoal production through not covered by this study can also lead to woodland depletion as deforestation tends to be rather intensive due to wide ranging search for suitable species [8,38]. Human use and resource depletion has also become a factor in the production of fine hardwoods. In Tanzania use of indigenous timber was declining along with supply [39]. In Zimbabwe, the share of indigenous woods from domestic sources in overall industrial wood supply fell from 534,600 m^3 to 108,500 m^3 between 1975 and 1990. This decline in the production of fine hardwoods reflected growing depletion of the stocks. In Tanzania, production in sawmills processing indigenous hardwoods dropped well below capacity in part due to depletion of wood resources in the vicinity of the mills [39].

4.3. Species and Size Class Selection for Building Poles

Poles, especially for house construction, were widely used as rafters covered in grass thatch of for walls coated with clay. More than 80% of the community in 1998 used poles with mud for walls and thatch for roofing. This number had declined to 50% in 2011. House building is not done annually except for new arrivals but repairs and extensions especially of granaries are done annually (89%, n = 600). Larger poles were needed for centre or corner posts or where additional strength was required. If the building was for public use such as a church or mosque, the roofing poles were longer than a private dwelling house. Wall poles took fewer but thicker poles than roofing poles. This was because the wall carries much of the weight of the building and requirements of desirable characteristics such as durability as in resistance to insect attack and rotting as well as straightness were critical. Rather than collect every species in equal quantities, communities select species because of their characteristics and labour costs involved. When species of desirable characteristics are lacking, in the settlement

zone, communities will disregard the rules restricting the cutting of trees in wildlife zone and National Park which indicates lack of legitimacy of the tenure systems in the eyes of the local people.

In general, the collection of building poles was not destructive to the woodland unless commercialized. This is because people were selective in their uses often going for durable and insect and moisture resistant species which prolong the replacement cycle after which considerable coppicing and regeneration would have taken place. When they return to harvest at a later time, they again select species and size classes according to human use type leaving others to grow. Additionally, each human use selects different species, a tree taken for firewood may sometimes be different from building pole and the same goes for girth size. However, it is argued that factors other than direct collection of plants affect abundance [8,40]. Increased disturbance would generally decrease diversity, yet it would increase abundance of those species which thrive in disturbed environments. *Diplorynchus condylocarpon* for instance which is fire tolerant thrives in areas burnt annually like in the settlement zone. However, he cautioned that human uses that target saplings and young trees should be closely monitored because these harvesting methods when combined with late fires may surprises regeneration, or for the latter, cause holes in trees and repeated fire resulting in death of the tree [41].

4.4. Methods of Collecting Building Poles

Regarding harvesting methods, the height at which a tree is cut is important because it affects the nature and rate of regeneration. The higher the tree is cut the less the biomass that is removed and the tree is likely to recover quickly [8,41,42]. Pollarding and lopping can be applied in harvesting fuel wood, building poles or obtaining fibre rather than felling the whole tree. Lopped trees regenerate faster producing greater biomass over shorter periods of time than trees cut close to the ground [8]. In fact selective branch removal could yield wood and leaf products with much less disturbance to the overall woodland than the removal of the whole tree [8]. In clearing for agriculture, trees were either stumped or cut at breast or waist height. In this case re-growth stems or branches are invariably defective at the point of cutting. Consequently the pole and timber value of re-growth is low [7,8,40,43]. Cutting trees at waist/breast height are not suitable techniques for regeneration. In cutting trees for fuel wood, poles and timber the cutting is usually done close to the ground. In such cases woodland regeneration is mainly by root coppice. If fire does not damage the regeneration, straight poles are produced [11,41]. Selective thinning can be practiced to provide building poles in short term and in the long term saw logs. It was suggested that trees should be felled at 15 cm above ground and at an angle of 45° using a sharp axe or saw to encourage vigorous regrowth from the stump [11,41,44]. Deep ploughing on the other hand would increase the stocking rate by causing root suckers to develop but only when the fallow period is long [44,45].

4.5. Firewood

Results showed that seven species were most popular species for firewood. The reason for their preference is due to their particular properties rather thjan abundance. The species chosen give good embers, hot fire, and burn easily producing little smoke. The species selection may continue for some in the future for as long as sources are nearby. However, as population increases and commercial charcoal production increases there would be need for intervention.

4.5.1. Areas of Intervention in the Fuel Wood Cycle

The starting point in tackling the fuel wood problem must be the multiple roles that trees perform in the environment within which people live and work. Starting by establishing people's needs, energy planners can then go on to look at what trees would be suitable for particular areas and how long they take to grow. These decisions enable the woody biomass production system to be developed and managed. By understanding and incorporating the knowledge of the people, they can make available a rich tool kit for the management of the woody biomass and explore new avenues for research and development programmes led by the needs of the people.

Simply getting trees into the ground may not slow deforestation or solve local fuel problems. Building on local traditional knowledge about uses of trees would be the most effective way of stimulating new tree growing initiatives. Where people do not generally plant trees as in the study area, management of coppicing would ensure a steady supply. The strong traditions on coppicing recorded in this study can be used to improve harvesting. Both coppicing and pollarding can be effective methods of obtaining a sustained yield of small diameter wood from stands of trees over long period of time.

4.5.2. Fuel Wood Consumption

Regarding firewood consumption per head per year, results from other parts of the world reveal similar trends as the ones observed in the study area. For example, a survey of 518 small farm families in different parts of Nicaragua where most cooked with fuel wood, showed that consumption ranged between 1100 kg - 2865 kg per person per year. In north eastern Tanzania, in Kwemzitu village, which is near a forest reserve from which people are allowed to collect fuel wood, the annual consumption

was found to be in the range of 1636 kg - 2605 kg per head per year. The 2366 kg consumption per head per year obtained in this study is therefore in the average upper range an indication that fuel wood is not yet a limiting factor for households in this area.

Availability of fuel wood was a key determinant factor of how people use it [8]. In areas where it is plentiful, people may use as much as 2000 kg per head per year or more. Where scarcity has forced to switch to fuels such as straw and animal dung, fuel wood consumption may drop to zero. In villages in Lesotho and South Africa's Transkei, where wood is scarce, the annual per capita fuel wood consumption was 288 kg and 271 kg supplemented 271 kg and 260 kg of crop residue and dung respectively. But in a village in Kwazulu in South Africa where wood was abundant, consumption was 1124 kg per head per year. In areas where wood is scarcer, consumption is lower. Survey of six villages in Mali and Niger found people using only 440 kg - 660 kg per head per year. Similar consumption patterns were recorded in from a survey of 17 villages in Tamil Nadu, India where a range of 344 - 676 kg per head per year with an average of 481 kg was recorded by [8]. In five other villages of Orissa, India, consumption ranged from 509 - 826 kg per head averaging 680 kg.

Family size also influenced per capita fuel requirements. Larger families cooking bigger meals can be much more efficient in their use of fuel wood. In one village in Nepal, families with between 1 - 4 members used an average of 890 kg of firewood per person each year. Families with 9 members or more used less than half this amount per person, about 340 kg. Based on the above experiences it was assumed that the average family size of 9.4 (minimum = 1, maximum = 23, mean = 9.4, SE = 4.4) in the study area, which is on the larger size, it is likely that fuel wood usage would be economical, but again this depends on whether firewood is commercialized or not [8].

4.5.3. Seasonal Use of Firewood

The second consumption of wood is dictated by weather. In the study area, an average 26 kg bundle of firewood was consumed in three days time in cold weather and five days in warm weather. Cold weather also coincided with harvesting of crops and this was also the time when many traditional rituals that require beer brewing and also weeding parties are done at this time which collectively increases firewood consumption. For as long as these traditional rituals do not attract outside interests the fire wood demand is unlikely to outstrip supply.

4.6. Bee Honey Collection

Bee honey hunters usually debark trees usually *Brachystegia* or *Julbernardia* to make a container for carrying honey home. To what extent this would be damaging to woody plants needs to be examined. There are however, some alternatives particularly when the community is encouraged to start bee keeping schemes. Calabashes, damaged clay pots, and bee hives made from exotic timber and many others can be used, hence the frequency of debarking is likely to be low. Currently the collection method is such that when a bee colony is located, especially if it is above average human height, a forked ladder is used is made to enable the collector reach the hive or the tree is felled down. A fire is made and quite often old pieces of cloth, old tyres are burnt to lessen stinging. If the tree is not cut down a hole is made in the tree. After collecting the honey, honey hunters rarely extinguish the fire. During the August-November honey collection season, honey collectors were major causes of wild fires as reported by 20% (n = 180) respondents that honey collection caused about 80% of fire incidences. This figure could not be verified as fires from fields as they being cleared in readiness for cultivation were also another major source. Honey collection therefore, as a type of human use of woody plants is damaging to large trees while fires left behind may cause damage to saplings and regenerating plants.

4.7. Fibre

Results of the PRA showed that firewood collectors often re-used the old fibre by soaking it in water to soften it. The frequency with which new fibre is collected is likely to be low. Other means of fire wood transportation such as by wheelbarrow do not require tying firewood in bundles. Fibre used in construction and basket/mat making on the other hand, is treated in black clay soil to strengthen it. By doing so, its life span is increased and therefore the replacement period is often long. Although data was not collected about quantities of fibre collected, it was assumed that the overall demand for new fibre is generally low [8].

4.8. Household Items, Tools, Species Used and Replacement Period

Utensils and tools are of immense value to the rural communities' daily lives and agricultural work; pounding food, firewood and building poles collection, all depend on locally made tools and utensils derived from wood.

In the study area, many households are made from woody plants. It was difficult in certain circumstances to take accurate girth measurements of the tree species from which the tool came from because most of them involve complete modification of the natural shape. On the whole, utensils represent low level exploitation because of the species selection and replacement cycles. Some tool making methods are however, destructive. For example,

axe handles require that the tree be uprooted to make use of the bulge at the base of the tree which has crossed grains to withstand the cutting impact, but again it is only destructive when commercialized as the replacement cycle for axe handles is approximately four years. Other items such as the motar or drums use larger trees of girth greater than 100 cm to allow for carving out a hollow, while a pestle only requires debarking and smoothening the ends, but these have a long replacement cycle of over 20 years.

4.9. Fruits

Indigenous fruits may not look as appetizing as exotics, but are often nutritious and healthier due to increasing use of insecticides and growth promoters in exotic species. Apart from yielding fruit for human consumption, they are also a source of income to small holders [46-49]. The extent to which indigenous fruit is marketed was not covered in this survey, but in other studies [48,49] it was noted that marketing depended on availability, shelf life, demand patterns and access to market, sale price and effectiveness of social controls regulating the sale of these products. Quite often the availability of exotic fruits such as mango takes precedence over indigenous fruits. Low indigenous sales, as low as 1%, but high sales of up to 56% for exotic fruits have been recorded in Zimbabwe [46-49] recorded very low. Although we did not gather information on income from wild fruits in this study, income from the sale of wild fruits is expected to contribute significantly to women's and children's income. This was based on the fact that most of the trees conserved in the fields were fruit trees. These observations also agree with those obtained in Mutoko Communal Area in Zimbabwe where most of the indigenous trees conserved in home fields were edible fruit bearing species [49]. It was also noted two indigenous trees *Ziziphus mucronata* and *Uapaca kirkiana*, both of which are edible, being planted by villagers [49].

4.10. Suggested Future Research Needs

The area of sustainable use of woody plants is an area of particular interest as local communities undergo rapid social, economic and political change. The transaction of establishing and centralizing institutional mechanisms for local resource management may be high to the point where institutions such as Forestry Department and Zambia Wildlife Authority may only be able to maintain high value and quick return resources such as game animals. In light of the above, we suggest future research activities as follows:

1) Examine the chances for local management can utilize low value resources such as woody plant resources on a sustained yield basis.

2) Identify simple and practicable mechanisms for strengthening local institutions and to develop methodologies to assess the comparative value of woodland resources especially in relation to other land uses.

3) Examine institutional mechanisms which ensure that local communities have control over resources of their landscape in the face of possible appropriation by outside elite groups.

4) Examine how scientific/ecological recommendations of using fire, coppicing and others can be feasibly be used as tools by local people to manage woodlands on traditional land.

5) Policy development that empowers local people to manage woodlands in their traditional landscape where law permits.

6) Investigates how empowerment would best be achieved, for what products and in what circumstances; so far the Community Based Natural Resources Management programmes may not be wholly applied to woodland products because of their low value compared to game animals; devolution per se is no panacea to sustainable use but it may be crucial first step, which needs to be investigated in detail by future research.

7) As for the ability for the woodland to coppice after human use, it is at the moment not clear whether the ability to produce coppice shoots and stump survival rate is correlated with girth size or age; it is important that future research activities thoroughly address this question.

8) The other topic of importance is that while attractive in theory, indigenous forest management faces serious technical challenges; little is known about feasible rotation patterns and how best to incorporate controlled grazing and wood cutting into indigenous forest management.

9) Develop a simple monitoring system for assessing human related threats to the ecology of the area.

10) Develop a mechanism for relating the extent of such threats to human related factors of society, population pressure, socio-economic standing, source of livelihood, contemporary and traditional law enforcement structures and mechanisms, incentives, environmental awareness and politics.

11) Develop a mechanism for using such information for planning and management as they relate to developing the potential of sustainable use of renewable woody plant resources.

12) In this survey, we excluded woody plants of dbh of less than 10 cm; this was a weakness as has been the case with other similar vegetation studies, as this implies excluding plants in the pre-sapling phase, which probably form a large reservoir for future trees; future research should consider saplings and that shoot growth among suppressed saplings is low while these plants accumulate

a relatively large below ground biomass constituting perennating tissue that regenerates new shoots following repeated shoot die back in the dry season.

5. ACKNOWLEDGEMENTS

Field assistants L. Mboote, L. Musa (late), P. Shamalundu, C. Malisawa, R. Chisangu, W. Sakala and M. Cheleka conducted vegetation surveys and identified tree species encountered and also assisted in the translation of names from vernacular to botanical names particularly for species used in tool making and implements. Enumerators; Jane Shimwense, Veronica Muyanda, Wambulwawaye Mubita and Joyce Mwenda Namukombo were excellent in questionnaire surveys. We highly appreciate the field skills which made this project a success.

REFERENCES

[1] Campbell, B., Frost, P. and Byron, N. (1996) Miombo woodlands and their use: Overview and key issues. In: Campbell, B.N., Ed., *The Miombo in Transition: Woodlands and Welfare in Africa*, Centre for International Forestry Research, Bogor, 1-5.

[2] Campbell, B. and Byron, N. (1996) Miombo woodlands and rural livelihoods: Overview and key issues. In: Campbell, B.N., Ed., *The Miombo in Transition: Woodlands and Welfare in Africa*, Centre for International Forestry Research, Bogor, 221-263.

[3] Clarke, J., Cavendish, W. and Coote, C. (1996) Rural households and miombo woodlands, use value and management. In: Campbell, B.N., Ed., *The Miombo in Transition: Woodlands and Welfare in Africa*, Centre for International Forestry Research, Bogor, 101-135.

[4] Bradley, P.N. and Dewees, P.A. (1993) Indigenous woodlands, agricultural production and household economy in communal areas. In: Bradley, P.N. and McNamara, Eds., *Living with Trees*: *Policies for Forestry Management in Zimbabwe*, World Bank Technical Paper 210, World Bank, Washington, 63-157.

[5] Fischer, F.U. (1993) Bee-keeping in the subsistence economy of the miombo savannah woodlands of south central Africa. Rural Development Forestry Network Paper, Overseas Development Institute, London, 1-2.

[6] Birchenoughl, L. (1993) The impact of natural resource utlisation on miombo woodland in Zambia. Unpublished MSc. Thesis University College of London, 1-96.

[7] Hosier, R.H. (1993) Charcoal production and environmental degradation. *Energy Policy*, **21**, 491-505.

[8] Chansa, W. (2000) Utilisation of woody plant species by local communities in Mumbwa Game Management Areas, Zambia. Unpublished MSc. Thesis, University of Zimbabwe, Harare, 1-99.

[9] Nahonyo, C.L., Mwasumbi, I. and Bayona, D.G. (1998) Survey of the vegetation communities and distribution of woody plant species in Mbomipa Project area. Report No. 1 MCRI Iringa, 1-81.

[10] Dove, M. (1993) A revisionist view of tropical deforesta-

tion and development. *Environmental Conservation*, **20**, 17-25.

[11] Chidumayo, E.N. (1987) Woodland structure, destruction and conservation in the Copperbelt areas of Zambia. *Biological Conservation*, **40**, 89-100.

[12] Chidumayo, E.N. (1989) A land use, deforestation and reforestation in the Zambian Copperbelt. *Land Rehabilitation*, **1**, 209-216.

[13] Price, L. (1998) Farmers do conserve and plant trees in communal areas: A case study in Mutoko communal area. *The Zimbabwe Science News*, **32**, 4-6.

[14] Hoffman, M.T. and Cowling, R.M. (1980) Vegetation changes in semi arid and eastern Karoo over the last two hundred years: An expanding Karoo? *South African Journal of Science*, **118**, 286-294.

[15] Burley, J. (1979) Choice of tree species and possibility for genetic improvement for small holder and community forests. *Commonwealth Forestry Review*, **59**, 3-7.

[16] Clarke, C. and William, C. (1976) Maintenance of agriculture and human habitats within Tropical Forest Ecosystems. *Human Ecology*, **4**, 217-259.

[17] Lusigi, W.J. (1981) New approaches to wildlife conservation in Kenya. *Ambio*, **10**, 2-3.

[18] Mackinnon, J., Mackinnon, K., Child, G. and Thorsell, J. (1986) Managing protected areas in the tropics. The World Conservation Union (IUCN), Gland, 99-117.

[19] Ehrlich, R. and Paul, R. (1982) Human carrying capacity, extinction and nature reserve. *Biosc*, **32**, 321-326.

[20] Nhira, C. and Fortman, L. (1993). Local woodland management: realities of the grass roots. In: Bradley, P.N. and McNamara, Eds., *Living with Trees: Policies for Forestry Management in Zimbabwe*, World Bank Technical Paper 210, World Bank, Washington DC, 139-156.

[21] Cunningham, A.B. (1990) The regional distribution, marketing and economic value of the palm wine trade in Ingwavuma District. *South African Journal of Botany*, **56**, 191-198.

[22] Cunningham, A.B. (1991) Development of a conservation policy on commercially exploited medicinal plants: A case study from southern Africa. In: Akerele, Heywood and Synge, Eds., *Conservation of Medicinal Plants*, Cambridge University Press, 337-358.

[23] Talukdar, S. (1993) The conservation of *Aloe polyphylla* endemic to Lesotho. *Bothaia*, **14**, 985-989.

[24] Lawton, R.M. (1978) A study of the dynamic ecology of Zambian vegetation. *Journal of Ecology*, **66**, 175-198.

[25] Trapnell, C.G. (1959) Ecological results of woodland burning experiments in Northern Rhodesia. *Journal of Ecology*, **47**, 129-168.

[26] Fanshawe, D.B. (1972) Useful trees of Zambia for the agriculturalist. Ministry of Lands and Natural Resources. Government Printer, Lusaka, 3-127.

[27] Cole, M.M. (1963) Vegetation and geomorphology in Northern Rhodesia: An aspect of the distribution of the savannah of central Africa. *Geology Journal*, **129**, 290-305.

[28] Stohlgren, T.J., Falker, M.B. and Schell, L.D. (1995) A modified Whittaker nested vegetation sampling method. *Vegetatio*, **117**, 113-121.

[29] Mueller-Dumbois, D. and Ellenberg, H. (1974) Aims and methods of vegetation ecology. John Wiley and Sons, London, 45-58.

[30] Walker, B.H. (1970) An evaluation of eight methods of botanical analysis in Rhodesia. *Journal of Applied Ecology*, **7**, 3-12.

[31] Prance, G.T., Balee, W., Boom, B.M. and Carneiro, R.L. (1987) Quantitative ethnobotany and the case for conservation in Amazonia. *Conservation Biology*, **1**, 4-10.

[32] Johns, A.D. (1981) The effects of selective logging on the social structure of resident primates. *Malaysian Applied Biology*, **10**, 221-226.

[33] Keith, G. (1984) The relationship between adjacent land and protected areas: Issues of concern for the protected area manager. In McNeely and Miller, Eds., *National Parks, Conservation and Development: The role of Protected Areas in Sustaining Society*, Smithsonian Institution Press, Washington DC, 65-71.

[34] Lewis, D., Kaweche, G.B. and Mwenya, A.N. (1990) Wildlife conservation outside protected areas, lessons from an experiment in Zambia. *Conservation Biology*, **4**, 171-180.

[35] Frost, P. (1996) The ecology of miombo woodlands. In Campbell, Ed., *Miombo in Transition: Woodlands and Welfare in Africa*, Centre for International Forestry Research, Bogor, 11-55.

[36] Lawton, R.M. (1982) Natural resources of the miombo woodlands and recent changes in agricultural land use practices. *Ambio*, **14**, 362-365.

[37] Dove, M. (1993) A revisionist view of tropical deforestation and development. *Environmental Conservation*, **20**, 17-25.

[38] Monela, G.C., O'Kting'ati, A. and Kiwele, P.M. (1993) Socio-economic aspects of charcoal consumption and environmental consequences along the Dar es Salaam-Morogoro highway. *Forest Ecology and Management*, **58**, 249-258.

[39] Kowero, G.S. and Temu, A.B. (1985) Some observations on implementing village forestry programmes in Tanzania. *The International Tree Crops Journal*, **3**, 1-12.

[40] Chidumayo, E.N. (1988) Estimating fuelwood production and yield in re-growth dry miombo woodland in Zambia. *Forest Ecology and Management*, **24**, 59-66.

[41] Frost, P. and Chidumayo, E.N. (1996) Population biology of miombo trees. In: Campbell, Ed., *Miombo in Transition: Woodlands and Welfare in Africa*, Centre for International Forestry research, Bogor, 59-71.

[42] Strang, R.M. (1974) Some man made changes in successional trends on the Rhodesian Highveld. *Journal of Applied Ecology*, **111**, 249-263.

[43] Chidumayo, E.N., Gambiza, J. and Grundy, I. (1996) Managing miombo woodlands. In: Campbell, Ed., *Miombo in Transition*: *Woodlands and Welfare in Africa*, Centre for International Forestry research, Bogor, 175-193.

[44] Winkworth, R.E. (1990) The use of point quadrats for analysis of heath land. *Australian Journal of Botany*, **3**, 68-81.

[45] Strang, R.M. (1966) The spread and establishment of *Brachystegia spiciformis* and *Julbernardia globiflora* in the Rhodesian high veld. *Commonwealth Forest Review*, 253-256.

[46] Gumbo, D.J., Mukamuri, B.B., Muzondo, M.I. and Scoones, I.C. (1990) Indigenous and exotic fruit trees; why do people grow them? In: Prinsley, Ed., *Agro Forestry for Sustainable Production*: *Economic Implications*, Common Wealth Scientific Council, London, 185-214.

[47] Ridker, R.G.M. and Cecelski, E.W. (1979) Resource, environment and population: The nature of future limits. Population Bulletin 34, Washington DC., 2-12.

[48] Gumbo, D.J. Makamuri, B.B., Muzondo, M.I. and Scoones, I.C. (1990) Indigenous and exotic fruit trees; why do people plant them? In: Prinsley, Ed., *Agroforestry for Sustainable Production*: *Economic Implications*, Commonwealth Scientific Council, London, 185-214.

[49] Wilson, K.B. (1989) Trees in fields in southern Zimbabwe. *Journal of Southern African Studies*, **15**, 369-383.

17

Phytoecological and phytoedaphological characterization of steppe plant communities in the south of Tlemcen (western Algeria)

Bahae-Ddine Ghezlaoui[*], Noury Benabadji, Nedjwa Benabadji

Department of Biology, Faculty of Natural Sciences and Life Sciences and Earth and the Universe, Pôle la Rocade No. 2, University of Tlemcen, Algeria; [*]Corresponding Author

ABSTRACT

In Algeria, the steppe areas of southern Sebdou between Tlemcen, El-Aricha and Mecheria are the scene of an adverse and continuous ecological imbalance often caused by the strong support of human pressure in these ecosystems. In arid and semi-arid regions, salinity remains a constraint for the development of plants. This study focuses on the realization of floristic surveys, where attention was paid to areas occupied by the dominant perennial species (*Tamarix gallica* L.). A correspondence analysis by Minitab 15 software has allowed us to individualize the groups of species attracted by some parameters (edaphic, nitrates, humidity), and ecological gradients appear to affect the distribution of these taxa. The study of plant diversity shows the dominance of biological type of therophytes (41%) and morphological type of the Chenopodiaceaes with (25%). For geographical types, the strict Mediterranean and circum-Mediterranean element predominates.

Keywords: Halophytes; Tamaricacées; Salinity; Human Impact; Algeria

1. INTRODUCTION

The flora of the arid, especially of high plains Algerian steppe is discontinuous. Plants use mainly the locations where water is a little more accessible than elsewhere. The vegetation of the steppe regions is relatively homogeneous and Mediterranean penetrations are common [1,2]. Endemism is high in these vast spaces [3]. A human action such as the action of man and his herd is a major factor that catalyzes the degradation of plant cover in these ecosystems.

The apparent homogeneity of the vegetation in xeric and halophytes hides considerable heterogeneity that is related to the diversity of microclimates, topography and varying degrees of human pressure [4]. Algerian steppe is an expanse of 20 million hectares covered with low and sparse vegetation [5].

The overgrazing and long dry periods are among the factors that have contributed to increasing the fragility of these ecosystems and reducing their ability to regenerate, namely a decrease in their potential production [6]. The floristic analysis of different plant communities and their biological characteristics and chronologic could allow us to evaluate the critical value necessary for conservation management.

The objective of this work attaches great importance to the spatial distribution of plant taxa that occupy these places. This distribution follows a very irregular distribution of different plant species according to some environmental variations including climate, soil and human impact. This study highlights the influence of these parameters on the vector shown in factorial designs.

2. MATERIALS AND METHODS

2.1. Location and Choice of Stations

The region, as shown in **Figure 1**, is located in the western part of western Algeria. It is part of an area called "High Plains". The north side opposes a clear way for its richness and variety of its landscapes to aridity and the relative monotony of the vast plains that stretch to the South.

It is close to Dayet El Ferd, bordered by mountains to the northwest by the hills of Sidi Djilali (Jebel Tenouchfi, 1840 m). It is bordered in the southwest by El-Abed Mountains (1450 m) and south-east by the mountains of El Gor (Djebel El-Hariga, 1600 m).

Figure 1. Location of study site.

The area is crossed by the National Road No. 22 connecting the cities of South Tlemcen (Sebdou, El-Aricha, Méchéria and Bechar). It lies to the north with a latitude of 34°29' north and longitude 1°16' West. It is a central part of the steppe zone of Tlemcen. The occupied area is about 73,700 ha.

2.2. Climate Summary

We took into account recent rainfall and temperatures, which are spread over almost fourteen years (1998-2012), taken at a weather station near El Aricha (**Table 1**).

This synthesis was based on a number of parameters such as climate regimes seasonal rainfall, the drought index, extreme temperatures, minimum values, and finally Q_2 of Emberger. A positioning on pluviothermic climagramme of Emberger was made to clarify the bioclimatic zone.

2.3. Soil Variables

Along a transect, we conducted soil samples corresponding to the horizon searched by rooting foster taxa [7,8]. The locations of these samples to the total number of six were selected in floristically homogeneous areas,

representative of the station studied. The samples with a shovel were brought to the laboratory in bags to carry out their analysis. Using a sieve of 2 mm in diameter, we recovered useful for the realization of physicochemical analyzes fine soil.

The measured parameters were: particle size, pH, salinity and organic matter, the rate of limestone and Munsell color (**Table 2**).

2.4. Floristic Diversity

In the arid conditions of the study area, the vegetation remains highly sensitive to climatic and mechanical influences, such as grazing or clearing [9].

In these areas, the kind *Tamarix* is among the taxa that play an important role in soil conservation against different physical erosion. Hardiness acquired resistance to drought, gives it a special interest in the floristic composition of the study area. The paths of these regions are also characterized by unfavorable conditions, mainly related to abiotic stresses (drought, salinity).

As a result of their location-related schemes aridity and well-defined salinity, halophytes are divided into groups arranged in zones around continental saline depressions or along the seashore [10].

Table 1. Average monthly rainfall and temperatures in the station of El-Aricha during the period 1998-2012.

Station	Months	J.	F.	M.	A	Ma	Ju	Jul.	Au.	S.	O.	N.	D.	Annual rainfall (mm)	
El-Aricha	Annual rainfall (mm)	21.3	16.4	24.4	21.2	19.6	5.1	3.1	7.8	15.4	19.9	21.5	12.0	188.1	
	Temperature (°C)	4.3	6.2	9.5	9.8	17.4	22.2	27.4	27	20.7	14.7	8.8	5.5	m (°C) -2.2	M (°C) 33.1

Table 2. Results of physic-chemical analyzes of soil.

Station	Sample. No	Soil texture (%)				pH	$CaCO_3$ %	O.M %	C (‰)	E.C mS/cm	Color Munsell
		A	L	S	T						
Dayet El-Ferd	1	10	17	31	S.l	8.9	21.8	1.7	10.2	1.2	5YR5/4
	2	14	15	32	S.l	8.9	23.6	1.8	10.4	7.3	5YR5/6
	3	18	34	26	L.s	8.7	17.2	2.2	13.2	1.2	5YR5/6
	4	23	40	23	L	8.7	16.3	3.5	20.6	4.4	5YR6/6
	5	10	23	29	S.l	8.8	19	1.9	11.5	13.6	5YR5/6
	6	12	39	24	L.s	8.6	14.5	1.9	11.1	14.5	5YR4/8

A: Clay; L: Silt; S: Sand; S.l: Sandy loam; L.s: Loamy sand; T: Soil Texture; $CaCO_3$: Limestone; O.M: Organic mater; C: Organic carbon; E.C: Electrical conductivity.

The method used for sampling the vegetation was that of Braun-Blanquet (1951) and Guinochet (1973) called stigmatiste [11,12].

The surface of the record must be at least equal to the minimum area [13], containing almost all of the species present [14]. The study area was divided into 3 zones with a minimum area measured at 64 m².

We made the statements accompanied recording stationers characters (location, altitude, exposure, substrate, geomorphology, slope and recovery rate).

This station is located near the inland basin Dayet El Ferd. We considered in our records only the presence and absence of the species, by setting the first number 1 and the second the number 0.

Analysis of floristic surveys was made from minitab 15 software. This bio-statistical analysis led to a hierarchy of environmental factors determining the diversity of vegetation. The variables were introduced in the form of codes to facilitate the reading of factorial designs. These codes were represented by lowercase letters from the vernacular name of the taxa presented and identified from the flora Quezel & Santa (1962) [15]. For example, it affects the species *Eurucaria uncata*, code (eu) (**Table 3**). Indices presence and absence were retained in the statistical treatment by correspondence analysis (A.F.C).

The determination of biological types of the species recorded was based on the Raunkiaer method (1934) [16].

3. RESULTS

3.1. Climate

The dry season lasts from May until October, or 6 months of drought with a drought index equal to 0.49.

The rainfall amounts type is P.A.H.E. The value of Q_2 was 18.51.

According climagramme of Emberger, the station is positioned at the upper arid bioclimatic with cold winter (**Figure 2**). The bioclimatic study allowed us to deduce that the two weather variables (temperature and rainfall) contributed to the ecological bio-components, such as salinity; flooding... These factors were considered in determining the spatio-temporal distribution different plant species of halophytes.

3.2. Soil Variables

Texture sandy loam to loamy sand is dominant. The limestone varies between 16.3 and 23.6. An alkaline pH exceeding 8 at all samples.

The electrical conductivity varies from 1.2 to 14.5 mS/cm, was increasing gradually as one approaches the Daya.

The level of soluble salts in the salt-affected soils was a function of the depth of the water table salt, texture, evapotranspiration and the humidity profile [17].

Organic carbon varied between 10.2‰ and 20.6‰. The following Munsell color varied between samples (5YR5/4 5YR5/YR6 6.5/6 and 5YR4/8).

The edaphic analysis revealed some specific pre-salt steppe depressions colonized by halophyte vegetation tree on a very loose substrate texture more or less fine (silt and sand).

Other tree formations were generally excluded from this type of environment. These soils were characterized by the presence of shallow salt water and were subjected

Table 3. Floristic inventories.

SITES	El-Aouedj										
SLOPE	0%										
RECOVERY LEVEL	5à10%										
SURFACE AREA	64 m^2										
SUBSTRATE	Lime store										
NUMEBER OF INVENTORIES	P	1	2	3	4	5	6	7	8	9	10
SPECIES											
Stipa tenacissima	**10**	1	1	1	1	1	1	1	1	1	1
Atractylis humilis	**5**	1	0	1	1	0	0	1	1	0	0
Cistus villosus	**2**	1	0	1	0	0	0	0	0	0	0
Tamarix gallica	**2**	0	1	0	0	0	0	1	0	0	0
Helianthemum rubellum	**1**	0	0	1	0	0	0	0	0	0	0
Thymus ciliatus	**2**	1	0	0	1	0	0	0	0	0	0
Schismus barbatus	**8**	0	1	1	0	1	1	1	1	1	1
Herniaria hirsuta	**8**	0	1	1	1	0	1	1	1	1	1
Adonis dentata	**6**	0	0	1	0	1	1	1	0	1	1
Plantago albicans	**6**	0	0	0	1	1	0	1	1	1	1
Bromus rubens	**5**	1	0	0	1	0	1	0	0	1	1
Poa bulbosa	**5**	1	0	0	1	0	1	0	0	1	1
Salvia verbenaca	**5**	0	1	0	0	0	1	1	1	0	1
Muricaria prostrata	**5**	0	1	0	0	0	1	1	0	1	1
Koeleria phleoides	**5**	0	1	0	0	0	1	1	1	0	1
Euphorbia falcata	**4**	1	0	1	0	0	0	0	0	1	1
Erodium moschatum	**4**	0	0	1	0	0	1	0	1	0	1
Ceratocephalus falcatus	**3**	0	0	0	0	0	1	0	0	1	1
Matthiola longipetala	**3**	0	0	0	0	0	1	0	0	1	1
Hordeum murinum	**2**	0	0	0	0	0	1	1	0	0	0
Peganum harmala	**2**	0	0	0	0	0	1	1	0	0	0
Sisymbrium runcinatum	**1**	1	0	0	0	0	0	0	0	0	0
Koeleria pubescens	**1**	1	0	0	0	0	0	0	0	0	0
Eruca vesicaria	**1**	1	0	0	0	0	0	0	0	0	0
Aegilops triuncialis	**1**	0	1	0	0	0	0	0	0	0	0
Marrubium vulgare	**1**	0	0	0	0	1	0	0	0	0	0
Paronychia argentea	**1**	0	0	0	0	0	1	0	0	0	0
Micropus bombycinus	**7**	1	1	1	1	0	0	1	1	0	1
Alyssum campestre	**6**	0	0	1	1	0	1	1	0	1	1
Astragalus epiglottis	**5**	0	0	1	0	0	1	0	1	0	1
Brachypodium distachyum	**4**	0	0	0	0	0	1	1	1	0	1
Trigonella polycerata	**3**	0	0	0	1	1	0	0	0	1	0
Echinaria capitata	**3**	0	0	0	0	0	1	0	0	1	1
Medicago truncatula	**2**	1	1	0	0	0	0	0	0	0	0
Astragalus pentaglottis	**2**	0	0	0	0	0	0	0	0	1	1
Teucrium pseudo-chamaepitys	**1**	1	0	0	0	0	0	0	0	0	0
Bupleurum semi-compositum	**1**	0	0	1	0	0	0	0	0	0	0
Euphorbia exigua	**1**	0	0	0	0	1	0	0	0	0	0
Evax pigmaea	**1**	0	0	0	0	0	0	0	0	0	1
Medicago minima	**7**	0	1	1	1	1	0	1	1	1	0
Malva aegyptiaca	**7**	0	1	0	1	0	1	1	1	1	1
Noaea mucronata	**5**	0	1	0	0	0	1	1	1	1	0

Continued

Filago spathulata	**5**	0	0	0	1	1	0	1	0	1	1
Atractylis serratuloides	**2**	1	1	0	0	0	0	0	0	0	0
Echium pycnanthum	**2**	0	1	0	1	0	0	0	0	0	0
Scorzonera undulata	**1**	1	0	0	0	0	0	0	0	0	0
Helianthemum apertum	**1**	1	0	0	0	0	0	0	0	0	0
Helianthemum hirtum	**1**	1	0	0	0	0	0	0	0	0	0
Launea residifolia	**1**	0	1	0	0	0	0	0	0	0	0
Salsola vermiculata	**1**	0	0	1	0	0	0	0	0	0	0
Sanguisorba minor	**1**	0	0	0	0	0	0	0	1	0	0

P: Presence.

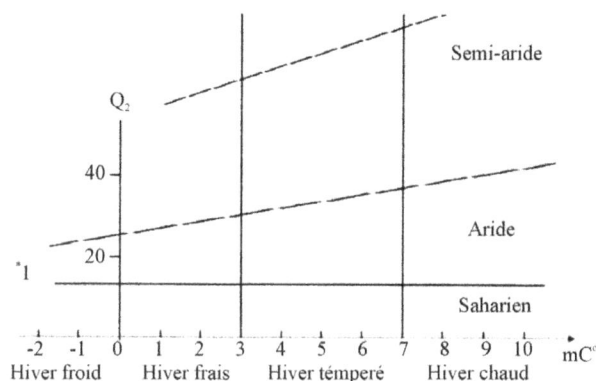

Figure 2. Positioning of the study area on the Emberger climagramme.

since their implementation in terms of salinity and large water logging with variable intensity [18].

3.3. Discrimination by Correspondence Analysis (A.F.C)

The A.F.C has allowed us to individualize sets of species with similar ecological affinities; it clarifies the structures of differentiated vegetation in these stands [19]. Therefore, it can highlight the strong cash contribution in factorial designs and understand the distribution of these. This distribution is undeniably dependent on environmental factors.

Floristic surveys at this station show the presence of a population dominated by halophyte species *Tamarix gallica* L. introduced into the area by reforestation in the 1970 around the periphery of the Daya. This taxon is reproduced perfectly alongside therophytes species (*Brachypodium distachyum, Micropus bombicinus...*).

The minitab 15 software processes sets of species were distributed along the factorial designs formed by the Axes (1 and 2) with strong contributions. We tried to understand the distribution function of gathering structures of vegetation governed by environmental factors identified by gradients whose meaning obeys the location

of the cash contribution of strong factorial designs [20].

Each species represented by a code took its place on factorial designs.

The rest of the taxa included in the (**Table 4**) which are represented, their contact with the values on Axis 1 and 2. These values allowed us to easily identify the species of the factorial design.

Interpretation of Factorial Designs

- (Zone 1). Axis 1 - 2 (**Figure 3**). Value = 4.73. Inertia ratio = 47%. This axis contains in the positive side species indicator gradients of salinization *Tamarix gallica* (2.0) of psammophitisation *Schismus barbatus* (1.9) and overgrazing *Noaea mucronata* (0.9). Tese gradients replaced limestone gradient indicated by the species *Poa bulbosa* (0.3). We also found some post-crop species *Malva aegyptiaca* (1.7) and *Muricaria prostrata* (1.0)...

On the negative side position, we found steppe species of Mediterranean dry grasslands, *Thymus ciliatus* (−0.9) and *Koeleria pubescens* (−1.2)... The gradients presented are those that show the human impact and therophitisation *Astragalus pentaglottis* (−0, 1) and *Marrubium vulgare* (−0.7)...

- (Zone 2). Axis 1 - 2 (**Figure 4**). Value = 4.63. Inertia ratio = 46%.

At this axis we find halophytes taxa *Tamarix gallica* (0.2), *Brachypodium distachyum* (0.6) creating a positive and increasing salinity gradient.

The gradient was also represented by psammophitisation, *Schismus barbatus* (0.4) alongside other gradients like overgrazing and steppe, *Noaea mucronata* (1.2), *Micropus bombycinus* (1.8).

On the negative side appeared species of the steppes and dry pre-steppes, like *Adonis dentata* (−1.4) *Filago spathulata* (−1.8), *Astragalus Pentaglottis* (−1.4) generating gradients steppe and human impact.

- (Zone 3). Axis 1 - 2 (**Figure 5**). Value = 2.70. Inertia ratio = 27%.

This line seemed to be managed by the same environmental factors that these precedents.

Table 4. Coordinates of species.

SPIECES	Code	Indiv.	Axis: 1	Axis: 2
Tamarix gallica	tg	1	2.06176483	1.82268615
Atractylis humilis	ahu	2	0.18746175	1.94189319
Cistus villosus	civ	3	−1.03007347	0.82624634
Helianthemum rubellum	hru	3	−0.78923077	0.09870069
Thymus ciliatus	tc	3	−0.9316649	1.32637496
Schismus barbatus	sb	4	1.96419167	0.09521696
Herniaria hirsuta	hh	5	1.98563516	0.18977741
Adonis dentata	ade	6	1.30057226	−0.05221344
Plantago albicans	pa	7	1.37906351	1.32570799
Bromus rubens	br	7	0.33587037	0.18333438
Poa bulbosa	pb	7	0.33587037	0.18333438
Salvia verbenaca	sav	7	0.97244641	−1.40852751
Muricaria prostrata	mp	7	1.03338802	−1.33890099
Koeleria phleoides	kph	7	0.97244641	−1.40852751
Euphorbia falcata	ef	7	−0.1562795	0.53203863
Erodium moschatum	emo	7	0.4173629	−1.11396642
Ceratocephalus falcatus	cf	7	0.23829721	−1.54413482
Matthiola longipetala	ml	7	0.23829721	−1.54413482
Hordeum murinum	hm	7	−0.13020138	−1.16316374
Peganum harmala	ph	7	−0.13020138	−1.16316374
Sisymbrium runcinatum	sr	3	−1.27008077	0.32645142
Koeleria pubescens	kp	3	−1.27008077	0.32645142
Eruca vesicaria	eve	3	−1.27008077	0.32645142
Aegilops triuncialis	at	3	−0.73944263	−0.28262378
Marrubium vulgare	mv	7	−0.71226568	0.50426885
Paronychia argentea	pa	7	−0.63549676	−1.24992711
Micropus bombycinus	mb	2	0.91628556	1.66756947
Alyssum campestre	alc	6	1.32201575	0.04234701
Astragalus epiglottis	ae	7	0.4173629	−1.11396642
Brachypodium distachyum	bd	7	0.68265098	−1.52699798
Trigonella polycerata	tp	7	0.06091577	1.60277886
Echinaria capitata	ec	7	0.23829721	−1.54413482
Medicago truncatula	mt	3	−0.98028533	0.44492188
Astragalus pentaglottis	ap	7	−0.15544409	−0.69530195
Teucrium pseudo-chamaepitys	tpc	3	−1.27008077	0.32645142
Bupleurum semi-compositum	bsc	3	−0.78923077	0.09870069
Euphorbia exigua	ee	7	−0.71226568	0.50426885
Evax pigmaea	ep	7	−0.59020968	−0.79388841
Medicago minima	mm	8	1.46983785	2.33676754
Malva aegyptiaca	ma	5	1.74562787	−0.31001751
Noaea mucronata	nm	7	0.96818362	−0.91714688
Filago spathulata	fs	7	1.00523953	1.29674805
Atractylis serratuloides	ats	3	−0.98028533	0.44492188
Echium pycnanthum	ep	3	−0.40102676	0.71729977

Phytoecological and phytoedaphological characterization of steppe plant communities in the south of Tlemcen
(western Algeria)

153

Continued

Scorzonera undulata	su	3	−1.27008077	0.32645142
Helianthemum apertum	hap	3	−1.27008077	0.32645142
Helianthemum hirtum	hhi	3	−1.27008077	0.32645142
Launea residifolia	Lr	3	−0.73944263	−0.28262378
Salsola vermiculata	sv	3	−0.78923077	0.09870069
Sanguisorba minor	sm	7	−0.65541408	−0.3721343
Stipa tenacissima	stt	7	−0.65541408	−0.3721343
Artemisia herba-alba	aha	7	−0.65541408	−0.3721343

Indiv.: Individuel.

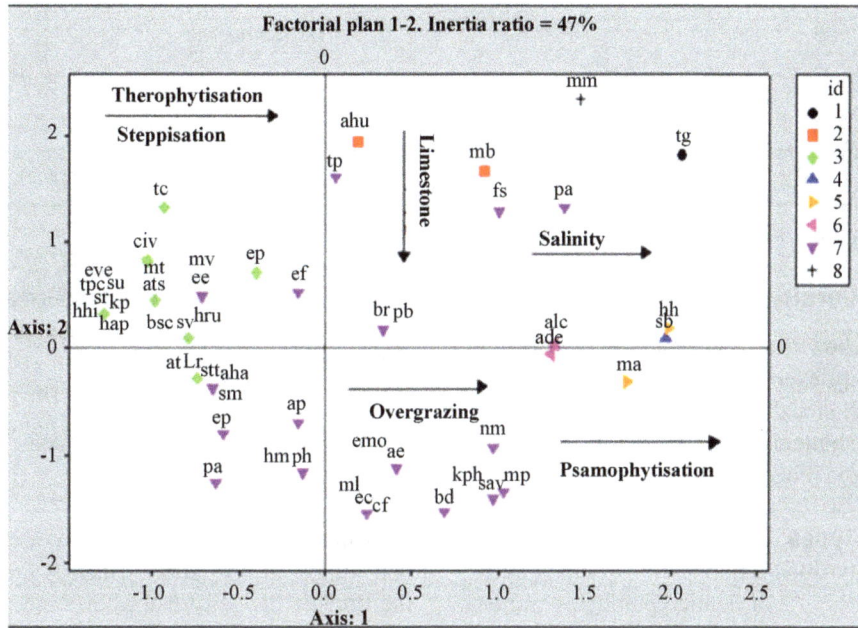

Figure 3. Factorial plan of taxa: Axis 1 - 2 (Zone 1).

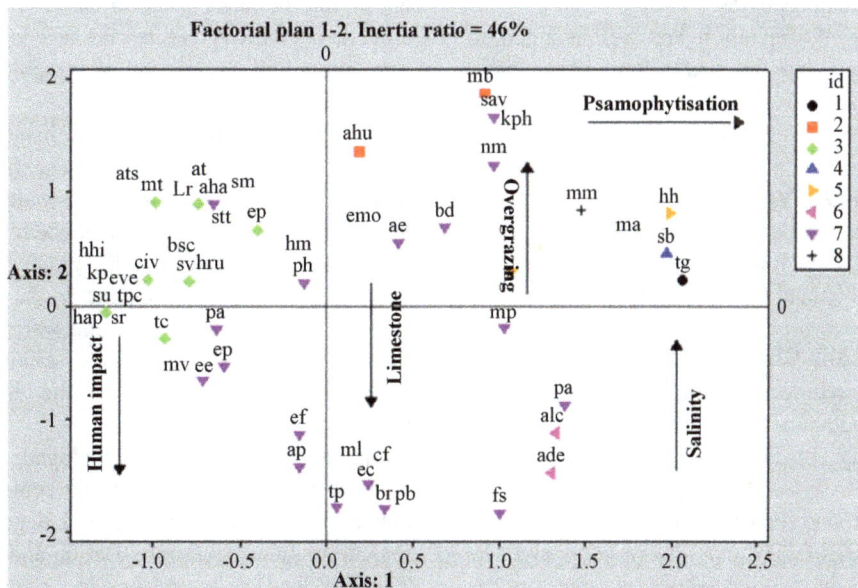

Figure 4. Factorial plan of taxa: Axis 1 - 2 (Zone 2).

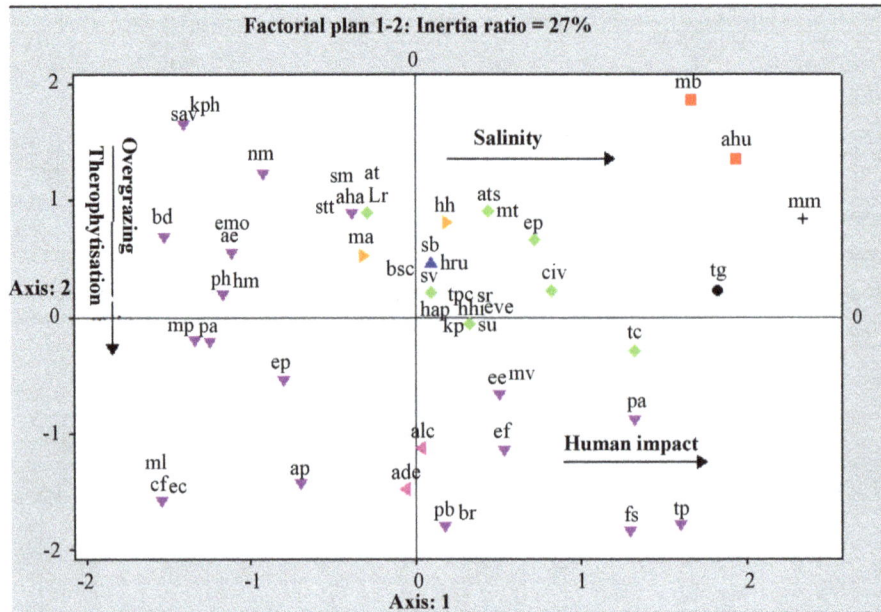

Figure 5. Factorial plan of taxa: Axis 1 - 2 (Zone 3).

3.4. Floristic Diversity (Figure 6)

Many examples show that changes in the composition and reduced diversity exert some influence on the biological characterization and especially the distribution of different classes of biological and morphological types and the systematic and biogeographical distribution.

3.4.1. Biological Types

The floristic composition contains 41% of therophytes, 38% of chamaephytes, 9% of hemicryptophytes and 6% of geophytes and phanerophytes. It seems again that the plant communities identified also give great importance to therophytes and chamaephytes.

These two types of vegetation are well adapted to steppe zones. Overgrazing and aridity have consistently led the development of taxa such as, *Noaea mucronata*, *Atractylis serratuloides*, *Peganum harmala*…

3.4.2. Morphological Types

Woody perennials types took the first dominant position with a rate of 37% followed by perennials herbaceous with 34% and annual herbs with 29%.

3.4.3. Systematically Characterization

This area is characterized by the dominance of taxa belonging to the botanical family of Chenopodiaceaes (25%). This category consists mainly of halophytes subjects.

The Asteraceaes and Poaceaes occupy second and third place with rates of 21% and 12% respectively. The rest is accounted for by families whose rate does not exceed 5%.

3.4.4. Biogeography Characterization

The strict Mediterranean element and circum-Méditerraneen occupy the first place with 30%, followed by the cosmopolitan species with 13%. The rest goes to the Sahara element with a rate that does not exceed 7%.

4. DISCUSSION

In this work, bioclimatic study area clearly indicates that we are in arid zone, which is a reduction factor of the floristic diversity that allows the installation of salt-tolerant vegetation. This halophytic vegetation has developed around the inland basin of the Daya creating a humid microclimate in dry areas [21]. At this scale, bioclimatic asymmetry creates fairness in the distribution of vegetation will depend on orographic and topographic contrasts.

On the substrate, the predominance of fine texture gives it a sandy-loam texture. The fine silts and clays react primarily to the dynamics of salts [22]. Indeed during the dry season, they favor the ascending soluble salts by capillary rise and salinity gradient is then directed towards the surface movement. In the rainy season, they promote stagnation of water and water logging of superficial horizon. Both soil factors (salinity and texture) are fundamentally responsible for the distribution of salt-tolerant species in our area.

The dynamic schemes developed from the factorial plans have managed to highlight ecological gradients of human impact (overgrazing and degradation) and biological gradients (therophytisation and psamophitisation). This link leads to a tendency to contiguity wealth therophytes which seems to be a corollary to the degradation

Phytoecological and phytoedaphological characterization of steppe plant communities in the south of Tlemcen
(western Algeria)

155

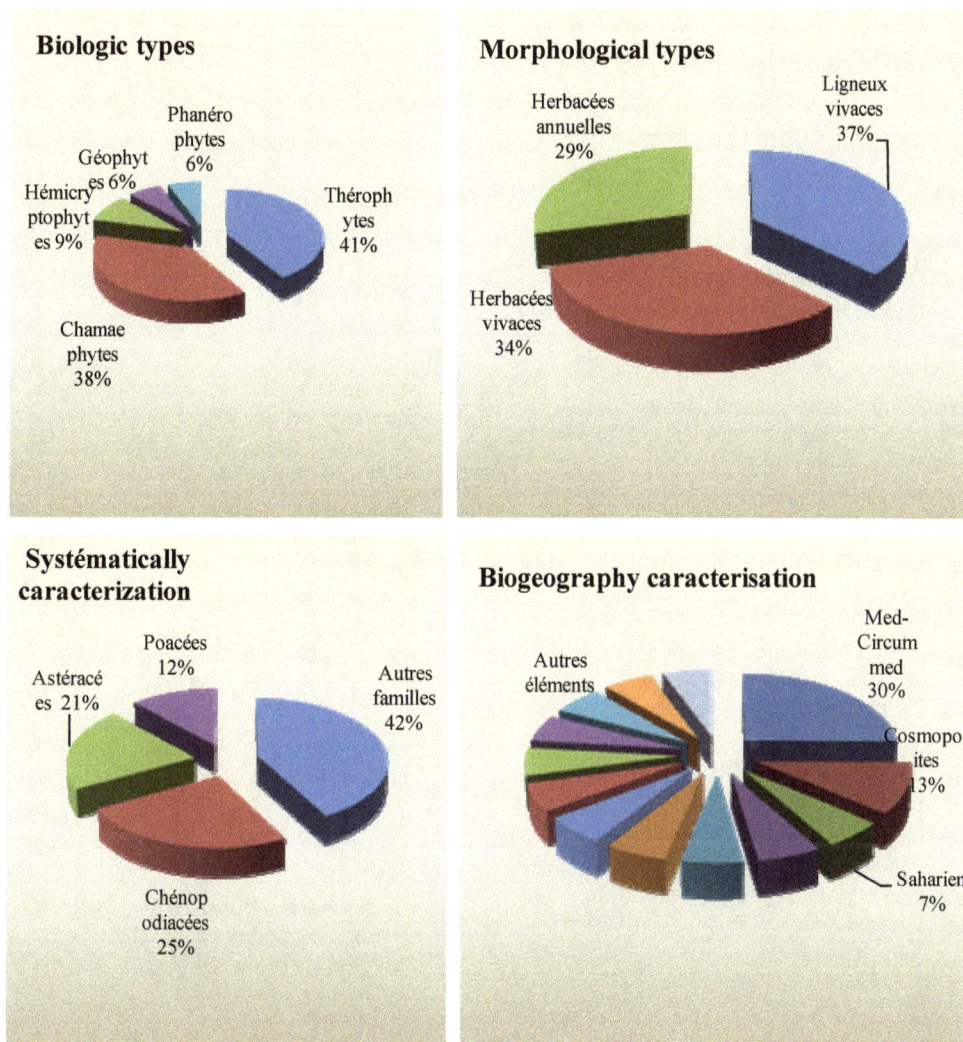

Figure 6. Representation areas of biological forms.

and desertification. We are in front of an adaptation strategy to lower canopy and soil resources. In any case, the rate of therophytes in plant communities naturally increases with aridity [23].

5. CONCLUSIONS

Algerian steppe was exposed for a long time to intensive exploitation of these resources (uprooting, grazing...). It is in the process of losing its potential resilience that ensured its restoration by a simple rational management. A state of floristic and ecological places could be very beneficial any time to initiate conservation programs.

The study of the vegetation in the steppe ecosystem sparked a demonstration of a number of gradients that are managed by different environmental factors. These gradients were positive or negative with respect to the center axes. The origin of these factors has a link with soil (salinity, lime...), human impact (overgrazing, deg-

radation) and biological (therophytisation, psamophitisation).

The plant communities of this area are dominated by chamaephytes and characterized by a low representation of hemicryptophytes and geophytes. On the phytochorique map the Mediterranean element (30%) dominates the other.

As for biodiversity beyond the local extinction of many species of pastoral taxa considered like "macro species", and lost their most important representation in terms of productive potential. Under the weight of permanent grazing, their gene pool is heavily eroded. This calls for further study of some endemic taxa to assess if possible the state of their regression.

The identification of these ecological gradients could be used to reorganize the research areas in these ecosystems. A triggering of a global approach will be based on a participatory conservation management device [24].

Finally, the recommendation for the preservation of

these populations in vulnerable areas could save a threatened biodiversity, and make sure even the breeding of livestock in the pastoral areas [25].

REFERENCES

[1]   Higazy, M.A., Shehata, M. and Allam, A. (1995) Free proline relation to salinity tolerance of three sugar beet varieties. Egypt. *Journal of Agriculture*, **73**, 175-189.

[2]   Lauchli, L. and Epstein, E. (1990) Plant response to saline conditions. *Agricultural Salinity Assensmentand Management*, **15**, 113-137.

[3]   Tremblin, G. (2000) Comportement auto-écologique de *Halopeplis amplexicaulis* plante pionnière des Sebkha de l'Ouest algérien. *Rev Sech*. **11**, 109-116.

[4]   Ozenda, P. (1991) Flore et végétation du sahara. C.N.R.S. 3rd Edition, T.III, Paris, 660 p.

[5]   Le Houerou, H.N. (2000) Use of fodder trees and shrubs (trubs) in the arid and semi-arid zones of west Asia and North Africa. *Proceeding of Workshop on Native and Exotic fodder Shrubs in arid and Semi-arid Zones*, **1**, 9-53.

[6]   Zohary, D. (1974) Domestication of pulses in the old world. *Rev Scien*, **182**, 887-994.

[7]   Viano, J. (1963) Etude phytosociologie et écologique de la région de Fès. Univ. Marseille III. Thèse 3ème Cycle Ecologie, 122 p.

[8]   Babinot, M. (1982) Promotoire oriental du grand Rhône (embouchure). Etude de la végétation et cartographie écologique des aires culicidogènes à Aedes. Caspius en milieu instable. Thèse Doct. Sci. St Jérôme, Aix-Marseille III.

[9]   Le Houerou, H.N. (1993) Land degradation in méditerranean Europe: Can agroforestery be a part of the solution? *Agroforestry Systems*, **21**, 43-61.

[10]   Lemee, G. (1978) Précis d'écologie végétale. Masson et Cie, Paris.

[11]   Braun Blanquet, J. (1951) PF Lauzensoziologie, Grundzuge, Vegetations Kunde Ed. 2 Springer, Vienne Autriche, 631 p.

[12]   Guinochet, M. (1973) Phytosociologie. Masson et Cie, Paris, 177 p.

[13]   Godron, M., Daget, P.H. and Emberger, L. (1983) Code pour le relevé méthodique de la végétation et du milieu. C.N.R.S., Paris, 296 p.

[14]   Chaabane, A. (1993) Etude de la végétation du littoral septentrional de la Tunisie: Typologie: Syntaxonomie et éléments d'aménagement. Thèse Doct Es-Sci, Univ-Aix-Marseille III. Fac-Sci et Tech. St Jerome, 216 p.

[15]   Quezel, P. and Santa, S. (1962) Nouvelle flore de l'Algérie et des régions méridionales CNRS. Tome 1 et 2, 1190 p.

[16]   Raunkiaer, C. (1934) The life form of plants and statistical plant geography. Clarendon Press, Oxford.

[17]   Bendaanoun, M. (1981) Etude synécologique et dynamique de la végétation halophile et hydro-halophile et de l'estuaire de Bou-Reg-Reg (Atlantique du Maroc). Application et perspectives d'aménagement. Thèse Doct. Ing Univ. St Jérome, Aix-Marseille III, 221 p.

[18]   Hanotiaux, G., Lancla, C. and Mathieu, L. (1976) Un exemple d'évaluation des sols salins suite à la mise en valeur par la rizière en Camargue. *Ann. Inst. Nat. Agron. EL-Harrach*, **6**, 259-318.

[19]   Aboura, R., Benmansour, D. and Benabadji, N. (2006) Comparaison phytoecologique des Atriplexaies en Oranie (Algérie). *Ecol Med*, **32**, 73-84.

[20]   El-Hamrouni, A. (1992) Végétation forestière et préforestière de la Tunisie. Typologie et éléments pour la gestion. Thèse Doct-état, Univ. Marseille, 220 p.

[21]   Aidoud, A., Le Floc'h, E. and Le Houerou, H.N. (2006) Les steppes arides du nord de l'Afrique. *Rev Sech*, **17**, 19-30.

[22]   Benabadji, N., Bouazza, M., Metge, G. and Loisel, R. (1996) Description et aspect des sols en régions semi-arides et arides au sud de Sebdou (Oranie, Algérie). *Bull Inst Sci*, **20**, 77-86.

[23]   Maire, R. (1962) Contribution à l'étude de la flore de l'Afrique du Nord. *Bull Soc Afr Nord*, **26**, 186-196.

[24]   Nedjraoui D., Hirche, A. and Boughani, A.F. (1999) Désertification par télédétection des hautes plaines steppiques du Sud-Ouest oranais. U.R.B.T. et I.N.C. Alger, **17**, 10-22.

[25]   Mostefa, S.A. and Benariad, M. (1999) Suivi diachronique des processus de désertification *in situ* et par télédétection des hautes plaines steppiques du Sud-Ouest oranais. U.R.B.T. et I.N.C. Alger, **16**, 9-15.

18

Bryophyte mass to stem length ratio: A potential metric for eco-physiological response to land use

Jason A. Hubbart[1*], Elliott Kellner[2]

[1]Department of Forestry and Department of Soils, Environmental and Atmospheric Sciences, University of Missouri, Columbia, USA; *Corresponding Author
[2]Department of Forestry, University of Missouri, Columbia, USA

ABSTRACT

Methods of analysis are needed that quantitatively characterize the response of organisms to anthropogenic disturbance. Herein a method is presented that characterizes bryophyte morphological variability in response to timber harvest treatments (clearcut and partial cut). Samples (n = 6196) of the semi-aquatic bryophyte *Brachythecium frigidum* were collected from clearcut, partial cut and full forest stream reaches between August 2003 and October 2005 and analyzed to obtain mass to stem length ratios (M:SL). Results show that relative to a full forest (*i.e.* full canopy cover condition), average M:SL ratios were reduced approximately 18% in the partial cut and 37% in the clearcut, indicating a decrease in biomass per unit stem length with increasing harvest intensities. Increased light intensities and higher air temperatures resulting from decreased canopy cover in the harvest treatments corresponded to lower M:SL ratios (0.31 and 0.24 for the partial cut and clearcut, respectively). Results quantify the morphological response of *B. frigidum* to habitat perturbation, thereby validating the method as a useful assessment of anthropogenic disturbance in post-timber harvest environments. Additional work should be conducted to test the method in other physiographic regions and to isolate bryophyte response to alterations of distinct environmental variables.

Keywords: Semi Aquatic Bryophytes; Timber Harvest; Microclimate; Morphological Growth Response

1. INTRODUCTION

Multiple authors have identified the need for studies that investigate bryophyte response to post-timber harvest environments [1-6]. Bryophytes are widely distributed among terrestrial and aquatic environments [7,8] and serve a number of important roles in forested ecosystems by contributing greatly to net primary productivity, nutrient retention, and by providing habitat and a food base for invertebrates [1,3,9]. Binkley and Graham [10] found that bryophytes made up an average of 5% net primary productivity, 20% of the understory biomass, and 95% of the understory photosynthetic tissue in an old-growth Douglas-fir forest of the pacific northwest, USA. Conceivably, semi-aquatic bryophytes may provide the most appropriate indices for anthropogenic change in many ecosystems since they are located at the terrestrial-aquatic interface, and are therefore influenced by changes occurring in both terrestrial and aquatic ecosystems. Timber harvest effects that may prove detrimental to bryophyte health include increased erosion and sediment, altered stream flow regimes, microclimate (including light availability and air and stream temperature) and nutrient regimes [11-21]. Given bryophyte sensitivity to disturbance, forest management including clearcutting or thinning (*i.e.* partial cutting) may result in nearly immediate changes to bryophyte health.

Bryophytes are known to have a narrow range of tolerance for many microclimatic conditions [8]. For example, low light intensities, low temperatures, and high relative humidity typical of closed canopy forests are assumed to be optimal for bryophytes [3,22]. Altered microclimate in post-harvest forests was shown to have detrimental effects on mosses in Acadian-mixed wood [2], boreal [16,23-25], mixed-conifer [3,19], and hemlock-northern hardwood forests [26]. For example, Nelson and Halpern [3] reported a 90% decline in bryophyte species richness within one year of harvest treatments. Likewise, Shields *et al.* [26] found an approximately 44% reduction in bryophyte cover in forest openings compared to undisturbed full forest plots.

A thorough review of recent primary literature concerning the effect of climatic alterations and/or timber harvest impacts on bryophytes suggested that the majority of studies have utilized community characteristics such as percentage cover, abundance, and various parameters of species richness and composition as measures of bryophyte response [4,6,18-21,24,25,27-33]. Of the studies that have addressed physiological and morphological responses of bryophyte species to altered climate conditions, many have focused on increased temperatures and radiation levels in Artic and sub-Artic ecosystems [34-36], or the response of bryophytes to increased levels of nutrients such as Nitrogen and Phosphorus [37,38]. Few studies have focused on physiological and morphological metrics of bryophyte growth response to timber harvest. Notable exceptions include Sollows *et al.* [22], who found that a species of liverwort (*Bazzania trilobata*) showed signs of severe desiccation stress, by means of decreased photosynthetic activity and reduced green pigmentation due to changes in microclimate (*i.e.* increased temperature, increased light intensity, and decreased relative humidity) that followed clearcutting. When closed canopy conditions were simulated in the laboratory, liverworts were capable of resuming photosynthesis and growth, but at rates considerably lower than those observed prior to canopy removal. Dynesius *et al.* [16] experimentally analyzed the effect of canopy sheltering on bryophyte transplants. Treatment included placement of transplants in positions of contrasting exposure and covering 50% of sample individuals with spruce branches. Response was evaluated by estimating proportion of living shoots and measuring radial transplant growth. Results showed a statistically significant (p < 0.0001) negative relationship between light intensity and bryophyte vitality. The results of these two studies indicate that clearcutting may affect bryophyte growth morphology, reduce recovery abilities, and significantly alter bryophyte density following harvest. Understanding how the effects of timber harvest influence bryophyte physiology and morphology will help scientists and land managers better understand how management practices influence riparian biota and water quality in complex forested ecosystems.

The overall objectives of this study were to 1) quantify the physiological/morphological response of a semi-aquatic bryophyte (*Brachythecium frigidum*) to anthropogenic disturbance by means of a new method, specifically the comparison of mass to stem length ratios (M:SL), and 2) characterize the potential for the method to predict aquatic ecosystem alteration in a post-timber harvest riparian ecosystem.

2. METHODS

2.1. Study Site

The study took place from August 2003 through October 2005 in the Mica Creek Experimental Watershed (MCEW) located in Shoshone County, northern Idaho, approximately 25 km southeast of St. Maries, Idaho (47.17°N latitude, and 116.25°W longitude) (**Figures 1** and **2**). The Mica Creek Experimental Watershed is a paired and nested catchment study basin that is privately owned and operated by Potlatch Corporation. Both continental and maritime weather patterns influence the watershed. The continental/maritime climate region, which includes northern Idaho, is a unique transitional climate influenced by both maritime and continental weather patterns [14,39]. Continental climates are characterized by large seasonal variations in air temperature and strong influences by relatively dry polar air masses, while maritime climates are generally wetter and influenced by the prevailing westerly winds with cool summers and mild winters. Given the distinct climate regime of the region, and the economic importance of the local timber industry, studies are warranted that address impacts of anthropogenic environmental change subsequent to timber harvest. The experimental watershed encompasses the Mica Creek and the West Fork of Mica Creek (2700 hectares) catchments, which are tributaries to the St. Joe River. The watershed varies in elevation from 1000 to 1585 meters at the headwaters, and receives approximately 1400 mm annual precipitation. The average annual temperature is approximately 4.5°C. The majority of precipitation falls from November to May, with at least 70% of all precipitation falling as snow [14].

Vegetation of the MCEW consists of 75- to 85-year-old naturally regenerated conifer stands, while remnant old-growth (pre-harvest) red cedar (*Thuja plicata*) remains along the upper tributaries of West Fork Mica Creek. Dominant canopy vegetation within the watershed includes red cedar (*Thuja plicata*), western larch (*Larix occidentalis*), grand fir (*Abies grandis*), western white pine (*Pinus monticola*), western hemlock (*Tsuga heterophylla*) and Engelmann spruce (*Picea engelmannii*).

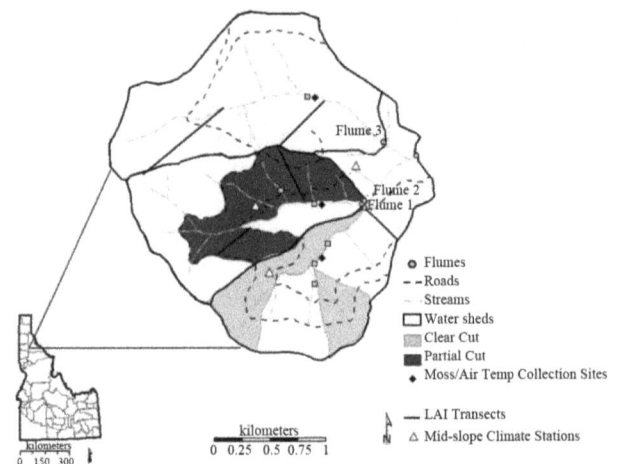

Figure 1. The Mica Creek Experimental Watershed (MCEW) located in the panhandle of northern Idaho, USA.

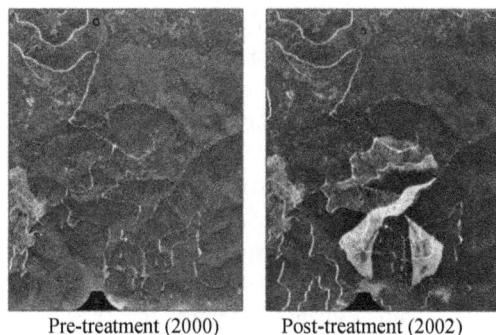

Pre-treatment (2000)   Post-treatment (2002)

Figure 2. Pre- and post-treatment aerial photos of the Mica Creek Experimental Watershed, northern Idaho, USA.

Understory vegetation is largely comprised of grasses, forbs, and shrubs. Stream riparian zones are dominated by alder (*Alnus* spp.) and dogwood (*Cornus* spp.) and many stream reaches are well populated with bracken fern (*Pteridium acquilinum*) undergrowth [14].

2.2. Forest Harvest

Timber harvest took place in the late summer and fall of 2001. In the clearcut, whole canopy harvest took place in 50% of the treatment area, and in the partial cut, 50% of the canopy was harvested in 50% of the treatment area. The clearcut was broadcast burned and replanted in May 2003 [15,17].

Timber harvest followed all Idaho Forest Practices Act regulations for Stream Protection Zone (SPZ) requirements at that time for the protection of in-stream and riparian biota and human beneficial uses. The state of Idaho has set SPZ requirements according to two stream classifications: class I and class II streams. Class I streams are fish bearing and are used for domestic water supply. The SPZ for class I streams must be at least 75 feet (22.9 m) wide on either side of the high water mark. Harvesting is still permitted in this zone, but 75% of the original shade must be retained. There are also target leave tree requirements per 1000 linear feet, depending on stream width. Class II streams are non-fish bearing, first order streams. In Idaho, equipment is excluded from a 30 foot (9.1 m) SPZ on either side of the high water mark. There are no shade requirements or leave tree requirements for SPZ's, but skidding logs in or through class II streams is not permitted [40].

2.3. Canopy Cover

Canopy cover was estimated using the ceptometer method, which estimates leaf area index (LAI) as per the methods of White *et al.* [41] and Keane *et al.* [42]. Using an AccuPAR ceptometer (Decagon Devices, Inc., Pullman, WA), radiometric readings were collected to estimate unimpeded incoming photosynthetically active ra-

diation (PAR) in the open and also along two 300 m transects in the partial cut harvest treatment and full forest (**Figures 1** and **2**). The sky was clear of clouds during the data collection for this work. Radiometric measurements were collected in the four cardinal directions every 20 m and then averaged to account for zenith angle, and foliar interception variability. LAI was computed using **Eq.1**.

$$LAI = \frac{-\ln\left(Q_i/Q_o\right)}{k} \quad (1)$$

where Q_i is the amount of radiation at the base of the canopy, Q_o is the amount of incoming radiation above the canopy, and k is a radiation extinction constant (normally assumed, $k = 0.5$) [43,44]. Limitations of the ceptometer method can include synchronizing changing light conditions between below and above canopy readings, and adjusting for solar zenith angle [42].

2.4. Air Temperature

Climate data were collected from stations installed at mid-slope positions within each treatment type (**Table 1**) to capture most representative climate variables in each treatment and avoid treatment induced edge effects. Air temperature was monitored with combination temperature/humidity probes in Gill radiation shields (Vaisala, HMP45C, Campbell Scientific, Inc., Logan, UT) in the valley bottom riparian zone in the nested catchment (**Figure 1**). Air temperature data were simultaneously collected at each riparian sampling site at hourly intervals 3 m above the water surface with a Thermochron iButton temperature data logger [45]. The device collects time series temperature data within a temperature range of −20°C to +85°C at 0.5°C resolution, with a manufacturer stated accuracy of ±1°C. Hubbart *et al.* [45] showed that with a sample size of 61 iButton temperature data loggers, the accuracy of the sensors was well within manufacturer stated specifications of ±1.0°C with a collective temperature variance of ±0.21°C. The radiation shield utilized for this work was comprised of a 6- and 8-oz white funnel assemblage with a 1-cm space created between the outermost and innermost perforated assemblage allowing passive airflow to the sensor, and thus accurate ambient temperatures as per Hubbart [46]. Stream temperature was not measured at sampling sites. In a peripheral study in the MCEW, Gravelle and Link [13] found that the best management practices (BMP's) employed in the harvest treatments, explained above, effectively mitigated the impact of canopy removal on stream temperature.

2.5. Semi-Aquatic Bryophyte

The species of semi-aquatic moss studied in this work was *Brachythecium frigidum* (C. Muell.) Besch. *Brachy-*

Table 1. Local topographic characteristics for mid-slope climate stations located within the Mica Creek Experimental Watershed, northern Idaho, USA.

Treatment	Elevation (m)	Slope (%)	Aspect (deg)	Aspect Categorized
Clearcut	1365	44	46	NE
Patial Cut	1340	20	117	ESE
Full Forest	1299	25	82	E

thecium frigidum is one of thirty species in this genus that occur in North America. It is found in mesic ecosystems from Alaska to California, Nevada, Utah, Montana, and Wyoming. A study by Jonsson [47] in the western Cascades of Oregon, found *B. frigidum* to occur along a wide elevation gradient (420 - 1250 m) on moist rocky substrates. It was most abundant at higher elevations, but was also common in lower elevation floodplain areas [48-50]. The abundance of this species of bryophyte makes it ideal for this type of study since results are transferable to other regions where *B. frigidum* is also likely to occur.

Bracythecium frigidum samples were collected in three, second order streams of the West Fork of Mica Creek from August 2003 to October 2005, on a quarterly basis (~3 month intervals). Each tributary was influenced by a different harvest condition: clearcut, thinned (partial cut), or fully forested control (**Figures 1** and **2**). Complete specimens were collected within the riparian zone at the terrestrial-aquatic interface (*i.e.* water surface) to measure changes occurring where anthropogenic effects are strongest. Sample collection was randomized, with no preference for size and presumably age (n = 6196). Moss samples were promptly placed in properly labeled brown paper bags as per methods described by Jönsson [51], and were allowed to air dry in the bags and stored in the same manner for later analyses. Samples were reconstituted in the lab in tap water for 24 hours in order to rinse out soil and woody debris without damaging the specimen as per methods described by Vargha *et al.* [52]. Moss samples were then blotted dry with a paper towel to remove any dripping water from samples before being weighed in order to standardize moisture content between samples. After weighing, each sample was dissected into individual stems and spread out on plastic transparency sheets for computer scanning. ImageJ 1.34s software (National Institutes of Health®) was used to trace the scanned images to measure total stem length of each sample. Subsequent to scanning, each sample was weighed again to account for any additional debris that was removed during the dissection process. Post-dissection sample weights were then compared to initial sample weights via regression analysis which indicated that before and after dissection, weights were statistically

identical (p < 0.01). On this basis, only pre-dissection weights were included in further analyses. Data were used to create a mass to stem length ratio (M:SL) for each sample, which provided a standardized measure of biomass per unit stem growth for unequally sized samples. Further, the M:SL ratio served as a measure of bryophyte morphological response, which mirrored the bryophyte physiological response to changing land use and environmental variables. One-way ANOVA was utilized to compare means between sites (p < 0.05) [53, 54]. Statistical analyses were conducted and graphs were generated utilizing Origin software (Origin Lab Corporation® 2010).

3. RESULTS AND DISCUSSION

3.1. Climate during Study

Meteorological data collected at mid-slope climate stations within each treatment from 1 Nov. 2003 to 14 Oct. 2005, showed many differences between the clearcut, partial cut, and full forest treatments (**Table 2**, **Figure 3**). Average mid-slope ambient air temperatures were 6.60°C, 6.09°C, and 6.12°C in clearcut, partial cut and full forest, respectively. Average ambient air temperatures in the riparian zone (Ta Riparian) were 4.94°C, 4.43°C and 4.36°C in the clearcut, partial cut and full forest, respectively.

These results show that average air temperatures in the riparian zones were 1.66°C, 1.66°C, and 1.76°C less than air temperatures at mid-slope climate stations in the clearcut, partial cut, and full forest, respectively. In addition, air temperatures measured in the riparian zone displayed a narrower range of variation than air temperatures measured at the mid-slope climate stations (**Figure 4**). In the riparian zone, average air temperatures ranged from

Table 2. Air temperature (Ta) descriptive statistics from 1 Nov. 2003 to 4 Oct. 2005, at mid-slope and riparian climate stations in clearcut, partial cut, and full forest within the Mica Creek Experimental Watershed, northern Idaho, USA. Min = Minimum, Max = Maximum, SD = Standard Deviation.

	Air Temperature: Mid-Slope (°C)			
	Min	Mean	Max	SD
Clearcut	−7.21	6.60	21.30	±7.48
Partial Cut	−7.27	6.09	20.13	±7.21
Control	−7.12	6.12	20.14	±7.19
	Air Temperature Riparian (°C)			
	Min	Mean	Max	SD
Clearcut	−4.28	4.94	16.56	±6.40
Partial Cut	−4.55	4.43	16.08	±6.46
Control	−5.03	4.36	16.36	±6.55

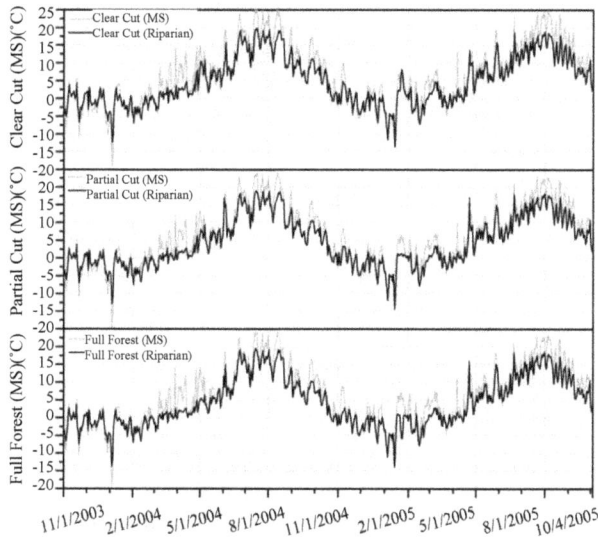

Figure 3. Average mid-slope and riparian air temperatures in clearcut, partial cut, and full forest catchments from 1 Nov. 2003 to 4 Oct. 2005, within the Mica Creek Experimental Watershed in northern Idaho, USA.

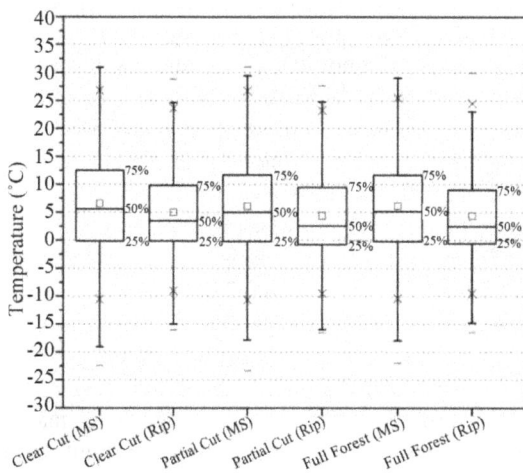

Figure 4. Average air temperature (Ta) box and whisker plots from 1 Nov. 2003 to 4 Oct. 2005, at mid-slope (MS) and riparian (Rip) climate stations in clearcut, partial cut, and full forest within the Mica Creek Experimental Watershed, northern Idaho, USA.

−4.28°C to 16.56°C in the clearcut, −4.55°C to 16.08°C in the partial cut, and −5.03°C to 16.36°C in the full forest, compared to average air temperatures at the mid-slope climate stations, which ranged from −7.21°C to 21.30°C in the clearcut, −7.27°C to 20.13°C in the partial cut, and −7.12°C to 20.14°C in the full forest.

Air temperature data collected from both the riparian zones and the mid-slope climate stations (**Figure 1**) showed only slight differences in average air temperature (<1°C) between the clearcut, partial cut, and full forest treatments (**Table 2**, **Figure 3**). However, average air temperatures in the riparian zone were ~25%, ~27%, and

~29% lower than average temperatures measured at the climate stations in the clearcut, partial cut and full forest, respectively. The clearcut treatment corresponded with the highest observed air temperatures in both the riparian zone and the treatment plot (**Table 2**). These results suggest that forest harvest practice had a warming effect on air temperature, even in the adjacent riparian areas. The range of air temperatures observed in the riparian zone was also narrower in comparison to air temperatures observed at the climate stations. In the riparian zone, the range of average air temperatures were ~27%, ~25%, and ~22% smaller than the range of air temperatures measured at mid-slope climate stations in the clearcut partial cut, and full forest, respectively (**Figure 4**).

In addition to shading, lower temperatures in the riparian zones may be partially due to cold air drainage and nocturnal temperature inversion processes that are known to occur in the MCEW [55]. Air temperature differences between the climate stations and the riparian zones also indicated that riparian zones may have a buffering effect against both increases in air temperature and fluctuations in air temperature occurring in adjacent harvest treatments. These results further demonstrate the need to maintain adequate riparian buffer zones in harvested environments to reduce the negative effects of altered temperature regimes on bryophytes and stream ecosystems in general.

3.2. Bryophyte Morphometrics

Bryophyte mass to stem length (M:SL) ratios varied broadly between the three harvest treatments (clearcut, partial cut, full forest) following timber harvest (**Tables 3** and **4**, **Figure 5**). Dashes in **Table 3** indicate no data due to collection complications or sample unsuitability. The mean M:SL ratio was 0.38 in the full forest, 0.31 in the partial cut, and 0.24 in the clearcut. While not statistically significant ($p > 0.05$), these results show an 18.4% reduction of M:SL in the partial cut and 36.8% reduction of M:SL in the clearcut, relative to the full forest control.

Canopy cover was most dense in the full forest, with a mean leaf area index (LAI) of 8.24 (**Table 4**, **Figure 5**). The mean LAI in the partial cut was 2.79 (clearcut LAI = 0.0). Reduced canopy cover (*i.e.* lower LAI) in harvested treatments resulted in higher levels of radiation reaching bryophytes in the stream channel. The relatively dense canopy in the full forest corresponded to the highest M:SL ratio of all three treatments (0.38). With less canopy cover in the partial cut, the mean M:SL ratio decreased to 0.31; and under little to no canopy cover in the clearcut (LAI = 0.0), the mean M:SL ratio further decreased to 0.24. These results suggest a positive relationship between the LAI of the forest canopy and the M:SL ratio of *B. frigidum*. Regression analysis between average M:SL and LAI (n = 6) resulted in a high coefficient

Table 3. Mass to stem length ratios by date and treatment in the clearcut, partial cut, and full forest of the Mica Creek Experimental Watershed, northern Idaho, USA.

	Clearcut		Partial Cut		Control	
Date	M:SL	n	M:SL	n	M:SL	n
08/31/03	0.28	184	-	-	0.62	74
10/21/03	-	-	0.21	255	1.14	43
11/14/03	0.20	161	0.23	83	0.02	85
03/13/04	0.28	475	0.20	292	0.54	50
06/01/04	0.40	230	0.35	203	0.22	193
08/28/04	-	-	0.62	152	0.19	336
11/12/04	0.25	185	0.18	340	0.35	192
01/21/05	0.23	365	0.22	256	0.29	342
04/21/05	0.18	223	0.45	175	0.19	339
07/18/05	0.13	123	0.24	304	0.25	233
10/03/05	0.24	190	0.45	113	-	-

M:SL by Treatment and Collection Date

Dashed line indicates no samples; M:SL = Mass to Stem Length Ratio; n = sample size.

Table 4. Descriptive statistics for mass to stem length (M:SL) ratios, and leaf area index (LAI), in each treatment (clearcut, partial cut, and full forest) within the Mica Creek Experimental Watershed, northern Idaho, USA.

Mass to Stem Length Ratio				
Treatment	Min	Mean	Max	SD
Clearcut	0.13	0.24	0.40	±0.07
Partial Cut	0.18	0.31	0.62	±0.15
Full Forest	0.02	0.38	1.14	±0.32
Leaf Area Index				
Treatment	Min	Mean	Max	SD
Clearcut	0.00	0.00	0.00	0.00
Partial Cut	0.01	2.79	6.57	±2.29
Full Forest	4.24	8.24	11.54	±2.41

SD = Standard Deviation.

of determination (R^2) of 0.97. Given lack of study site replication more rigorous statistical analyses were not practical. Regardless, these relationships are compelling. Future replicated studies should be conducted to supply statistical rigor and validation to the method established in the current work.

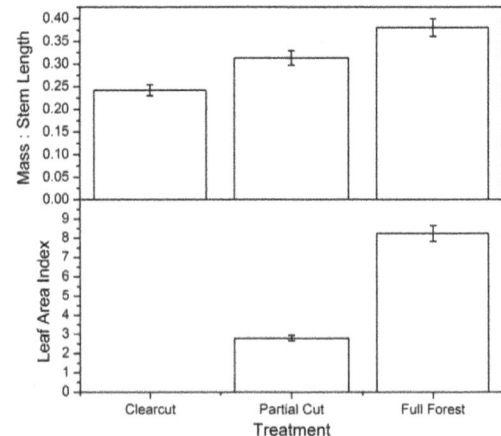

Figure 5. Mean mass to stem length (M:SL) ratios and leaf area index (LAI) in each treatment (clearcut, partial cut, and full forest) within the Mica Creek Experimental Watershed, northern Idaho, USA.

3.3. Timber Harvest, Microclimate and Bryophyte Morphology

Results indicated that altered environmental variables including leaf area index and air temperature subsequent to altered canopy conditions (*i.e.* clearcut and partial cut) corresponded to a decrease in biomass per unit stem length in *B. frigidum*. Additional replicated work should be conducted in other physiographic regions to test the method used in the current work and assess ecosystem response to perturbation. Mass to stem length ratios decreased from full forest, to partial cut, to clearcut (in that order), from 0.38, to 0.31, to 0.24, respectively (**Table 4**, **Figure 5**). This result is supported by the negative relationship between timber harvest and bryophyte physiological/morphological growth response reported by Dynesius *et al.* [16]. Stem elongation in bryophytes may be a response to elevated light, temperature, and nutrient levels observed in the partial cut and clearcut treatments but it is not necessarily an indication of healthier, more productive plants [56,57]. Longer stems often lacked secondary stem and leaf growth, giving them an etiolated appearance with less biomass per unit stem length (**Figure 6**). This observation is consistent with previous work by Furness and Grime [56,57] who showed that bryophytes growing at above optimum temperatures exhibited elongated stem growth and a reduction in overlapping leaves. They suggested that bryophytes may be capable of exploiting short-term seasonal increases in temperature and light conditions. However, prolonged exposure to those conditions leads to changes in growth form that may be detrimental to bryophyte health, including reduced leaf density and stem etiolation. Further investigations on the physiology of *B. frigidum* are warranted to better explain relationships between moss growth form and environmental conditions.

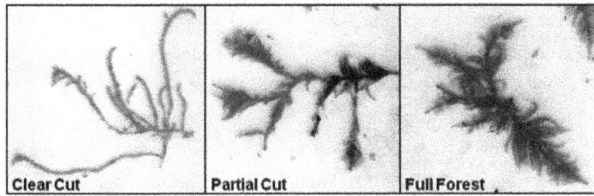

Figure 6. Representative moss (*Brachythecium frigidum*) samples from the clearcut, partial cut, and full forest treatments within the Mica Creek Experimental Watershed, northern Idaho, USA.

Meteorological data collected at the climate stations located mid-slope in both treatment catchments clearly illustrated microclimatic alterations subsequent to harvest (**Table 2**, **Figure 3**). There were slight differences (<1°C) in average air temperature between the clearcut, partial cut, and full forest throughout the study period, with the clearcut exhibiting the widest range of air temperatures (−7.21°C to 21.30°C) of all three treatments. Climate differences between treatments demonstrate how timber harvest leads to multiple environmental changes, many of which may alter bryophyte health, and thus stream ecosystem health.

Although riparian zones were protected by a 75 foot (22.9 m) buffer, harvesting was still permitted providing 75% of the original shade was retained. Regardless, in the steep sloped v-notch valleys of the MCEW, increased light in the adjacent harvest areas was also transmitted horizontally through the thinned riparian canopy to the stream channel, adding to the amount of light reaching stream bryophytes. Future research is warranted to better understand light attenuation through riparian stands and the impact of thinning on the stream light regime and primary production. Conceivably, any alteration to light may alter primary productivity and consequently the aquatic food web. The stream channel canopy gap should therefore be considered since, in harvested treatments, more radiation reaches the stream via horizontal transmittance following riparian thinning practices [58,59]. However, since bryophyte morphology is determined by multiple micrometeorological and nutrient criteria, further studies are needed to isolate the physiological response of this species to altered light regimes.

Bryophytes may exhibit a variable response to changing light intensities between partial cut and clearcut harvest treatments due to influences from other interacting variables (e.g. nutrient alterations, suspended sediment). Other published studies conducted in the MCEW confirm the effects of timber harvest on various environmental variables. For example, Hubbart *et al.* [14] reported significant (p < 0.01) increases in annual water yield from harvested plots. Koeniger *et al.* [60] discovered alterations in the isotopic composition of snow corresponding to harvest practices. Gravelle *et al.* [17]

found significant (p < 0.001) increases of nitrate + nitrite ($NO_3 + NO_2$) concentrations in water draining from the MCEW. Ultimately, the combined effects of altered environmental variables may not be merely cumulative, but rather have a non-linear effect on bryophyte morphology. Elevated light levels and stream nutrient concentrations in the partial cut (relative to the full forest) may have a smaller combined effect on bryophyte morphology than in the clearcut, where these variables may exhibit a greater deviation from the full forest. For example, the broadcast burning that took place in the clearcut following harvest may have contributed to the higher stream nitrate concentrations, and therefore, lower M:SL ratios in that treatment. Elucidating these interactions was beyond the scope of the current work but supplies impetus for future research.

Semi-aquatic bryophytes such as *B. frigidum* are understood to draw the majority of their nutrients from stream water [1], so it is possible that differences in nutrient availability subsequent to harvest treatments may affect bryophyte productivity. Future controlled experiments should seek to measure the response of semi-aquatic bryophytes to increased stream nitrate levels. One method of evaluating this response would be to culture bryophytes in a laboratory and measure the response. This work could easily be done in conjunction with varying light levels in order to measure both the individual and cumulative effects of these variables on bryophyte productivity.

Similar to Dynesius *et al.* [16], this study sought to determine the physiological/morphological growth response of bryophytes to climate alterations subsequent to timber harvest. However, in contrast to Dynesius *et al.* [16], this research was conducted on individuals sampled from their natural habitat, as opposed to experimental transplants. This study developed a practical method of assessing the relative response of bryophyte individuals, a method which can be utilized by land managers to evaluate the effect of harvest practices on aquatic/terrestrial ecosystem viability. While results of this work were not statistically significant, even minor responses subsequent to timber harvest, although statistically undetectable, should not be interpreted as ecologically benign. Results indicate that bryophytes respond proportionally to disturbance levels (in this case, harvest intensity). For example, this work showed that shoot lengthening was greater in the clearcut than in the partial cut, relative to the full forest. This suggests that perhaps riparian forests should be selectively harvested in order to maintain a stand density that will continue to protect riparian stream ecosystems and biological and water quality integrity in post-harvest environments. Future work is thus warranted to quantify these relationships and validate contemporary riparian forest buffer best management practices.

4. CONCLUSIONS

This work adds to the limited body of knowledge pertaining to the eco-physiological response of riparian bryophytes to anthropogenic disturbance, and presents an assessment method to measure that response. Average mass to stem length ratios (M:SL) were 18.4% less in a partial cut and 36.8% less in a clearcut relative to a fully forested control, indicating stem elongation and reduced leaf density per unit stem length in the harvested treatments. Canopy density in the full forest (mean LAI = 8.24) corresponded to the highest M:SL ratios (0.38) of all three treatments. Reduced canopy cover in the partial cut and clearcut (mean LAI = 2.79 and 0.0, respectively) corresponded to lower M:SL ratios (0.31 and 0.24, respectively). Regression analysis between mean average M:SL and LAI demonstrated a high coefficient of determination (R^2) of 0.97. Lower average air temperature in the full forest corresponded to the highest biomass per unit stem length (M:SL = 0.38), whereas higher air temperatures in the clearcut and partial cut corresponded to lower biomass per unit stem length (M:SL = 0.24 and 0.31, respectively). These results suggest a negative relationship between altered climate conditions subsequent to timber harvest and bryophyte growth response. Average air temperatures in the riparian zones were cooler than air temperatures measured at the mid-slope climate stations, which may be explained by persistent cold air drainage that is known to occur within the watershed. Lower air temperatures in the riparian zones also indicate that riparian zones may have a cooling effect on air temperatures relative to the adjacent harvest treatments.

Bryophytes are particularly sensitive to land use impacts, as they exhibit a narrow range of tolerance to environmental disturbance. As primary producers, they are quick to respond to alterations to the riparian environment. The practical assessment method presented here can be utilized by land managers to evaluate the effect of land use practices on aquatic/terrestrial ecosystem fitness. This holds important implications for minimizing disturbance to higher trophic level organisms such as macroinvertebrates. Results of this work should serve as a springboard to further investigations (both laboratory and field) that isolate the direct bryophyte morphological response to individual environmental variables, including light and nutrient availability, and stream and air temperature. Further research is warranted to measure bryophyte response to timber harvest and other land uses (e.g. urbanization) over greater spatial and temporal scales, and to investigate the rate of bryophyte recovery from disturbance.

5. ACKNOWLEDGEMENTS

This research was carried out with funding provided by the United States Department of Agriculture, and USFS Research Joint Venture Agreement #03JV-11222065-068, USDA-CSREES 2003-01264. A profound acknowledgment is due to Malia Volke for lab, analysis, and manuscript preparation assistance. Special acknowledgments are extended to Timothy Link, John Marshall, and John Gravelle, and multiple reviewers whose feedback greatly improved the article.

REFERENCES

[1] Stream Bryophyte Group (1999) Roles of bryophytes in stream ecosystems. *Journal of the North American Benthological Society*, **18**, 151-184.

[2] Fenton, N.J., Frego, K.A. and Sims, M.R. (2003) Changes in forest floor bryophyte (moss and liverwort) communities 4 years after forest harvest. *Canadian Journal of Botany*, **81**, 714-731.

[3] Nelson, C.R. and Halpern, C.B. (2005) Short-term effects of timber harvest and forest edges on ground-layer mosses and liverworts. *Canadian Journal of Botany*, **83**, 610-620.

[4] Perhans, K., Appelgren, L., Jonsson, F., Nordin, U., Soderstrom, B. and Gustafsson, L. (2009) Retention patches as potential refugia for bryophytes and lichens in managed forest landscapes. *Biological Conservation*, **142**, 1125-1133.

[5] Hardman, A. and McCune, B. (2010) Bryoid layer response to soil disturbance by fuel reduction treatments in a dry conifer forest. *The Bryologist*, **113**, 235-245.

[6] Root, H.T. and McCune, B. (2010) Forest floor lichen and bryophyte communities in thinned *Pseudotsuga menziesii-Tsuga heterophylla* forests. *The Bryologist*, **113**, 619-630.

[7] Cenci, R.M. (2000) The use of aquatic moss (*Fontinalis antipyretica*) as monitor of contamination in standing and running waters: Limits and advantages. *Journal of Limnology*, **60**, 53-61.

[8] Gignac, D. (2001) Bryophytes as indicators of climate change. *The Bryologist*, **104**, 410-420.

[9] Turetsky, M.R. (2003) The role of bryophytes in carbon and nitrogen cycling. *The Bryologist*, **106**, 395-409.

[10] Binkley, D. and Graham, R.L. (1981) Biomass, production, and nutrient cycling of mosses in an old-growth Douglas-fir forest. *Ecology*, **62**, 1387-1389.

[11] Tiedemann, A.R., Quigley, T.M. and Anderson, T.D. (1988) Effects of timber harvest on stream chemistry and dissolved nutrient losses in northeast Oregon. *Forest Science*, **34**, 344-358.

[12] Hutchens Jr., J.J., Batzer, D.P. and Reese, E. (2004) Bioassessment of silvicultural impacts in streams and wetlands of the eastern United States. *Water, Air, and Soil Pollution: Focus*, **4**, 37-53.

[13] Gravelle, J.A. and Link, T.E. (2007) Influence of timber harvesting on headwater peak stream temperatures in a northern Idaho watershed. *Forest Science*, **53**, 189-205.

[14] Hubbart, J.A., Link, T.E., Gravelle, J.A. and Elliot, W.J. (2007) Timber harvest impacts on hydrologic yield in the continental/maritime hydroclimatic region of the US. *Special Issue on Headwater Forest Streams, Forest Science*, **53**, 169-180.

[15] Karwan, D.L., Gravelle, J.A. and Hubbart, J.A. (2007) Effects of timber harvest on suspended sediment loads in Mica Creek, Idaho. *Forest Science*, **53**, 181-188.

[16] Dynesius, M., Åström, M. and Nilsson, C. (2008) Microclimatic buffering by logging residues and forest edges reduces clear-cutting impacts on forest bryophytes. *Applied Vegetation Science*, **11**, 345-354.

[17] Gravelle, J. A., Ice, G., Link, T. E. and Cook, D.L. (2009) Nutrient concentration dynamics in an inland Pacific Northwest watershed before and after timber harvest. *Forest Ecology and Management*, **257**, 1663-1675.

[18] Caners, R.T., MacDonald, S.E. and Belland, R.J. (2010) Responses of boreal epiphytic bryophytes to different levels of partial canopy harvest. *Botany*, **88**, 315-328.

[19] Dovciak, M., Halpern, C.B., Evans, S.A. and Heithecker, T.D. (2010) Forest management changes microclimate and bryophyte communities in the Cascade Mountains of western Washington. *The 95th ESA Annual Meeting*, Pittsburgh, 1-6 August 2010.

[20] Stehn, S.E., Webster, C.R., Glime, J.M. and Jenkins, M.A. (2010) Elevational gradients of bryophyte diversity, life forms, and community assemblage in the southern Appalachian Mountains. *Canadian Journal of Forest Research*, **40**, 2164-2174.

[21] Baldwin, L.K., Petersen, C.L., Bradfield, G.E., Jones, W.M., Black, S.T. and Karakatsoulis, J. (2012) Bryophyte response to forest canopy treatments within the riparian zone of high-elevation small streams. *Canadian Journal of Forest Research*, **42**, 141-156.

[22] Sollows, M.C., Frego, K.A. and Norfolk, C. (2001) Recovery of *Bazzania trilobata* following dessication. *The Bryologist*, **104**, 421-429.

[23] Dynesius, M. and Hylander, K. (2007) Resilience to bryophyte communities to clear-cutting of boreal stream-side forests. *Biological Conservation*, **135**, 423-434.

[24] Wozniewski, R. and Diekmann, M. (2009) Bryophyte vegetation on decaying spruce logs in wind throws. *Forstarchiv*, **80**, 173-180.

[25] Arseneault, J., Fenton, N.J. and Bergeron, Y. (2012) Effects of variable canopy retention harvest on epixylic bryophytes in boreal black spruce—Feathermoss forests. *Canadian Journal of Forest Research*, **42**, 1467-1476.

[26] Shields, J.M., Webster, C.R. and Glime, J.M. (2007) Bryophyte community response to silvicultural opening size in a managed northern hardwood forest. *Forest Ecology and Management*, **252**, 222-229.

[27] McGee, G.G. and Kimmerer, R.W. (2002) Forest age and management effects on epiphytic bryophyte communities in Adirondack northern hardwood forests, New York, USA. *Canadian Journal of Forest Research*, **32**, 1562-1576.

[28] Cleavitt, N.L., Eschtruth, A.K., Battles, J.J. and Fahey, T.J. (2008) Bryophyte response to eastern hemlock decline caused by hemlock woolly adelgid infestation. *Journal of the Torrey Botanical Society*, **135**, 12-25.

[29] Mathieson, K. and Frego, K. (2008) Bryophyte-substrate associations and relationships to forest management responses in the Acadian Forest of southern New Brunswick. *Eastern CANUSA Forest Science Conference Proceedings*, Orono, 17-18 October 2008.

[30] Witkowski, A. D. and Frego, K.A. (2008) The response of bryophytes to pre-commercial thinning in the Acadian Forest. *Eastern CANUSA Forest Science Conference Proceedings*, Orono, 17-18 October 2008.

[31] Dynesius, M., Hylander, K. and Nilsson, C. (2009) High resilience of bryophyte assemblages in streamside compared to upland forests. *Ecology*, **90**, 1042-1054.

[32] Hudson, J.M.G. and Henry, G.H.R. (2009) Increased plant biomass in a High Arctic heath community from 1981 to 2008. *Ecology*, **90**, 2657-2663.

[33] Lang, S.I., Cornelissen, J.H.C., Shaver, G.R., Ahrens, M., Callaghan, T.V., Molau, U., Ter Braak, C.J.F., Holzer, A. and Aerts, R. (2012) Arctic warming on two continents has consistent negative effects on lichen diversity and mixed effects on bryophyte diversity. *Global Change Biology*, **18**, 1096-1107.

[34] Hudson, J.M.G. and Henry, G.H.R. (2010) High Arctic plant community resists 15 years of experimental warming. *Journal of Ecology*, **98**, 1035-1041.

[35] Arroniz-Crespo, M., Gwynn-Jones, D., Callaghan, T.V., Nunez-Olivera, E., Martınez-Abaigar, J., Horton, P. and Phoenix, G.K. (2011) Impacts of long-term enhanced UV-B radiation on bryophytes in two sub-Arctic heathland sites of contrasting water availability. *Annals of Botany*, **108**, 557-565.

[36] Bjerke, J.W., Bokhorst, S., Zielke, M., Callaghan, T.V., Bowles, F.W. and Phoenix, G.K. (2011) Contrasting sensitivity to extreme winter warming events of dominant sub-Arctic heathland bryophyte and lichen species. *Journal of Ecology*, **99**, 1481-1488.

[37] Arróniz-Crespo, M., Leake, J.R., Horton, P. and Phoenix, G.K. (2008) Bryophyte physiological responses to, and recovery from, long-term nitrogen deposition and phosphorus fertilisation in acidic grassland. *New Phytologist*, **180**, 864-874.

[38] Bu, Z.J., Rydin, H. and Chen, X. (2011) Direct and inter-action-mediated effects of environmental changes on peat-land bryophytes. *Oecologia*, **166**, 555-563.

[39] Hubbart, J.A., Link, T. and Elliot, W.J. (2011) Implemen-tation strategies to improve WEPP snowmelt simulations in mountainous terrain. *American Society of Agricultural and Biological Engineers*, **54**, 1333-1345.

[40] Idaho Department of Lands (IDL) (2000) Forest practices cumulative watershed effects process for Idaho. Idaho Department of Lands, Boise.

[41] White, J.D., Running, S.W., Nemani, R., Keane, R.E. and Ryan, K.C. (1997) Measurement and remote sensing of LAI in Rocky Mountain montane ecosystems. *Canadian Journal of Forest Research*, **27**, 1714-1727.

[42] Keane, R.E., Reinhardt, E.D., Scott, J., Gray, K. and Reardon, J. (2005) Estimating forest canopy bulk density using six indirect methods. *Canadian Journal of Forest Research*, **35**, 724-739.

[43] Marshall, J.D. and Waring, R.H. (1986) Comparison of methods of estimating leaf-area index in old-growth Douglas-fir. *Ecology*, **64**, 975-979.

[44] Duursma, R.A., Marshall, J.D. and Robinson, A.P. (2003) Leaf area index inferred from solar beam transmission in mixed conifer forests on complex terrain. *Agricultural and Forest Meteorology*, **118**, 221-236.

[45] Hubbart, J.A., Link, T.E., Campbell, C. and Cobos, D. (2005) An evaluation of a low-cost air temperature meas-urement system. *Hydrological Processes*, **19**, 1517-1523.

[46] Hubbart, J.A. (2011) An inexpensive alternative solar radiation shield for ambient air temperature micro-sen-sors. *Journal of Natural and Environmental Sciences*, **2**, 9-14.

[47] Jonsson, B.G. (1996) Riparian bryophytes of the H.J. Andrews experimental forest in the western Cascades, Oregon. *The Bryologist*, **99**, 226-235.

[48] Conrad, H.S. and Redfearn, P.L. (1979) How to know the mosses and liverworts. Wm. C. Brown Company Pub-lishers, New York.

[49] Dhindsa, R.S. and Matowe, W. (1981) Drought tolerance

in 2 mosses correlated with enzymatic defense against lipid per oxidation. *Journal of Experimental Botany*, **32**, 79-92.

[50] Martinez-Abaigar, J., Nunez-Olivera, E. and Beaucourt, N. (2002) Moss communities in the irrigation channels of the river Iregua basin (La Rioja, northern Spain). *Cryp-togamie Bryologie*, **23**, 97-117.

[51] Jönsson, K.I. (2003) Population density and species com-position of moss-loving tardigrades in a boreo-nemoral forest. *Ecography*, **26**, 356-364.

[52] Vargha, B., Ötvös, E. and Tuba, Z. (2002) Investigations on ecological effects of heavy metal pollution in Hungary by moss-dwelling water bears (*Tardigrada*), as bioindi-cators. *Annals of Agriculture and Environmental Medi-cine*, **9**, 141-146.

[53] Sokal, R.R. and Rohlf, F.J. (1981) Biometry: The prince-ples and practice of statistics in biological research. Free-man, New York.

[54] Zar J.H. (1999) Biostatistical analysis. Prentice Hall, Upper Saddle River.

[55] Hubbart, J.A., Kavanagh, K.L., Pangle, R., Link, T.E. and Schotzko, A. (2007) Cold air drainage and modeled noc-turnal leaf water potential in complex forested terrain. *Tree Physiology*, **27**, 631-639.

[56] Furness, S.B. and Grime, J.P. (1982) Growth rate and temperature responses in bryophytes: I. An investigation of *Brachythecium Rutabulum*. *The Journal of Ecology*, **70**, 513-523.

[57] Furness, S.B. and Grime, J.P. (1982) Growth rate and temperature responses in bryophytes: II. A comparative study of species of contrasted ecology. *The Journal of Ecology*, **70**, 525-536.

[58] Bonan, G. (2002) Ecological climatology: Concepts and applications. Cambridge University Press, Cambridge.

[59] Monteith, J.L. (1973) Principles of environmental physics. Edward Arnold Limited, London.

[60] Koeniger, P., Hubbart, J.A., Link, T. and Marshall, J.D. (2008) Isotopic variation of snow cover and streamflow in response to changes in canopy structure in a snow-dominated mountain catchment. Hydrological Processes, 1-10.

19

Genotype and task influence stinging response thresholds of honeybee (*Apis mellifera* L.) workers of African and European descent

Jose L. Uribe-Rubio[1], Tatiana Petukhova[2], Ernesto Guzman-Novoa[3*]

[1]CENIDFA-INIFAP, Animal Physiology Research Center, Ajuchitlán, Mexico
[2]Department of Statistics, University of Guelph, Guelph, Canada
[3]School of Environmental Sciences, University of Guelph, Guelph, Canada; *Corresponding Author

ABSTRACT

The stinging response thresholds of individual European and Africanized worker honeybees (*Apis mellifera* L.) were analyzed. Workers of each genotype performing defense (guard and soldier bees) and non-defense (nest and forager bees) associated tasks were collected and exposed to an electric stimulus of 0.5 mA, and the time they took to sting a leather substrate was recorded. Africanized bees had significant lower thresholds of response than European bees. Guards and soldiers were faster to sting than nest and forager bees for the Africanized genotype, whereas for the European genotype, guards stung significantly faster than bees of the other three task groups. This is the first study that shows that individual bees specialized in two defensive tasks also have a lower response threshold for stinging. Our results fit a model of division of labor based on differences in response thresholds to stimuli among workers of different genotypes and task groups.

Keywords: *Apis mellifera*; Africanized Honeybees; Genotypic Effects; Defensive Behavior; Response Thresholds; Division of Labor

1. INTRODUCTION

Division of labor in insect societies can be explained by behavioral threshold variance among individual members of a colony [1,2]. In this model, it is assumed that individuals respond to stimuli on the basis of their response thresholds. Members with similar sets of thresholds will tend to perform similar tasks and thus end up in the same behavioral roles. Defensive behavior is an example of a highly advantageous behavior for the evolutionary and ecological success of honeybees [3]. Defensive tasks are carried out by guards and soldiers, individuals that specialize in either guarding the colony entrance or in stinging potential intruders [4]. Guards are middle-age bees (one to three weeks old) that patrol the entrance of their colony and inspect incoming individuals to identify them as nest mates or non-nest mates. Guards exclude bees (or other invertebrates) that are foreign to their colony, and alert other colony workers about intruders. By releasing pheromones, they recruit bees from the interior of the colony and some of the recruited bees (the soldiers) fly out, detect, pursue and sting vertebrate intruders [5].

Numerous studies have demonstrated that guarding and stinging are specialized tasks that are performed by few, genetically predisposed individuals [3] and loci associated with guarding and stinging behavior have been mapped and confirmed [6-9]. Thus, it is clear that guards and soldiers play an important role in colony defense. Genotypic effects on the defensive behavior of honeybees are also evident in nature. Africanized bees are an example of a defensive genotype; they are significantly more defensive than European bees [10-12]. There are other groups of bees that do not perform defensive tasks in honey bee colonies, but that specialize in nest-associated tasks such as cleaning cells, feeding larvae, tending the queen, and building comb. Foraging is also another non-defensive task performed typically by older individuals [13].

Individual tests are important to study the defensive behavior of honeybees because group measures may be subject to nonlinear effects owing to interactions among workers inside colonies. Electric stimulation assays reliably detect variability between genotypes for defensive response thresholds using individual honeybees as has been repeatedly demonstrated in previous studies [14-17], but very few studies have demonstrated genotype by task interactions for behavioral traits of honeybees. In the case of defensive behavior, it is not known how much the thresholds of response of individual bees are affected by specialized tasks. A recent study demonstrated that guard bees of Africanized colonies have a lower threshold of response to electrical stimulation than bees performing nest tasks [17], but such evaluations have never been conducted to compare bees performing other defensive and non-defensive tasks, which is critical to further strengthen the notion that response thresholds to defense stimuli are influenced by both genotype and behavioral tasks performed by individuals.

Here we report results of stinging response-thresholds to electric stimulation of two non-defensive groups of young and old workers (nest and forager bees) and two defensive groups of middle age and presumably older workers (guard and soldier bees) taken from European and Africanized honeybee colonies.

2. MATERIALS AND METHODS

2.1. Study Area and Genotypes

This study was conducted at Mexico's Animal Physiology Research Center in Ajuchiltlan, Queretaro, Mexico (21°N, 22°W). Experimental honeybees were obtained from five Africanized and five European bee colonies. The Africanized colonies were derived from queens of swarms captured locally. The European colonies were derived from queens of Carniolan descent that were previously imported from Ontario, Canada. Morphometric and mitochondrial DNA analyses [18] of the queens' progeny confirmed the origin of the source colonies.

2.2. Bioassay and Data Collection

The stinging response thresholds to an electric stimulus of bees performing two distinct defensive tasks [4] and two distinct non-defensive tasks (nest and forager bees) [13] were determined. To collect bees of the above behavioral categories from the source colonies, different procedures were followed. Nest bees were individually captured from brood combs in the interior of the colonies by gently placing an inverted 15-ml plastic tube over the bee on the comb, whereas bees performing guarding bouts were captured at the hives' entrances. Guard bees were detected by observing their "typical" behavior [4] for a period of 5 min before taking the sample. Guard bees stand with their forelegs off the ground and their antennae held forwards to touch and inspect incoming bees. Once identified, guards were captured by gently placing a plastic tube over the bee on the landing board of the hive. To capture foragers, a wire mesh was used to

blockade the hive entrance, which forced returning foragers to land on it. Foragers were identified as bees returning to the hive with pollen loads and were collected following the same procedure used to capture guard bees. Soldiers were collected as previously described [19]. Briefly, a pyramidal-shaped net trap was made to fit the dimensions of the hive with four triangular pieces of white wood (two 55.5 × 90 cm and two 45.5 × 90 cm) that were covered with mosquito net. A flag (10 × 8 cm black paper patch attached to a 100 cm long wooden stick) was inserted through the tip of the net. Each experimental colony was opened and the triangular net placed on top of the hive, trying not to disturb the bees. Then a flag was presented ca. 10 cm above the top of the combs in the brood chamber. The flag was moved in zigzag manner across all frames at a rate of 1 circuit/s for 20 s to present a quick stimulus to attract the most sensitive bees to try to sting the paper. The trap was withdrawn and closed to prevent the soldiers inside it from escaping and was immediately transported to the laboratory. In the laboratory, bees were individually transferred from the trap into plastic containers as above. Bees of all task groups contained in tubes were tested within 1 h after collection. A total of 125 bees per task group were tested for each genotype, and all colonies were equally represented.

A constant-current stimulator (Isostim 360, model A 320 R-E, World Precision Instruments, Sarasota, Fla., USA) was used to provide a constant stimulus regardless of variation in resistance caused by the manner a bee touched the wires, or the size of the bee. A fixed electric stimulus of 0.5 mA was applied, with a pulse interval of 250 ms and an amplitude of 2.0 ms. These conditions were set based on previous studies that tested the effects of varying pulse rate, interval, voltage or current to induce stinging. A constant current of 0.5 mA showed good separation between European and Africanized workers captured at the hive entrance [17]. The electric stimulator was connected to a wire grid composed of parallel stainless steel wires (13 cm × 2 mm) with a separation of 3.5 mm between them [15]. The grid wires were assembled in an alternate electric manner (+ and −) so that the circuit was closed by the bee touching two adjacent wires.

Individual bees were placed unrestrained above the device's grid with the aid of entomological forceps. Then, the individual was covered with an inverted plastic Petri dish (57 mm diameter) that restrained the bee within the grid, but allowed her to freely walk on it. Then, the bee was left undisturbed for 5 min to allow her to acclimate before being stimulated. Each bee was then subjected to the electric stimulation and the time taken by the individual from the application of the electric stimulus to the stinging of a black suede leather patch placed underneath the wire grid, was measured with a precision chronometer (Leonidas trackmaster, model 8042, Bern, Switzerland). The black suede was used to provide a more realistic target to sting. Defenders localize potential intruders in the field by visual cues and odor and are more responsive to dark colored objects with a mammalian smell [3]. A new suede patch was used for each bee and the wire grid was cleaned with a solution of water and neutral soap after each test to remove possible residues of alarm pheromones left by the previous bee.

2.3. Statistical Analysis

The data did not show a normal distribution based on histogram visual assessments; thus, they were subjected to the Box-Cox method to power-transform them and stabilize the variance. A linear mixed effects model was used to analyze colony and task effects on time to sting within each genotype because colonies are nested within genotype groups and as such are not independent. Since colony effects were not significant after this initial analysis, the data of both genotypes were combined and subjected to factorial analyses of variance to detect genotype and genotype x task interaction effects. When significant differences were found, means were compared with Tukey tests.

3. RESULTS

Time to sting was not significantly affected by colony effects neither for European bees ($F_{4,495} = 1.27$, $P = 0.283$), nor for Africanized bees ($F_{4,495} = 2.02$, $P = 0.090$). However, Africanized bees responded significantly faster to the electric stimulus than European bees ($F_{1,998} = 108.15$, $P < 0.0001$) with mean stinging times of 2.32 ± 0.15 and 3.06 ± 0.18 s, respectively. Africanized bees performing both defense activities (guards and soldiers) stung significantly faster than bees not performing defense tasks (nest and forager bees) ($F_{3,496} = 70.81$, $P < 0.0001$; **Figure 1**), but in European bees, no differences were detected for time to sting among three task groups (nest, forager and soldier bees), although guards stung significantly faster than forager and nest bees ($F_{3,496} = 5.92$, $P < 0.001$; **Figure 2**). Additionally, interaction effects between genotype and task were evident ($F_{3,992} = 27.42$, $P < 0.0001$), with Africanized bees being faster responders specifically for the defense associated tasks.

4. DISCUSSION

Africanized bees responded faster than Europeans to the electric stimulus used in this study. The genotypic effects reported here agree with studies that have shown that Africanized bees are more defensive than European bees [3]. These results also confirm those of a previous study showing that Africanized bees have a lower threshold of stinging response than European bees to

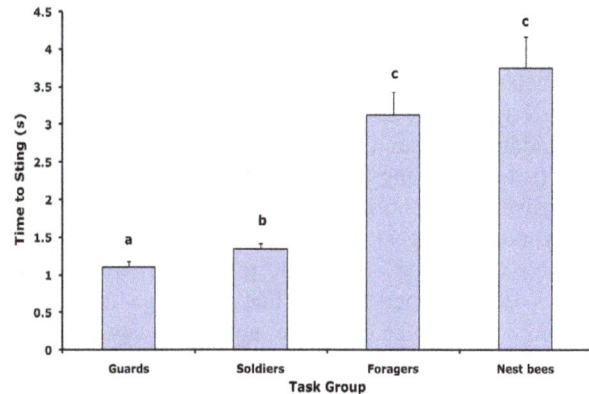

Figure 1. Effect of task on mean time to sting (s ± s.e.) for individual Africanized honeybees performing defensive tasks (guards and soldiers) and non-defensive tasks (nest and foragers). Worker bees (n = 500) collected from five colonies were subjected to a 0.5 mA electric stimulus and the time they took to sting a leather substrate was recorded. Different literals indicate significant differences of means, based on analysis of variance and Tukey tests of Box-Cox transformed-data. Actual non-transformed values are shown.

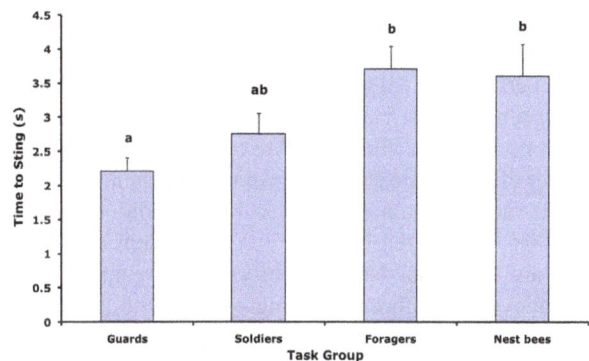

Figure 2. Effect of task on mean time to sting (s ± s.e.) for individual European honeybees performing defensive tasks (guards and soldiers) and non-defensive tasks (nest and foragers). Worker bees (n = 500) collected from five colonies were subjected to a 0.5 mA electric stimulus and the time they took to sting a leather substrate was recorded. Different literals indicate significant differences of means, based on analysis of variance and Tukey tests of Box-Cox transformed-data. Actual non-transformed values are shown.

electric stimulation [17] and support the validity and reliability of using an electric stimulator to study defensive responses and to test threshold models in honeybee studies. Additionally, results of this study provide further evidence of variability for stinging response thresholds depending on the task performed by the bees. Overall, bees performing defensive tasks responded faster than bees performing non-defensive tasks. This variability associated to defensive and non-defensive tasks was more evident in Africanized workers than in European workers, where only one task group (guards) showed significantly faster responses than the other three groups. This is the first time that evidence is generated showing that geno-type and task specialization, rather than age, influence thresholds of stinging response in Africanized bees. Africanized bees specialized in defensive tasks performed by middle-age bees (guards) and by presumably older bees (soldiers), stung faster than bees specialized in non-defensive tasks performed by both, young (nest bees) and old workers (foragers). In European bees, young and old workers did not differ for time to sting. Therefore, our results do not support the idea that as bees get older, they are more prone to be defensive, unless they specialize in defensive tasks. This is particularly clear in Africanized bees.

Only one study has previously compared the stinging response to electric stimulation of Africanized and European bees [17], but this is the first time that two non-defensive and two defensive-associated tasks are evaluated in the two bee types. This study provides evidence indicating that individuals specialized in two different defense tasks, also have a lower response threshold for stinging, which supports the hypothesis that defensive bees may have lowered response thresholds to defensive stimuli [20].

Perhaps differences were found between Africanized bees performing defensive tasks and Africanized bees performing non-defensive tasks in the colonies of this study because the alleles of African origin may have a greater effect on the stimulus response threshold than the allelic variation within European bees. Additionally, the fact that Africanized bees are more persistent as guards than are European bees reinforces this argument [21]. The above speculations are in agreement with the evolutionary histories of African and European subspecies of honeybees, in which bees of African ancestry show a more evolved level of defense specialization than bees of European origin [22].

Our results indicate that genotype and task specialization influence the stinging response thresholds of honeybees, particularly in workers of African ancestry. The identification of these components associated with the variability in stinging behavior of individual honeybees is consistent with a response-threshold model that explains the division of labor in the colony as the result of variability for the sensitivity to various stimuli among its members [1,2].

5. ACKNOWLEDGEMENTS

We thank Eusebio Pedroza for technical support and bee management received during the course of the experiments.

REFERENCES

[1] Robinson, G.E. (1992) Regulation of division of labor in insect societies. *Annual Review of Entomology*, **37**, 637-665.

[2] Page, R.E. and Erber, J. (2002) Levels of behavioral organization and the evolution of division of labor. *Naturwissenschaften*, **89**, 91-106.

[3] Breed, M.D., Guzmán-Novoa, E. and Hunt, G.J. (2004) Defensive behavior of honey bees: Organization, genetics, and comparisons with other bees. *Annual Review of Entomology*, **49**, 271-298.

[4] Breed, M.D., Robinson, G.E. and Page, R.E. (1990) Division of labor during honey bee colony defense. *Behavioural Ecology and Sociobiology*, **27**, 395-401.

[5] Guzmán-Novoa, E., Hunt, G.J., Uribe-Rubio, J.L. and Prieto-Merlos, D. (2004) Genotypic effects of honey bee (*Apis mellifera*) defensive behavior at the individual and colony levels: The relationship of guarding, pursuing and stinging. *Apidologie*, **35**, 15-24.

[6] Hunt, G.J., Guzmán-Novoa, E., Fondrk, M.K. and Page, R.E. (1998) Quantitative trait loci for honeybee stinging behavior and body size. *Genetics*, **148**, 1203-1213.

[7] Guzmán-Novoa, E., Hunt, G.J., Uribe-Rubio, J.L., Smith, C. and Arechavaleta-Velasco, M.E. (2002) Confirmation of QTL effects and evidence of genetic dominance of honeybee defensive behavior: Results of colony and individual behavioral assays. *Behavioral Genetics*, **32**, 95-102.

[8] Arechavaleta-Velasco, M.E., Hunt, G.J. and Emore, C. (2003) Quantitative trait loci that influence the expression of guarding and stinging behaviors of individual honey bees. *Behavioral Genetics*, **33**, 357-364.

[9] Arechavaleta-Velasco, M.E. and Hunt, G.J. (2004) Binary trait loci that influence honey bee guarding behavior. *Annals of the Entomological Society of America*, **97**, 177-183.

[10] Collins, A.M., Rinderer, T.E., Harbo, J.R. and Bolten, A.B. (1982) Colony defense by Africanized and European honeybees. *Science*, **218**, 72-74.

[11] Guzmán-Novoa, E. and Page, R.E. (1994) Genetic dominance and worker interactions affect honeybee colony defense. *Behavioral Ecology*, **5**, 91-97.

[12] Uribe-Rubio, J.L., Guzmán-Novoa, E., Hunt, G.J., Correa-Benitez, A. and Zozaya, R.J.A. (2003) Effect of Africanization on honey production, defensive behavior and size in honey bees (*Apis mellifera* L.) of the Mexican high plateau. *Veterinaria Mexico*, **34**, 47-59.

[13] Winston, M.L. (1987) The biology of the Honeybee. Harvard University Press, Cambridge.

[14] Kolmes, S.A. and Fergusson-Kolmes, L.A. (1989) Measurements of stinging behaviour in individual worker honeybees (*Apis mellifera* L.). *Journal of Apicultural Research*, **28**, 71-78.

[15] Paxton, R.J., Sakamoto, C.H. and Rugiga, F.C.N. (1994) Modification of honey bee (*Apis mellifera* L.) stinging behaviour by within-colony environment and age. *Journal of Apicultural Research*, **33**, 75-82.

[16] Lenoir, J.C., Laloi, D., Dechaume-Moncharmont, F.X., Solignac, M. and Pham, M.H. (2006) Intra-colonial variation of the sting extension response in the honey bee *Apis mellifera*. *Insectes Sociaux*, **53**, 80-85.

[17] Uribe-Rubio, J.L., Guzmán-Novoa, E., Vázquez-Peláez, C. and Hunt, G.J. (2008) Genotype, task specialization, and nest environment influence the stinging response thresholds of individual Africanized and European honeybees to electrical stimulation. *Behavioral Genetics*, **38**, 93-100.

[18] Nielsen, D.I., Ebert, P.R., Hunt, G.J., Guzmán-Novoa, E., Kinnee, S.A. and Page, R.E. (1999) Identification of Africanized honey bees (hymenoptera: Apidae) incorporating morphometrics and an improved PCR mitotyping procedure. *Annals of the Entomological Society of America*, **92**, 167-174.

[19] Alaux, C., Sinha, S., Hasadsri, L., Hunt, G.J., Guzman-Novoa, E., DeGrandi-Hoffman, G., Uribe-Rubio, J.L., Southey, B.R., Rodriguez-Zas, S. and Robinson, G.E. (2009) Honey bee aggression supports a link between gene regulation and behavioral evolution. *Proceedings of the National Academy of Sciences USA*, **106**, 15400-15405.

[20] Robinson, G.E. and Page, R.E. (1988) Genetic determination of guarding and undertaking in honey-bee colonies. *Nature*, **333**, 356-358.

[21] Hunt, G.J., Guzmán-Novoa, E., Uribe-Rubio, J.L. and Prieto-Merlos, D. (2003) Genotype by environment interactions in honey bee guarding behavior. *Animal Behavior*, **66**, 469-477.

[22] Guzman-Novoa, E., Correa-Benítez, A., Guzmán, G. and Espinoza-Montaño, L.G. (2011) Colonization, impact and control of Africanized honey bees in Mexico. *Veterinaria Mexico*, **42**, 149-178.

20

Modification of soil properties by *Prosopis* L. in the Kalahari Desert, South-Western Botswana

Samuel Mosweu[1*], Christopher Munyati[2], Tibangayuka Kabanda[2]

[1]Department of Environmental Science, University of Botswana, Gaborone, Botswana;
[*]Corresponding Author
[2]Department of Geography and Environmental Science, North-West University, Mafikeng Campus, Mafikeng, South Africa

ABSTRACT

The aim of this research was to investigate the interactions between *Prosopis* plants and soils in the Kalahari area, south west of Botswana. The underlying assumptions of the research were that *Prosopis* plants significantly enhanced the nutrient content and improved the condition of soils in the study area, and that the height and canopy size of *Prosopis* plants affected the interactions between *Prosopis* plants and the soils. Firstly, soil samples were collected under 42 randomly selected *Prosopis* plant canopies and in the spaces between *Prosopis* plant canopies at the depth of 0 - 20 cm and 60 - 80 cm. Secondly, soil samples were collected under 45 randomly selected *Prosopis* plant canopies of three different categories of height and canopy size at the depth of 0 - 10 cm. The soil samples were analysed for soil organic carbon, pH, total nitrogen (N), electrical conductivity (EC), calcium (Ca), sodium (Na), potassium (K), and magnesium (Mg). Soil collected under *Prosopis* plant canopies and in the spaces between *Prosopis* plant canopies showed statistically significant difference in the soil organic carbon content ($F = 2.68$, $P = 0.05$, $\alpha = 0.05$), pH ($F = 44.81$, $P < 0.001$; $\alpha = 0.05$) and electrical conductivity (EC) ($F = 3.75$, $P = 0.01$, $\alpha = 0.05$). Statistically significant difference was also observed in the comparison of soils existing under Classes 1, 2 and 3 *Prosopis* plant canopies in relation to pH and EC ($F = 6.56$, $P = 0.01$ and $F = 4.77$, $P = 0.01$ respectively at $\alpha = 0.05$). Therefore, it was concluded that the fundamental assumptions of the study were valid.

Keywords: *Prosopis* Species; Soil Properties;

Kalahari Desert; Botswana

1. INTRODUCTION

Prosopis Linnaeus amended Burkart genus belongs to the family Leguminosae (Fabaceae), sub-family Mimosoideae [1]. The genus range covers arid and semi-arid regions in Africa, Asia, Central, Northern and Southern regions of America [1]. The best known and most widely spread *Prosopis* species is *Prosopis juliflora* [1]. [2] described *Prosopis* species as trees or shrubs of various sizes which are primarily xerophilous, aculeate, and spiny. The taxonomy of the *Prosopis* genus compiled by [2] included 44 *Prosopis* species and a number of varieties. The species are aggressive pioneers which predominate over other flora wherever they are introduced [1]. They are famous for their rapid growth and their resilience under harsh arid and semi-arid environments [1]. They also have the capacity to assimilate and store nutrients and moisture in their root systems. Consequently, they usually have relatively large root mass [3]. Since their introduction in Africa, they have aggressively invaded and continue to invade large areas of rangelands [1]. The rangelands in the south east of Botswana are amongst the areas that are seriously affected by the invasion of *Prosopis* species.

Previous studies on the interactions between *Prosopis* species and soil indicated that leguminous tree species modify the characteristics of soil on which they grow [4,5]. Elevated soil nutrient content and lower values of pH are generally associated with soils found under canopies of *Prosopis* plants compared to the inter-canopy areas in arid and semi-arid environments [1,4,5]. Empirical research has shown that the size of a tree, particularly canopy cover, may affect the condition of soil under tree canopy [6,7]. For instance, at earlier stages when a tree is young and its canopy size is small, organic matter may not be efficiently trapped under a small tree canopy within a relatively short period of time [8,9]. With advancing developmental stage of an individual tree, the canopy size and duration of nutrient accumulation may increase, leading to the improvement in understory soil conditions [8,9]. Therefore, this research aims to investigate the interactions between *Prosopis* plants and soils in the Kalahari area, south west of Botswana. The research specifically tests the assumptions that *Prosopis* plants significantly enhance the nutrient content and improve the condition of soils in the Kalahari area and that the height and canopy size of *Prosopis* plant affect the interactions between *Prosopis* plants and the soils.

2. MATERIALS AND METHODS

2.1. Description of the Study Area

The study focuses on four villages (Bokspits, Rappelspan, Vaalhoek and Struizendam) located in the Kalahari district in Botswana (**Figure 1**). The area lies within the Kalahari Desert; a vast area covered in sand stretching between the Orange River and the Zambezi River covering the western and central part of Botswana, eastern Namibia and North western regions of South Africa. The

area is mainly undulating plains with interspersed pans, rocky outcrops, dry river valleys and dune fields [10]. It is dominated by longitudinal dunes and some barchan or transverse dunes [11].

It is generally believed that the Nossob-Molopo River valley that exists in the area was part of the Orange River system [10]. The sand stone and quartz comprise the rocky outcrops in the study area with calcrete dominating the riparian zones along the Nossob-Molopo River. The area is also characterized by ephemeral and often relict closed basins of varying scales and origin [12] called pans. The vegetation of the area is generally open tree and grass savanna with sparse cover of tussock grasses. *Acacia erioloba, Acacia haematoxylon, Rhigozum trichotomum, Lycium namaquense, Monechma incanum, Prosopis chilensis, Prosopis velutina, Prosopis juliflora, Prosopis glandulosa*, hybrids of *P. juliflora* and *P. glandulosa, P. Juliflora* and *P. pallida, P. chilensis* and *P. glandulosa, P. glandulosa* and *P. pallida*, and *P. juliflora* and *Acacia karoo* comprise the main trees and shrubs found in the study area [13] while *Schmidtia pappophoroides*, and *Eragrostis* species are the main grass species growing in the area [10].

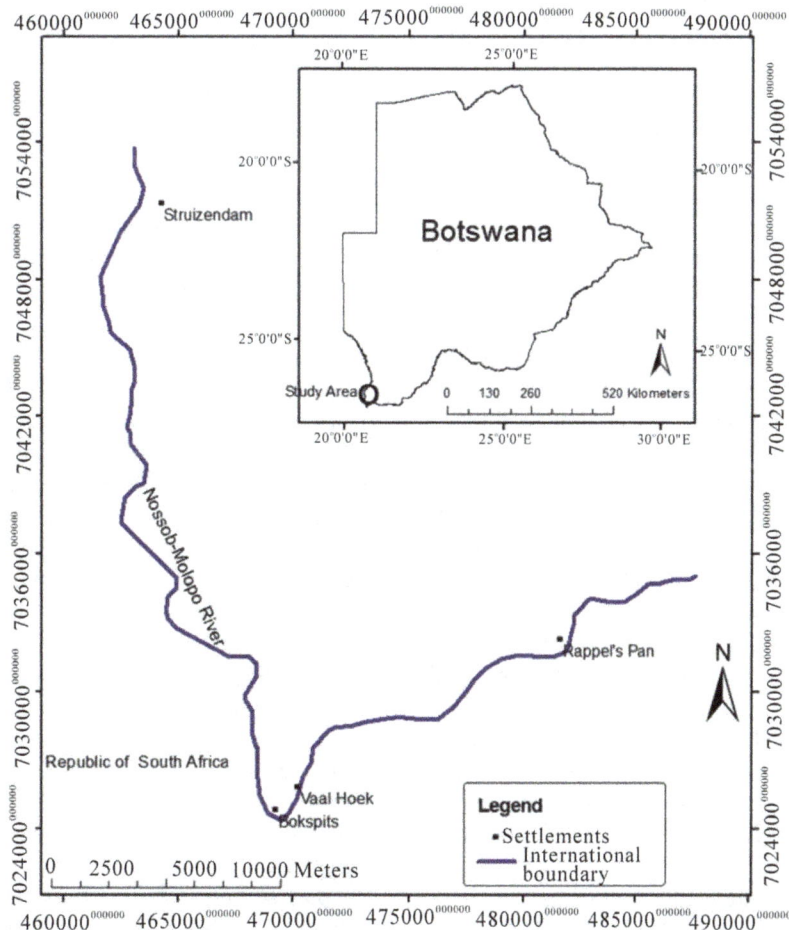

Figure 1. Location of the study area.

The study area forms part of the driest region of Botswana where the mean annual rainfall is 300 mm and the rainfall season is characterized by erratic rainfall pattern [14]. The period starting from November to April marks the season during which the area experiences about 80 per cent of the precipitation. The area experiences very high temperatures in summer which may reach up to over 40°C, while the winter temperatures are normally between 2 to 4°C [10].

2.2. Experimental Design and Laboratory Analyses

Firstly, soil samples were collected along three transects spaced equally and radiating from the main stem of 42 randomly selected mature ($\geq$2.5 m height and $\geq$20 m^2 canopy cover) Prosopis tree. Using an auger, soil cores of 5 cm in diameter and 20 cm in length were obtained: 1) under plant canopy or crown at 0.5 m of the radius of the plant canopy and 2) in the space between plant canopy (inter-space/inter-canopy) at a distance of 150 m from the nearest Prosopis plants to reduce the influence of Prosopis plants on the inter-canopy soil samples. A total of 504 soil samples were collected at the depth of 0 - 20 cm and 60 - 80 cm and pooled by depth and location resulting in a total of 168 soil samples which were prepared for laboratory analysis.

Secondly, Prosopis plants were categorized into three classes (i.e. Class 1: 0.3-1.5 m height and 1-9 m^2 canopy cover; Class 2: 1.6-2.5 m height and 10-19 m^2 canopy cover; Class 3: 2.5+ m height and 20+ m^2 canopy cover). Soil samples were then collected along three transects spaced equally and radiating from the main stem of 45 randomly selected Prosopis plants (15 per class) at 20 cm distance from the stem under plant canopy. The depth of sampling was 0 - 10 cm and sampling was conducted by the use of a hand trowel. A total of 270 soils samples was collected and pooled according to sampling location to make 90 composite samples for laboratory analysis.

All soil samples were prepared by air-drying and sieving through a 2 mm mesh sieve to remove plant material. The soil samples were analyzed for total nitrogen (N), soil organic carbon, pH, electrical conductivity (EC), exchangeable cations (calcium (Ca), sodium (Na), potassium (K), and magnesium (Mg)) as outlined below.
- Total nitrogen and organic carbon were determined using the LecoTruspec CN instrument.
- Soil pH and electrical conductivity (EC) were investigated using the 1:2 (soil: water) ratio extract method [15]. The pH meter was used to measure soil pH and the electrical conductivity meter was used to measure soil EC.
- Ca, Mg, K and Na were analysed through the silver thiourea method [16] and the Varian 220 FS Atomic Absorption Spectrophotometer (AAS).

3. RESULTS

The mean values of the selected soil properties were determined and presented in **Tables 1** and **2**. Analysis of Variance (one way ANOVA) showed that soil organic carbon at 0 - 20 cm under canopy, 60 - 80 cm under canopy, 0 - 20 cm inter-canopy and 60 - 80 cm inter-canopy soil depths were statistically significantly different ($F = 2.68$, $P = 0.05$, $\alpha = 0.05$). Student's t-Test (one tailed) indicated that soil at 0 - 20 cm depth had statistically significantly lower soil organic carbon content in comparison with soil at 60 - 80 cm depth under Prosopis plant canopies (**Table 3**). Additionally, soil at 0 - 20 cm depth in the inter-canopy had statistically significantly lower soil organic carbon compared to soil at 60 - 80 cm under canopy and 60 - 80 cm inter-canopy depth. Generally soil organic carbon at 0 - 20 cm depth was statistically significantly lower than at 60 - 80 cm depth.

It was observed through Analysis of Variance (one way ANOVA) that soil pH differed significantly ($F = 44.81$, $P < 0.001$; $\alpha = 0.05$) at 0 - 20 cm under canopy, 60 - 80 cm under canopy, 0 - 20 cm inter-canopy and 60 - 80 cm inter-canopy soil depths. Student's t-Test (one tailed) showed that soil pH was significantly lower at 0 - 20 cm under canopy depth in comparison with 60 - 80 cm under canopy, 0 - 20 cm inter-canopy and 60 - 80 cm inter-canopy depths (**Table 3**). The soil pH was also significantly lower at 0 - 20 cm inter-canopy compared to 60 - 80 cm under canopy and 60 - 80 cm inter-canopy soil depths. The general pattern in the soil pH showed lower pH levels at 0 - 20 cm soil depth as compared to 60 - 80 cm soil depth (**Table 3**). The observed values were above pH7, indicating that the soils in the study area were generally basic.

Analysis of Variance (one way ANOVA) revealed statistically significant difference ($F = 3.75$, $P = 0.01$, $\alpha = 0.05$) in the soil electrical conductivity (EC) at 0-20 cm under canopy, 60 - 80 cm under canopy, 0 - 20 cm inter-canopy and 60 - 80 cm inter-canopy depths. Student's t-Test (one tailed) showed that soil EC was significantly higher at 0 - 20 cm under canopy depth compared to 60 - 80 cm under canopy, 0 - 20 cm inter-canopy and 60-80 cm inter-canopy depths (**Table 3**). In addition, soil had significantly higher EC ($P \leq 0.001$, $\alpha = 0.05$) at 0 - 20 cm depth under Prosopis plant canopies. The content of soil Ca, Mg, K and Na and N did not show statistically significant difference (**Table 3**). Statistically significant difference was observed through one way ANOVA in the comparison of soils existing under Classes 1, 2 and 3 Prosopis plant canopies in relation to pH and EC ($F = 6.56$, $P = 0.01$ and $F = 4.77$, $P = 0.01$ respectively at $\alpha = 0.05$). Further, Student's t-Test (one tailed) revealed that soils found under Class 3 Prosopis plant canopies were statistically significantly different from soils found under Classes 1 and 2 *Prosopis* plant canopies with respect to pH and EC (**Table 4**), while soils existing under Classes 1 and 2 *Prosopis* plant canopies were not statistically significantly different.

4. DISCUSSION

Studies similar to the current research have indicated that *Prosopis* plants significantly modify the characteristics of soils on which they grow (e.g.[5,17]). Such studies showed that increased soil nutrient content (e.g. soil organic matter, Ca, Mg, K and Na and N) and lower values of pH levels are normally associated with soils found under the canopies of *Prosopis* plants compared to the inter-canopy areas in arid and semi-arid environments [1,17]. However, the Ca, Mg, K, Na and N in the soils were low and the soils under *Prosopis* plant canopies and the inter-canopy areas were not statistically significantly different in relation to the afore-mentioned soil nutrients in the study area. This suggested that *Prosopis* plants did not significantly enhance the Ca, Mg, K, Na and N content of soils in the study area.

The general pattern in the soil pH indicated significant lower pH levels at 0 - 20 cm soil depth as compared to 60 - 80 cm soil depth. *Prosopis* plants growing in the study area produced considerable amount of litter fall as it is usually the case in other habitats where *Prosopis* plants grow. For instance, [21] observed that a 4 to 6 year old *Prosopis juliflorast* and produced 5 to 8 tonnes per hector per year of dry leaf litter, while 8 year old *Pro-*

Table 1. Selected soil properties under *Prosopis* plant canopies and the spaces between tree canopies.

Sampling Site	pH	EC (µS/cm)	Ca (cmol/kg)	Mg (cmol/kg)	K (cmol/kg)	Na (cmol/kg)	C (%)	N (%)
Under Canopy 0 - 20 cm	7.75 ± 0.44	308.70 ± 10.21	1.11 ± 0.03	0.62 ± 0.01	0.11 ± 0.03	0.04 ± 0.01	0.344 ± 0.041	0.066 ± 0.002
Under Canopy 60 - 80 cm	8.70 ± 0.42	202.57 ± 8.37	1.29 ± 0.05	0.20 ± 0.03	0.09 ± 0.05	0.08 ± 0.01	0.541 ± 0.063	0.069 ± 0.001
Inter-canopy 0 - 20 cm	8.11 ± 0.43	223.72 ± 12.16	1.17 ± 0.07	0.17 ± 0.01	0.10 ± 0.01	0.05 ± 0.01	0.340 ± 0.022	0.043 ± 0.003
Inter-canopy 60 - 80 cm	8.19 ± 2.14	212.72 ± 9.43	1.21 ± 0.02	0.16 ± 0.02	0.09 ± 0.03	0.07 ± 0.01	0.500 ± 0.015	0.045 ± 0.002

Table 2. Selected soil properties under canopies of three categories of *Prosopis* plants.

Sampling Site	pH	EC (µS·cm^{-1})	Ca (cmol·kg^{-1})	Mg (cmol·kg^{-1})	K (cmol·kg^{-1})	Na (cmol·kg^{-1})	C (%)	N (%)
Class 1	8.28 ±0.27	158.48 ± 13.54	1.20 ± 0.42	0.12 ± 0.06	0.12 ± 0.05	0.15 ± 0.01	0.455 ± 0.023	0.058 ± 0.001
Class 2	8.19 ± 0.22	170.72 ± 9.47	1.11 ± 0.37	0.11 ± 0.05	0.12 ± 0.04	0.14 ± 0.01	0.461 ± 0.012	0.080 ± 0.003
Class 3	8.09 ± 0.20	218.30 ± 10.23	1.29 ± 0.44	0.13 ± 0.05	0.15 ± 0.06	0.15 ± 0.01	0.572 ± 0.014	0.053 ± 0.001

Table 3. Comparisons (t-Test *P*-values) of soil properties in the study area.

Sampling sites		pH	EC	Ca	Mg	K	Na	N	C
0 - 20 cm canopy	60 - 80 cm canopy	<0.001*	<0.001*	0.28	0.27	0.22	0.05*	0.11	0.05*
	0 - 20 cm inter-canopy	<0.001*	0.020*	0.69	0.87	0.42	0.67	0.06	0.95
	60 - 80 cm inter-canopy	<0.001*	0.021*	0.59	0.91	0.13	0.13	0.19	0.07
60 - 80 cm canopy	0 - 20 cm inter-canopy	<0.001*	0.610	0.47	0.34	0.63	0.07	0.08	0.04*
	60 - 80cm inter-canopy	0.861	0.472	0.42	0.26	0.77	0.78	0.09	0.71
0 - 20 cm inter-canopy	60 - 80 cm inter-canopy	<0.001*	0.871	0.92	0.80	0.13	0.20	0.60	0.05*

*The mean difference is significant at $\alpha = 0.05$.

Table 4. Comparison (t-Test *P*-values) of soils under canopies of three categories of *Prosopis* plants.

		pH	EC	Na	K	Ca	Mg	N	C
Class 1	Class 2	0.176	0.395	0.371	0.863	0.453	0.877	0.532	0.937
	Class 3	0.004*	0.004*	0.976	0.105	0.411	0.415	0.963	0.126
Class 2	Class 3	0.079	0.013*	0.397	0.134	0.112	0.311	0.423	0.121

*The mean difference is significant at the $\alpha = 0.05$ level.

sopis juliflora stand produced 7.4 tonnes per hector per year of dry leaf litter [17] in India. The dominant tree forms of *Prosopis* plants that grow in the study area are prostrate and decumbent, and these types of plant growth promote the accumulation of tree litter under the trees. The accumulation of tree litter under *Prosopis* plant canopies appeared to have led to the establishment of the conditions that promoted reduction in soil pH, such as low evaporation rates and initiation of biological activities [20,21]. Additionally, the accumulation of tree litter under *Prosopis* plant canopies promoted the accumulation of soil organic carbon in soil existing under *Prosopis* plant canopies. *Prosopis* plants evidently enhanced soil organic carbon content and also influenced the soil pH particularly under *Prosopis* plant canopies.

Tree growth and canopy development normally lead to increase in the period of nutrient accumulation and improvement in understory soil conditions [8,9]. For this reason, statistically significant differences in the soils existing under the canopies of Classes 1, 2 and 3 *Prosopis* plants in relation to the selected soil properties was expected. On the contrary, no statistically significant difference in the soils existing under the canopies of Classes 1, 2 and 3 *Prosopis* plants in relation to soil organic carbon, Ca, Mg, K, Na and N was observed, suggesting that the influence of height and canopy size of *Prosopis* plants on the soil properties was not significant. Statistically significant different observed between soils found under Class 3 *Prosopis* plant canopies and soils found under Classes 1 and 2 *Prosopis* plant canopies in relation to soil pH and EC indicated that mature *Prosopis* plants influenced the understory soil pH and EC.

5. CONCLUSION

This study specifically tested the assumptions that *Prosopis* plants significantly enhance the nutrient content and improve the condition of soils in the Kalahari area and that the height and canopy size of *Prosopis* plants affect the interactions between *Prosopis* plants and the soils. Empirical evidence from this study showed that *Prosopis* plants enhanced soil organic carbon content and also influenced soil pH and EC in soils existing under *Prosopis* plant canopies. In addition, the height and canopy size of *Prosopis* plant affected the interactions between *Prosopis* plants and the soils.

REFERENCES

[1]   Pasiecznik, N.M. (2001) The *Prosopis juliflora-Prosopis* pallid complex: A momograph. HDRA, Coventry.

[2]   Burkart, A. (1976) A monograph of the genus *Prosopis* (Leguminosae sub-fam. Mimosoideae): Part 1. and Part 2. Catalogue of the recognized species of *Prosopis*. *Journal of the Arnold Arboretum*, **57**, 219-249.

[3]   Felker, P. and Clark, P.R. (1981) Rooting of mesquite (*Prosopis*) cuttings. *Journal of Range Management*, **34**, 466-468.

[4]   Schlesinger, W.H., Raikes, J.A., Hartley, A.E. and Cross, A.F. (1996) On the spatial pattern of soil nutrients in desert ecosystem. *Ecology*, **77**, 364-374.

[5]   Geesing, D., Felker, P. and Bingham, R.L. (2000) Influence of mesquite (*Prosopis glandulosa*) on soil nitrogen and carbon development: Implications for global carbon sequestration. *Journal of Arid Environments*, **46**, 157-180.

[6]   Robinson, D. (1994) The responses of plants to non-uniform supplies of nutrients. *New Phytologist*, **127**, 635-674.

[7]   Maestre, F.T. and Reynolds, J.F. (2006) Small-scale spatial heterogeneity in the vertical distribution of soil nutrients has limited effects on the growth and development of *Prosopis glandulosa* seedlings. *Plant Ecology*, **183**, 65-75.

[8]   Ludwig, J.A., Reynolds, J.F. and Whitson, P.D. (1975) Size-biomass relations of several Chihuahuan Desert shrubs. *American Midland Naturalist*, **94**, 451-461.

[9]   Reynolds, J.F., Virgia, R.A., Kemp, P.R., De Soyza, A.G. and Tremmel, D.C. (1999) Impact of drought on desert shrubs: Effects of seasonality and degree of resource island development. *Ecological Monographs*, **69**, 69-106.

[10] Timberlake, J. (1980) Handbook of Botswana Acacias. Ministry of Agriculture, Gaborone.

[11] Lancaster, I.N. (1978) The pans of Southern Kalahari, Botswana. *Geographical Journal*, **144**, 81-98.

[12] Shaw, P.A. and Thomas, S.G. (1997) Pans, playas and salt lakes. In: Thomas, D.S.G., Ed., *Arid Zone Geomorphology; Process, Form and Change in Drylands*. John Wiley and Sons, New York, 293-312.

[13] Muzila, M., Setshogo, M.P., Moseki, B. and Morapedi, R. (2011) An assessment of *Prosopis* L. in the Bokspits area, south-western Botswana, based on morphology. *The African Journal of Plant Science and Biotechnology*, **5**, 75-80.

[14] Bhalotra, Y.P.R. (1985) Rainfall maps of Botswana. Department of Meteorological Services, Gaborone.

[15] Sonnevelt, C. and vandenEnde, J. (1971) Soil analysis by means of a 1:2 volume extract. *Plant Soil*, **35**, 505-506.

[16] Chhabra, R., Pleysier, J. and Cremers, A. (1975) The measurement of the cation exchange capacity and exchangeable cations in soils: A new method. *Proceedings of International Clay Conference*, Mexico City, 439-449.

[17] Garg, V.K. (1998) Interaction of tree crops with a sodic soil environment: Potential for rehabilitation of degraded environments. *Land Degradation a Development*, **9**, 81-93.

[18] Skarpe, C. (1986) Plant community structure in relation to grazing and environmental changes along a North South transect in Western Kalahari. *Vegetatio*, **68**, 3-18.

[19] Vanderpost, C., Ringrose, S., Matheson, W. and Arntzen, J. (2011) Satellite based long-term assessment of rangeland condition in semi-arid areas: An example from Botswana. *Journal of Arid Environments*, **75**, 383-389.

[20] Bhojvaid, P.P., Timmer, V.R. and Singh, G. (1996) Reclaiming sodic soils for wheat production by *Prosopi juliflora* (Swartz) DC afforestation in India. *Agroforestry Systems*, **34**, 139-150.

[21] Singh. G. (1996) The role of *Prosopis* in reclaiming high-pH soils and in meeting firewood and forage needs of small farmers. In: Felker, P. and Moss, J., Eds., *Prosopis*: *Semiarid Fuelwood and Forage Tree*; *Building Consensus for the Disenfranchised*, Center for Semi-Arid Forest Resources, Kingsville, 4.21-4.34.

Permissions

The contributors of this book come from diverse backgrounds, making this book a truly international effort. This book will bring forth new frontiers with its revolutionizing research information and detailed analysis of the nascent developments around the world.

We would like to thank all the contributing authors for lending their expertise to make the book truly unique. They have played a crucial role in the development of this book. Without their invaluable contributions this book wouldn't have been possible. They have made vital efforts to compile up to date information on the varied aspects of this subject to make this book a valuable addition to the collection of many professionals and students.

This book was conceptualized with the vision of imparting up-to-date information and advanced data in this field. To ensure the same, a matchless editorial board was set up. Every individual on the board went through rigorous rounds of assessment to prove their worth. After which they invested a large part of their time researching and compiling the most relevant data for our readers. Conferences and sessions were held from time to time between the editorial board and the contributing authors to present the data in the most comprehensible form. The editorial team has worked tirelessly to provide valuable and valid information to help people across the globe.

Every chapter published in this book has been scrutinized by our experts. Their significance has been extensively debated. The topics covered herein carry significant findings which will fuel the growth of the discipline. They may even be implemented as practical applications or may be referred to as a beginning point for another development. Chapters in this book were first published by Scientific Research Publishing Inc.; hereby published with permission under the Creative Commons Attribution License or equivalent.

The editorial board has been involved in producing this book since its inception. They have spent rigorous hours researching and exploring the diverse topics which have resulted in the successful publishing of this book. They have passed on their knowledge of decades through this book. To expedite this challenging task, the publisher supported the team at every step. A small team of assistant editors was also appointed to further simplify the editing procedure and attain best results for the readers.

Our editorial team has been hand-picked from every corner of the world. Their multi-ethnicity adds dynamic inputs to the discussions which result in innovative outcomes. These outcomes are then further discussed with the researchers and contributors who give their valuable feedback and opinion regarding the same. The feedback is then collaborated with the researches and they are edited in a comprehensive manner to aid the understanding of the subject.

Apart from the editorial board, the designing team has also invested a significant amount of their time in understanding the subject and creating the most relevant covers. They scrutinized every image to scout for the most suitable representation of the subject and create an appropriate cover for the book.

The publishing team has been involved in this book since its early stages. They were actively engaged in every process, be it collecting the data, connecting with the contributors or procuring relevant information. The team has been an ardent support to the editorial, designing and production team. Their endless efforts to recruit the best for this project, has resulted in the accomplishment of this book. They are a veteran in the field of academics and their pool of knowledge is as vast as their experience in printing. Their expertise and guidance has proved useful at every step. Their uncompromising quality standards have made this book an exceptional effort. Their encouragement from time to time has been an inspiration for everyone.

The publisher and the editorial board hope that this book will prove to be a valuable piece of knowledge for researchers, students, practitioners and scholars across the globe.

List of Contributors

Schweitzer Paul
Laboratoire d'Analyses et d'Ecologie Apicole Centre d'Etudes Techniques Apicole de Moselle, Guenange, France

Boussim I. Joseph
Institut des Sciences, Ouagadougou, Burkina Faso

Nombré Issa
Institut des Sciences, Ouagadougou, Burkina Faso
Laboratoire de Biologie et Ecologie Végétales UFR Science et Technique Université de Ouagadougou, Ouagadougou, Burkina Faso

Aidoo Kwamé
International Stingless Bee Centre, Department of Entomology and Wildlife, University of Cape Coast, Cape Coast, Ghana

Bruno Sampaio Sant'Anna, Erico Luis Hoshiba Takahashi and Gustavo Yomar Hattori
Federal University of Amazonas (Amazonas University), Institute of Exact Science and Technology (ICET), Itacoatiara, Brazil

Susanne Facchin
Departamento de Genética, Instituto de Ciências Biológicas, Universidade Federal de Minas Gerais, Belo Horizonte, Brazil
Phoneutria Biotecnologia e Serviços Ltda, Belo Horizonte, Brazil

Priscila Divina Diniz Alves and Júnia Maria Netto Victória
Phoneutria Biotecnologia e Serviços Ltda, Belo Horizonte, Brazil

Flávia de Faria Siqueira, Tatiana Moura Barroca and Evanguedes Kalapothakis
Departamento de Genética, Instituto de Ciências Biológicas, Universidade Federal de Minas Gerais, Belo Horizonte, Brazil

Sung-Ryong Kang
School of Renewable Natural Resources, Louisiana State University Agricultural Center, Baton Rouge, USA

Sammy L. King
U.S. Geological Survey, Louisiana Cooperative Fish and Wildlife Research Unit, School of Renewable Natural Resources, Louisiana State University Agricultural Center, Baton Rouge, USA

Chansa Chomba
School of Agriculture and Natural Resources, Disaster Management Training Centre, Mulungushi University, Kabwe, Zambia

Eneya M'simuko
School of Natural Resources, Copperbelt University, Kitwe, Zambia

Vincent Nyirenda
Department of Research, Zambia Wildlife Authority, Chilanga, Zambia

Olena V. Derev'yanko and Oleksandr I. Raichenko
Frantsevych Institute for Problems of Materials Science of NASU, Kyiv, Ukraine

Vladimir S. Mosienko, Vladimir O. Shlyakhovenko, Yuri V. Yanish and Olena V. Karnaushenko
Kavetsky Institute of Experimental Patology, Oncology and Radiobiology of NASU, Kyiv, Ukraine

Edward B. Mondor, Carl N. Keiser, Dustin E. Pendarvis and Morgan N. Vaughn
Department of Biology, Georgia Southern University, Statesboro, USA

Sung-Ryong Kang
School of Renewable Natural Resources, Louisiana State University Agricultural Center, Baton Rouge, USA

Sammy L. King
U.S. Geological Survey, Louisiana Cooperative Fish and Wildlife Research Unit, School of Renewable Natural Resources, Louisiana State University Agricultural Center, Baton Rouge, USA

Korehisa Kaneko
Ecosystem Conservation Society-Japan, Tokyo, Japan

Seiich Nohara
Center for Environmental Biology and Ecosystem Studies, National Institute for Environmental Studies, Ibaraki, Japan

Mark Thorne
Environmental Science Graduate Program, The Ohio State University, Columbus, USA

Landon Rhodes
Department of Plant Pathology, The Ohio State University, Columbus, USA

John Cardina
Department of Horticulture and Crop Science, Ohio Agricultural and Research Development Center, The Ohio State University, Columbus, USA

Bennett Sall
Department of Biological Sciences, California State University, Chico, USA

James Pushnik
Department of Biological Sciences, California State University, Chico, USA
Institute for Sustainable Development, California State University, Chico, USA

Michael W. Jenkins
Department of Ecology and Evolutionary Biology, University of California, Santa Cruz, USA

Brahim Babali, Abderrahmane Hasnaoui, Nadjat Medjati and Mohamed Bouazza
Laboratory of Ecology and Management of the Natural Ecosystems, Department of Ecology and Environment, Aboubakr Belkaid University, Tlemcen, Algeria

James T. Anderson
Division of Forestry and Natural Resources and Environmental Research Center, West Virginia University, Morgantown, USA

Andrew K. Zadnik
Wildlife and Fisheries Resources, Division of Forestry and Natural Resources, West Virginia University, Morgantown, USA

Petra Bohall Wood
US Geological Survey West Virginia Cooperative Fish and Wildlife Research Unit, West Virginia University, Morgantown, USA

Kerry Bledsoe
West Virginia Division of Natural Resources, Wildlife Resources Section, Fairmont, USA

Amos Enock Majule
Institute of Resource Assessment (IRA), University of Dar es Salaam, Dar es Salaam, Tanzania

Tao Zou
Beijing Tsinghua Tongheng Urban Planning and Design Institute, Beijing, China

Zhengnan Zhou
School of Architecture, Tsinghua University, Beijing, China

Chansa Chomba and Mitulo Silengo
School of Agriculture and Natural Resources, Disaster Management Training Centre, Mulungushi University, Kabwe, Zambia

Vincent Nyirenda
Zambia Wildlife Authority, Directorate of Research P/B 1 Chilanga, Zambia

Bahae-Ddine Ghezlaoui, Noury Benabadji and Nedjwa Benabadji
Department of Biology, Faculty of Natural Sciences and Life Sciences and Earth and the Universe, Pôle la Rocade No. 2, University of Tlemcen, Algeria

Jason A. Hubbart
Department of Forestry and Department of Soils, Environmental and Atmospheric Sciences, University of Missouri, Columbia, USA

Elliott Kellner
Department of Forestry, University of Missouri, Columbia, USA

Jose L. Uribe-Rubio
CENIDFA-INIFAP, Animal Physiology Research Center, Ajuchitlán, Mexico

Tatiana Petukhova
Department of Statistics, University of Guelph, Guelph, Canada

Ernesto Guzman-Novoa
School of Environmental Sciences, University of Guelph, Guelph, Canada

Samuel Mosweu
Department of Environmental Science, University of Botswana, Gaborone, Botswana

Christopher Munyati and Tibangayuka Kabanda
Department of Geography and Environmental Science, North-West University, Mafikeng Campus, Mafikeng, South Africa

www.ingramcontent.com/pod-product-compliance
Lightning Source LLC
Chambersburg PA
CBHW050459200326
41458CB00014B/5235

* 9 7 8 1 6 3 2 3 9 1 5 9 9 *